D1503547

ENERGY AND PACKAGING

I. BOUSTEAD, B.Sc., M.Sc., Ph.D.
and
G. F. HANCOCK, B.Sc., M.Sc., Ph.D.
Faculty of Technology
The Open University
Milton Keynes

ELLIS HORWOOD LIMITED
Publishers · Chichester

Halsted Press: a division of
JOHN WILEY & SONS
New York · Brisbane · Chichester · Toronto

First published in 1981 by

ELLIS HORWOOD LIMITED

Market Cross House, Cooper Street, Chichester, West Sussex, PO19 1EB, England

The publisher's colophon is reproduced from James Gillison's drawing of the ancient Market Cross, Chichester.

Distributors:

Australia, New Zealand, South-east Asia:
Jacaranda-Wiley Ltd., Jacaranda Press,
JOHN WILEY & SONS INC.,
G.P.O. Box 859, Brisbane, Queensland 40001, Australia

Canada:
JOHN WILEY & SONS CANADA LIMITED
22 Worcester Road, Rexdale, Ontario, Canada.

Europe, Africa:
JOHN WILEY & SONS LIMITED
Baffins Lane, Chichester, West Sussex, England.

North and South America and the rest of the world:
Halsted Press: a division of
JOHN WILEY & SONS
605 Third Avenue, New York, N.Y. 10016, U.S.A.

© I. Boustead and G. F. Hancock/Ellis Horwood Ltd.

British Library Cataloguing in Publication Data
Boustead, I.
   1. Beverages – Packaging
   2. Containers
   3. Energy consumption
   I. Title   II. Hancock, G. F.
   663   TP506
ISBN 0-85312-206-7 (Ellis Horwood Limited)
ISBN 0-470-27269-4 (Halsted Press)

Typeset in Press Roman by Ellis Horwood Limited
Printed in England by R. J. Acford, Chichester

Table of Contents

Table of Contents

Author's Preface

In 1977, the Packaging and Containers Working Party of the Waste Management
Advisory Council, a body jointly sponsored by the Departments of Industry
and the Environment, began work to consider the environmental and economic
impact of the various types of container used to package beer, cider and carbonated soft drinks in the U.K. Very early during the work of this Committee,
it was realised that no accurate, quantitative data were available for the energy
and raw materials required in the production and use of these containers. As a
result we were asked by the committee to carry out an examination of such
systems with the aim of providing a representative picture of practices in the
U.K. The results of this work were reported to the Committee at the end of
1979.

Since that time there has been a considerable interest from industrial and
other workers in the detailed results of the study and it is in response to this
interest that this book has been prepared.

Essentially the book is based upon the report presented to the Waste
Management Advisory Council. To reduce the text to manageable proportions,
it has been edited and many of the original tables have been aggregated. However, no important information in the original report has been removed and new
material has been added in two important areas. First, more detailed information
has been obtained on the collection and disposal of domestic refuse and so
the whole of Chapter 27 has been rewritten using this new data. Secondly, at
the time the original report was being prepared, the plastic (PET) bottle was
just making its appearance in the U.K. market, and this was primarily in the
1.5 litre size. Since then, this market has grown and the 2.0 litre size has been
introduced. The data on PET bottles has therefore been updated to include
this change.

This book will be of direct interest to all concerned with the many different
aspects of beverage packing since it provides the first extensive treatment of this
subject based on U.K. practices. Much of the information is however of much
wider application. Anyone who has worked in the field of energy analysis will

be only too well aware of the grave shortage of detailed numerical information based on actual operating practice. This book should be invaluable to all such workers seeking such information since it provides detailed energy requirements for the production of a large number of different products.

Finally we would like to thank the many industries who have helped us in this work and those companies and organisations who are willing to be identified are listed in the acknowledgements. We would also like to thank our publishers, Ellis Horwood Ltd., for their help in the preparation of this book and we must thank Pat Brittin for her work in the preparation of the manuscript. We hope that the text is free from serious errors, but if any are discovered, we would apologise and accept responsibility.

I. Boustead and G. F. Hancock
The Open University
1981

Acknowledgements

A significant part of the work reported in this book is taken from a report prepared for and financed by the Secretary of State, Department of Industry, London. The views expressed are those of the authors and do not necessarily coincide with those of the Secretary of State.

We would like to thank the following companies and organisations who kindly supplied information: Air Products Ltd., Alcan Aluminium (UK) Ltd., Alcoa of Great Britain Ltd., Allied Breweries Ltd., The Aluminium Federation, Appleby Calumite Ltd., Avon Tyres Ltd., A G Barr & Co. Ltd., Bass Charrington Ltd., Batchelor Robinson Metals and Chemicals Ltd., Beatson Clark & Co. Ltd., Beecham (Food and Drinks) Ltd., Wm Blythe & Co. Ltd., Bowater Packaging Ltd., The Brewers Society, British Aluminium Co. Ltd., British Gas Corporation, British Gypsum Ltd., British Industrial Sand Ltd., British Oxygen Co. Ltd., British Petroleum Ltd., British Rail, British Road Services, British Steel Corporation, British Soft Drinks Council, Britvic Ltd., H P Bulmer Ltd., Canning Town Glass Ltd., Cantrell & Cochrane Ltd., Chloride Industrial Batteries Ltd., Coca-Cola Southern Bottlers Ltd., Courage Ltd., Crown Cork Co. Ltd., Dayla Soft Drinks Ltd., Distillers Company Ltd., Dunlop Ltd., Emhardt UK Ltd., Electricity Council, Esso Petroleum Ltd., Fell & Briant Ltd., Firestone Tyre and Rubber Co. Ltd., Glass Manufacturers Federation, W R Grace Ltd., David Hughes & Co. Ltd., Imperial Chemical Industries Ltd., Institute of Grocery Distribution, International Alloys Ltd., Irwell Minerals & Chemicals Ltd., Keymarkets Ltd., Laporte Industries Ltd., Lojo Kalverk (Finland), Longcliffe Quarries, Material Recovery Ltd., Metal Box Ltd., Mobil Oil Ltd., National Association of Cider Makers, National Association of Soft Drinks Manufacturers, National Coal Board, National Freight Corporation. P.L.M. Plastic (Holland), R Posnett & Sons Ltd., C E Ramsden & Co. Ltd., Reads Ltd., Redfearn National Glass Ltd., Reed Engineering & Development Services Ltd., Reynolds Aluminium (USA), Rockware Glass Ltd., G. Ruddle & Co. Ltd., J Sainsbury Ltd., Schweppes International Ltd., Scottish & Newcastle Breweries Ltd., Shell Oil Co. Ltd., Showerings Ltd., Tarmac Ltd., Tesco Ltd., Tilgate Pallets Ltd., James Townsend & Sons Ltd.,

United Glass Ltd., Vickers Ltd., Victoria Wine Ltd., Waitrose Ltd., Warren Spring Laboratories, Watney Mann Ltd., Barry Wehmiller Ltd., Whitbread Ltd., R White & Sons Ltd., White Bros. (Derby) Ltd.

We would also like to thank the following Councils for supplying information on their collection and disposal practices: Bath City, Blackpool Borough, Boston District, Bradford Metropolitan City, Cambridge City, Cambridgeshire County, Chelmsford District, Chesire County, Chesterfield Borough, Cleveland County, Corby District, Coventry City District, Croydon Borough, Cumbernauld and Kilsyth District, Darlington Borough, Derby City, Dorset County, Dundee District, Dunfermline District, East Kilbride District, East Sussex County, Edinburgh City District, Essex County, Exeter City, Gosport Borough, Great Grimsby Borough, Greater Manchester County, Guildford Borough, Hartlepool Borough, Hereford City, Horsham District, Isle of Wight County, Kent County, Kingston upon Thames Royal Borough, Kirklees Metropolitan, Lancaster City, Leicester City, Lincolnshire County, Liverpool City, Luton Borough, Macclesfield Borough, Manchester City, Mendip District, Merseyside County, Middlesbrough Borough, Midlothian District, Newcastle under Lyme Borough, North Yorkshire County, Northampton Borough, Northamptonshire County, Nottinghamshire County, Nuneaton Borough, Oadby and Wigston Borough, Oldham Metropolitan Borough, Rugby Borough, St. Albans City and District, St. Helens Metropolitan Borough, Scunthorpe District, South Lakeland District, South Yorkshire County, Staffordshire County, Stirling District, Stockport Metropolitan Borough, Stoke on Trent City, Suffolk County, Tyne and Wear County, Watford Borough, West Lothian District, West Sussex County, Wiltshire County, Wolverhampton Metropolitan Borough and York City.

Crown copyright subsists in the following Tables which are reproduced by kind permission of the Controller of Her Majesty's Stationery Office: Tables 7.1–7.16, 8.2–8.6, 8.9, 10.2–10.15, 11.1, 11.2, 11.6, 11.8, 11.12, 11.13, 11.15, 11.19, 11.22, 11.24, 11.25, 11.27, 11.30–11.32, 12.1, 13.1–13.13, 14.1–14.9, 16.1–16.4, 17.1–17.4, 18.1–18.19, 19.1–19.9, 20.1–20.4, 21.1, 22.1–22.18, 23.1–23.12, 24.1–24.23, 25.1–25.6, 26.1, 26.2, 28.1–28,7 and 29.2–29.13.

CHAPTER 1

Introduction and Methodology

1.1 GENERAL INTRODUCTION

Two essential inputs to manufacturing industry are fuels and raw materials, the cost of which contribute to total production costs. In many industries however, this proportion of the total cost is small in comparison with other costs such as labour and capital. Nevertheless, one aim of industry is to reduce the consumption of raw materials wherever possible and devising the means of achieving such savings has been the job of engineers for decades. The essential difference between the type of analysis traditionally performed by the engineer and that only recently attempted, and used in this work, lies in the size of the systems analysed. The engineer's task is normally confined to an examination of operations within a company and more usually, within a single factory. In contrast, the type of analysis reported here extends over many companies and, as will be seen later is bounded on the one hand by the extraction from the earth of the raw materials and fuels necessary to support the system and on the other by the return to the earth of the system products at the end of their useful life.

Quite apart from the economic need to improve industrial efficiency, recent years have seen increasing awareness of the implications of the continued and unrestrained use of all forms of raw materials, especially fossil fuels. Again, it is important to recognise that such problems have always existed. In the early seventeenth century for example, it was feared that coal would become scarce because the coal mines of north-east England would have to cease production due to flooding and the impossibility, at that time, of removing water from deep mines. More recently, alarming predictions have been made of the date at which known reserves of fossil fuels and some minerals will be exhausted. Whilst arguments can be produced to justify a whole range of predicted lifetimes it remains true that fossil fuels and rich mineral deposits are present in the earth in strictly limited quantities and with continued use, they will one day be exhausted. Although efficient recycling of materials can reduce the demand for minerals, the use of fossil fuels as energy sources represents an irreversible consumption; once used they are gone forever. There is therefore good reason to use them prudently and efficiently but until we know in detail how they are used at present, it is difficult to indicate where the most effective savings might be made. One of the principal aims of the type of analysis reported here is to provide just such a description, not only of existing systems, but also of projected modifications.

Much of the impetus for this type of work arose from the dramatic increase in oil prices in 1973 which led to a serious reconsideration not only of oil consumption but of consumption of all fuels in an attempt to reduce fuel costs. The result was the introduction of energy saving measures which in many industries were long overdue and which were simply the result of improving the 'good housekeeping' aspect of their operations. At the same time, investigations were started in many companies to seek further energy saving schemes by detailed analysis of current practices.

The degree of control which an industry can exercise is often misunderstood by proposers of conservation schemes. Within industry, a company has direct control only over its own operations. It may attempt to change practices in the suppliers of its raw materials, in users of its products or in its competitors, by persuasion, but it can in no way guarantee controlled behaviour outside its own factories. For example, the manufacturer of non-returnable glass bottles may claim that such bottles need only a rinse and not a hot wash before filling. Nevertheless, many fillers do wash these bottles for a variety of reasons. For example, their equipment may not be adapted for rinsing only or they may have found that occasionally some of the bottles did require washing because damaged packaging during storage allowed contamination which could not be removed by rinsing alone. This is a relatively simple example, but it illustrates the problems likely to occur when schemes involving a number of operations are proposed.

With these considerations in mind, the current position appears to be that most industrial operators and Government see a need for some form of action to reduce the consumption of energy and raw materials. Few specific workable schemes have yet been devised for extended systems largely because of the grave shortage of reliable quantitative data concerning existing practices in the U.K. Such data as do exist are the subject of more controversy than reliability. Despite this, it is commonly thought that the greatest savings will accrue from modifications to extended processing systems. The present work is therefore an attempt to provide satisfactory quantitative data for the beverage packaging industry so that this sector of industry will have some base from which to work in the future. It should also be added that much of the base data described in this book is of direct relevance to other sectors of industry since they take products from some of the industries described.

1.2 METHODOLOGY OF ENERGY ANALYSIS

The methodology of energy analysis of industrial processes has been described in detail elsewhere [1] and the short description given here is intended only as an outline guide. The basis of the procedure is the assumption that all materials processing operations, irrespective of size, may be considered as systems whose sole function is the transformation of input raw materials to output products with the consumption of energy. Fig. 1.1 represents such a system. The box represents a system boundary which encloses all operations of interest. Raw materials of mass M_i enter the system and output products of mass M_o leave. To sustain this throughput of materials, energy E_i must be fed in. There will also be an energy output, E_o, usually as waste heat, which is seldom measured.

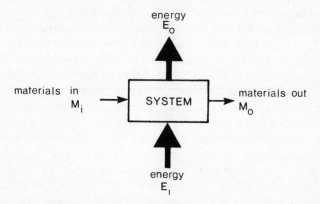

Fig. 1.1 – Schematic representation of an industrial operation as a system.

For the system of Fig. 1.1, the law of conservation of mass applies so that

$$M_i = M_o \tag{1.1}$$

Once the system of Fig. 1.1 is defined, it may be described quantitatively by specifying the input mass M_i, the output mass M_o and the energy input E_i. In general, the magnitude of E_i depends upon the magnitude of the materials throughput, a dependence which can be eliminated by normalising the energy requirement with respect to some parameter which is proportional to materials throughput. In this instance, parameters M_i or M_o could be used to give normalised system energy requirements E_s or E_s', defined by the equations:

$$E_s = E_i/M_o \tag{1.2}$$

$$E_s' = E_i/M_i \tag{1.3}$$

In the special case of Fig. 1.1,

$$E_s = E_s' \tag{1.4}$$

This normalisation procedure confers no additional significance on the energy requirement and is simply a convenient way of eliminating dependence on throughput. Moreover, mass is not the only normalising parameter that may be chosen. For fluids, volume may be a more convenient measure and in other operations, such as paint spraying, surface area may be more appropriate.

The crucial point to appreciate is that this technique applies to *systems* and the specification of the system energy such as E_s or E_s' must not be divorced from the definition of the system. Unless the system is properly defined, the system energy requirement is meaningless. It is commonly supposed that the specification of the system output is a sufficient definition of the system but this is not so. A simple example demonstrates the point. Suppose that the output of a system was a steel casting. Using the product alone as the definition of the system gives no indication of the nature of the input which could be iron ore in the ground, ore at the blast furnace, pig iron at the converter or even scrap at the consumer. Hence the input as well as the output must be closely defined since different processing sequences are needed to convert the different inputs to a steel casting, each requiring a different energy expenditure to effect the change. Furthermore, even a close specification of both inputs and outputs may be inadequate. Suppose in the steel example, that the input was specified as pig iron at the steel furnace. The function of the system is the conversion of this pig iron to the steel casting; but what type of furnace is to be used? It could be a basic oxygen furnace, an open earth furnace or even an electric arc furnace and each will exhibit a different energy requirement. It should be clear that the definition of the system is of the utmost importance if misinterpretation of the results is to be avoided. In other words, this definition is an integral part of the statement of the energy requirement.

It follows from the above discussion that the calculated energy requirement of any defined system represents the energy of the *system* and not the energy of the products. Questions are commonly posed in relation to a specific product but energy analysis can only evaluate the energy consumed by a system which produces the product. Consequently the questioner must ensure that he adequately specifies the system so that it yields the product of interest.

When a system has been defined and numerical value obtained for the inputs and outputs of materials and energy, calculation of a normalised energy requirement is straightforward as shown by (1.2) and (1.3). For systems yielding only a single product, the system energy may reasonably be regarded as that needed to produce the product. This is as close as energy analysis can ever come to stating the energy requirement of a product. When however systems yield more than one saleable product and detailed monitoring of fuel consumption in different areas of the plant is not carried out, it is necessary to partition the total measured energy consumption between different product streams passing through the system. Such partitioning exercises can be complex and examples occur in the present work. A detailed discussion of the general procedures is given in [1].

1.3 CALCULATING THE ENERGY REQUIREMENT OF A SYSTEM

The overall system describing any industrial operation must be broken down into a series of sub-systems which match physical operations for which data are available. For any sub-system, the energy requirement is the sum of the contributions from **four** sources. These are:

(a) energy directly consumed as fuels,
(b) energy needed to produce these fuels from raw materials in the ground,
(c) energy needed to erect and maintain plant and machinery, and
(d) energy of labour.

Of these (a) and (b) usually account for well over 95% of the total energy requirement associated with any operation and the method of calculating the contributions due to fuel production and consumption is considered in Chapter 2. In this work we have neglected the contributions from (c) and (d); for detailed reasoning, see [1]. Other useful data are given in [3] and [5].

1.4 SOURCES OF DATA

Numerical data are needed to describe both the fuel consumption and the materials throughput of any operation. Such information may be derived from two main sources, published data and factory operators.

Most published energy analyses are based on previously published data — largely Government and Trade statistics. Although in the early years of energy analysis, use of these sources provided considerable information not readily available elesewhere, it is important to recognise the limitations in the calculations. These are:

1. The system must be defined to fit the available data. This limits the choice of systems available for analysis.
2. Much statistical information is recorded in monetary terms and not in terms of physical quantities. The conversion is usually effected by assuming a constant conversion factor from monetary to energy units. It is however seldom possible to ascertain the magnitude of this conversion factor with any accuracy so that significant errors may be introduced.
3. Statistical data are inevitably aggregated to provide manageable tables. It is seldom possible to disaggregate the data and hence it is usually impossible to identify specific products. Instead, these data usually relate to industrial sectors rather than to specific products.
4. There always remains the uncertainty, often unknown, introduced in the collection and processing of the raw data used to construct the statistical tables. Inaccurate and incomplete returns from manufacturers as well as unsuitable methods of scaling up by the compilers of the tables all lead to a reduction in the reliability of the statistics.

Other published information is often free from the more obvious uncertainties of statistical information but unless the analysis is reported in extensive detail, it is seldom useful as a satisfactory source of data. Even when detailed reporting occurs, some caution must be exercised because few industries are likely to permit publication of detailed analyses of current operations for commercial reasons. Consequently such publications tend to report historical or obsolescent processes which are no longer commercially sensitive.

For these reasons, published data are used in this study only when no satisfactory information could be obtained directly from manufacturers themselves, and only then when the contribution to the total system energy is so small that no significant errors are introduced. Wherever such data have been used, it is clearly indicated in the text.

The most satisfactory information is that derived directly from the process operator and this is the approach employed almost exclusively in this work. The major advantages are that it is possible to ascertain accurately the system to which the data refer and to check any anomalies directly with the originator. Most industries experience fluctuations in activity and this is particularly true of the beverage industries where the Christmas and summer peaks produce unusual operating conditions. To smooth out such variations, data have been obtained, wherever possible, for a 12 month period.

1.5 UNITS

The work is reported in S.I. (metric) units employing the kilogramme (kg) as the unit of mass and the metre (m) as the unit of length. For large masses we have used the metric tonne (t) equivalent to 1000 kg.

Energy is measured in joules (J) and to eliminate powers of 10 for industrial energy, the unit used throughout the report is the megajoule (MJ) where $1 \text{ MJ} = 10^6 \text{J}$.

Beverage volumes are commonly measured in fluid ounces (fl. oz.) but to facilitate comparisons, the metric units, the litre (l) and the millilitre (ml) are usually also given (1 fl. oz. = 28.41 ml). Container sizes are frequently referred to in terms of the capacity stated on the label when filled; thus a container of nominal capacity 10 fl. oz. would be referred to as 9.68 fl. oz. when filled with beer.

There are a number of non-metric units in widespread use in the U.K. and it is pointless to convert these to metric units for the sake of uniformity. These have therefore been retained in those instances where the retention is thought likely to permit a more ready understanding of the calculations. Of these non-metric units the most important are:

(a) the mile has been retained as the unit of transport distance,
(b) the imperial gallon has been retained as the measure in which some liquids are sold, and
(c) the therm has been retained as the measure of energy supplied in gas distribution.

CHAPTER 2

Fuels

2.1 PRIMARY AND SECONDARY FUELS

Materials from which energy is derived are usually referred to as fuels, a word which nowadays tends to include forms of energy such as electricity as well as metals such as uranium which produce energy by nuclear reaction. Two quite distinct groups of fuels can be recognised; **primary** and **secondary** fuels. Fig. 2.1 illustrates the distinction. The left hand side of the diagram shows the three main materials inputs extracted from the ground which when burnt in air produce useful heat energy. These are the primary fuels. A primary fuel may therefore be defined as a naturally occurring raw material which can be used as a technologically useful source of energy without modifying its chemical structure prior to the reaction which releases the energy.

The most obvious example of this type of fuel is coal. This is mined and

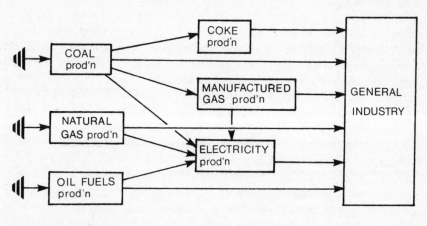

Primary Fuels   Secondary Fuels

Fig. 2.1 – Schematic flow diagram showing the flow of fuels within the fuel producing industries. For clarity not all flows have been shown.

apart from washing, removing stones and possibly crushing and grading by size, it is used in exactly the same chemical form as it is mined. Natural gas likewise is subjected to only minor treatment to remove corrosive impurities before it can be used. In contrast, crude oil is usually subjected to extensive treatment before use (refining), yet it can still be regarded as a primary fuel because the treatment is essentially the physical separation of components of different molecular mass. There is little or no change in the chemical structure of these components. Some workers regard oil as a secondary fuel because of the extensive processing and blending; the distinction is however of no great significance in energy analysis.

In all the above examples of primary fuels, energy is extracted from them by burning in air. That is, the energy release originates from the change which takes place in the chemical structure of the compounds making up the material. However, primary fuels do exist in which the energy release results from a change in some property other than the chemical structure. Uranium, for example, may liberate heat energy by nuclear fission and here it is a change in the structure of the atomic nucleus that is the source of the energy. Similarly, water in a mountain reservoir can also be regarded as a primary fuel; here it is a physical property, the mechanical potential energy, which is the source of energy.

Returning to Fig. 2.1, the three primary fuels shown take one of two routes after extraction and treatment. The most direct route is that leading directly to use by industry and the domestic sector, shown as the block on the right hand side of the diagram. The second route is to one of the three operations collectively referred to as secondary fuel production systems. In these operations, the energy content of the primary fuel is converted to some other form. We can therefore define a secondary fuel as a source of energy which has been derived from a primary fuel.

One very important feature of the fuel producing industries is the interchange of fuels between the different fuel producers. Coal mines, for example, use electricity for lighting and driving machinery and they also use oil fuels for transport. Moreover, a fuel producing industry may also consume part of its own output; electric lights are used in electricity generating stations. Thus the primary and secondary fuel producing industries present an interlinked network.

When a consumer receives a fuel, he is concerned with the maximum energy that he may potentially derive from it, that is the gross calorific value in the case of a combustible fuel. It must be remembered that commercially available fuels are not chemically pure compounds and the exact calorific value will vary from one sample to another. Where standards are laid down, they usually refer to a lower limit for calorific value although the manufacturer attempts to produce a fuel with a calorific value above this minimum in order that his product will not be rejected. However, as a result of this variation in calorific values for fuels of the same type, there will always be small fluctuations in the calculated values for the associated energy from time to time. Over a sufficiently long period however, these effects will be smoothed out so that the average calorific values given in Table 2.1 are reasonable representations of the total energy available per unit mass of fuel.

Table 2.1

Typical measured gross calorific values for a number of fuels and other materials of complex composition.

Material	Gross calorific value (MJ/kg)	Burning medium
Coal	29.5	oxygen
Coke	25.5	oxygen
Heavy fuel oil	42.6	oxygen
Diesel oil	44.8	oxygen
Liquified petroleum gas	50.0	air
Lignite	17.0	oxygen
Rubber	27.5	oxygen
Garbage (average)	6.0	oxygen

When an industrial operator uses a fuel, the gross calorific value represents the energy that is potentially available to him. The fact that he may not use the whole of this energy efficiently is an inherent problem of the operation but by taking in the fuel, he has essentially deprived any other user of its use. All of the fuel intake must therefore be attributed to the operation and not just that fraction which survives conversion to useful work. The energy associated with the consumption of fuel by a process is known as the **direct energy consumption** of the operation and is equal to the energy content of the fuel. From the above observation it should be clear that the direct energy consumption of an operation can be decreased by increasing the efficiency with which the available energy in the fuel is converted to useful work. The idea is usually summarised in an **overall thermal efficiency**, η, for the plant, defined as the ratio:

$$\eta = \frac{\text{Work output}}{\text{Fuel energy supplied}} . \tag{2.1}$$

2.2 ENERGY TO PRODUCE A FUEL

Additional energy must be expended to produce fuels in a usable form and to deliver them to a consumer. A fuel producing industry can be represented by a simple system of the type shown in Fig. 2.2 where all inputs and outputs are measured in terms of energy. The fuel input E_i and the energy required to process it, E_p, give an energy E_o at the consumer. As a consequence of the

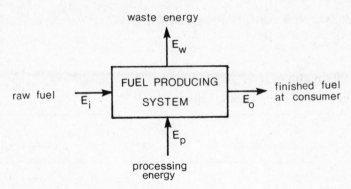

Fig. 2.2 – Schematic representation of a fuel producing operation with inputs and outputs represented in terms of energy.

second law of thermodynamics, there will also be a waste energy output E_w. The advantage of this simple representation is that it allows definition of an overall efficiency, η_e, for the production of the fuel as:

$$\eta_e = \frac{\text{Total delivered energy available to consumer}}{\text{Total input energy}} \qquad (2.2)$$

Using the parameters defined in Fig. 2.2, this gives:

$$\eta_e = \frac{E_o}{E_i + E_p} \,. \qquad (2.3)$$

The definition of an efficiency in this manner eliminates the need to consider further the parameter E_w, the waste energy, which is seldom known. The use of an energy efficiency of production is therefore convenient for expressing production energies for fuels.

For a secondary fuel, the overall production system is a two stage process. The first is the production of the primary fuel with an energy efficiency η_p and the second stage is the conversion of this primary fuel into the secondary fuel with an energy efficiency of conversion of η_s. If the overall system has an energy efficiency η_e, then

$$\eta_e = \eta_p \cdot \eta_s \,. \qquad (2.4)$$

Since both η_p and η_s will be less than unity, it follows that the overall production efficiency will always be less than both η_p and η_s alone. Hence the energy efficiency of production of a secondary fuel will always be less than that of the primary fuel from which it is made.

The calculation of η_e is central to energy analysis since all subsequent computations require a knowledge of this parameter. However, its evaluation is not simple and a large number of values exist in the published literature, as is discussed in some detail in [1].

Despite these problems, suitable values for production efficiencies of fuels are needed for energy analysis and the inevitable problem is that of choosing from the available range. Those selected for use throughout this work are given in Table 2.2 and the basis of the calculations leading to them are given in [1] and [6] to [13].

2.3 REPRESENTATION OF ENERGY REQUIREMENTS

The representation of the energy requirements of a system are best illustrated by a simple numerical example. Suppose that the production of 1 kg of a product requires the consumption of:

1 gallon of heavy fuel oil
1 gallon of diesel oil
1 kWh of electricity
1 therm of natural gas

Because the materials output of the system is 1 kg of product, these fuel inputs also represent the fuel requirements per unit mass output of the system, that is they are already normalised. Using the data of Table 2.2, these fuel requirements can be converted to energy requirements to give Table 2.3.

In this instance the detailed energy requirements of the system are described by the 15 entries in Table 2.3. However, in real systems there could be up to 16 fuel inputs (see Table 2.2) leading to a table containing 51 entries. To decide whether all of these numbers need separate identification consider the usefulness of the tabulated parameters.

One function of energy analysis, is the description of the total resource requirements of a process. In Table 2.3, the total of column 4 (559.10) provides such a description. This is a measure of the total energy that must be extracted from the ground to process 1 kg of material in the defined system.

The total of column 2 (466.78) is of immediate use to the plant operator since this represents the total fuel energy actually used within the processing plant. Irrespective of the mix of fuels employed, this is the energy directly consumed within the system and is therefore directly within his control. In the special case of the system defined to include only a single processing operation, this energy requirement may be used to compare plant performance with similar operations within the factory or with those of a competitor.

The entries in column 1 of Table 2.3 are derived from those of column 2 using the different fuel production efficiencies in Table 2.2. Because a variety of values exist for the production efficiency of any fuel particularly on an international basis, it is important to present the results of energy analysis in a form such that they may be readily revised to take account of different fuel

Table 2.2

Typical values for the total energy associated with fuels

Fuel	Quantity	Production and delivery energy/MJ	Energy content of fuel /MJ	Total energy /MJ	Energy efficiency of production /%
Coal	1 kg	1.39	28.01	29.40	95.0
Coke	1 kg	3.93	25.42	29.35	86.6
Electricity (grid)	1 kWh	10.80	3.60	14.40	25.0
Natural gas	1 therm	8.67	105.44	114.11	92.4
Manufactured gas	1 therm	41.21	105.44	146.65	71.9
Heavy fuel oil	1 imp gall	38.97	186.30	225.27	82.7
Medium fuel oil	1 imp gall	38.64	186.02	224.66	82.8
Light fuel oil	1 imp gall	37.68	182.66	220.34	82.9
Gas oil	1 imp gall	33.75	172.02	205.77	83.6
Kerosine	1 imp gall	31.63	166.07	197.70	84.0
Diesel	1 imp gall	33.88	171.44	205.32	83.5
Petrol (gasoline)	1 imp gall	31.15	163.53	194.68	84.0
LPG (propane)	1 kg	8.89	50.00	58.89	84.9
LPG (butane)	1 kg	8.89	49.30	58.19	84.7
Lubricating oil	1 imp gall	37.68	182.66	220.34	82.9
Grease	1 kg	8.89	42.60	51.49	82.7

Table 2.3

Energy requirements per unit output for the hypothetical system given in the text

Fuel type	Production and delivery energy/MJ	Energy content of fuel/MJ	Feedstock energy /MJ	Total energy /MJ
Heavy fuel oil	38.97	186.30	nil	225.27
Diesel oil	33.88	171.44	nil	205.32
Electricity	10.80	3.60	nil	14.40
Natural gas	8.67	105.44	nil	114.11
Totals/MJ	92.32	466.78	nil	559.10

production efficiencies. This implies that columns 1 and 2 in Table 2.3 be retained and moreover, that the different horizontal rows also be retained. However, the number of rows can be reduced because for most practical purposes, fuels can be divided into three groups: electricity, oil fuels and other fuels.

Electricity needs separate identification because, with very few exceptions, it is consumed by all industrial processes to a greater or lesser extent. It is also the fuel with the lowest production efficiency and is the one fuel most likely to experience a marked change in production efficiency in the foreseeable future. For these reasons it is separately identified by almost all energy analysts. The primary fuel used for electricity generation can be readily obtained from published sources [59].

All oil fuels are derived from the same source, crude oil. This is a fuel of special political sensitivity and, as can be seen from Table 2.2, the production efficiencies for the different types of oil fuels, all lie within a narrow well defined band. Consequently for most applications an average production efficiency of 83% can be assumed without introducing any serious errors into the calculations. For these reasons there is a strong case for keeping oil fuels separate in the results tables.

Of the remaining fuels: natural gas, manufactured gas, coal and coke, the last three are closely linked since both coke and manufactured gas are produced from coal. Furthermore, natural gas and manufactured gas, when used as fuels, are usually interchangeable. It is therefore reasonable to group these fuels together. Unlike oil fuels however, they have markedly different production efficiencies and so, although the energies associated with their use may be aggregated in the final table, calculations must be performed on the consumption of the individual fuel types before aggregation.

Adopting these procedures, Table 2.3 for example can be rewritten as shown in Table 2.4.

Table 2.4

Table 2.3 rewritten with the fuels arranged into three groups

Fuel type	Production and delivery energy/MJ	Energy content of fuel/MJ	Feedstock energy /MJ	Total energy /MJ
Electricity	10.80	3.60	nil	14.40
Oil fuels	72.85	357.74	nil	430.59
Other fuels	8.67	105.44	nil	114.11
Totals/MJ	92.32	466.78	nil	559.10

2.4 FEEDSTOCK ENERGY

Tables 2.3 and 2.4 contain a column labelled 'Feedstock' which has not so far been discussed. The materials inputs to any operation are often referred to as **feedstocks** especially in those industries which use organic materials. These materials, which include oil or natural gas feedstocks for the petrochemical industry or wood in paper making, are also conventional fuels. Consequently if the calculated energy requirement of a system is to describe the *total* energy that must be taken from the ground to support the system then it must include an energy contribution which reflects this use of potential fuels as materials. This is the reason for separate identification of the 'Feedstock' column.

A proportion of the feedstock material entering a processing system will eventually form part of the saleable output. The balance will appear as processing losses. Consequently the feedstock energy requirement of a system must take account of these losses and therefore will be the energy content of the *input* materials. This is the value appearing in the Tables. That proportion of the input material which appears in the product is often referred to as the **rolled-up feedstock energy**. Unlike fuels which are taken into the system and consumed, the rolled-up feedstock energy is potentially reclaimable, for example by burning paper or plastic. Although there is a clear connection between feedstock energy and the fuel content of the output products, they are not the same. A more detailed discussion is given in [1].

Table 2.5
Typical values of feedstock energies for some raw materials

Feedstock	Typical gross calorific values (MJ/kg)
Coal	28.5
Oil	44.5
Natural gas	50.0
Wood	12.5
Rubber	27.5

CHAPTER 3

Transport

3.1 ROAD TRANSPORT

Road transport is the major means of bulk commodity movement in the U.K. Road vehicles for long haul (trunking) movements and for local distribution are available in a variety of sizes ranging from small petrol driven pick-ups to large diesel powered articulated vehicles, with gross weights exceeding 32 t. Not surprisingly, the energy requirements of these different vehicles vary considerably.

The energy requirements for road transport can be considered as the sum of the energy requirements of a three component system. The main contributor is the fuel directly consumed by the vehicle on its journey with two other subsystems responsible for (a) the construction and maintenance of the vehicle and (b) the construction and maintenance of the roads. The energy requirement associated with the fuel consumption comprises some 60% of the total, construction and maintenance of vehicles has been estimated as a further 32% and construction and maintenance of routes as 7% [17]. The inclusion of road construction and maintenance energies has been much debated [18] and this contribution has been excluded from the calculations in this report.

Detailed calculations of road transport energies have been presented elsewhere [1] based on industrial data and published information [11, 14-21] and Table 3.1 summarises the energy requirements for the main types of vehicles employed in the U.K. when operating with maximum payloads. In general, when a journey is made but no load is carried, the requirements of Table 3.1 may be reduced by a factor of 0.7 [1, 19].

Table 3.1

Total energy per vehicle-mile for road vehicles of various types operating in the U.K. under full load conditions

Vehicle type	Electricity/MJ		Oil fuels/MJ			Other fuels/MJ			Total energy /MJ
	Fuel production and delivery	Energy content of fuel	Fuel production and delivery	Energy content of fuel	Feedstock energy	Fuel production and delivery	Energy content of fuel	Feedstock energy	
Less than 1 t rigid	0.57	0.19	2.20	11.58	0.02	0.07	0.84	—	15.47
1 – 2 t rigid	0.69	0.23	2.59	13.42	0.02	0.09	1.06	—	18.10
3 t rigid	0.72	0.24	2.88	15.06	0.03	0.10	1.27	—	20.30
4 t rigid	0.78	0.26	3.22	16.59	0.03	0.12	1.46	—	22.46
5 – 8 t rigid	0.87	0.29	3.72	19.10	0.03	0.15	1.77	—	25.93
9 t rigid	0.93	0.31	4.04	10.63	0.05	0.16	1.96	—	28.08
10 – 12 t rigid	0.93	0.31	4.12	21.05	0.05	0.17	2.01	—	28.64
13 – 20 t rigid	0.96	0.32	4.34	22.27	0.05	0.18	2.17	—	30.29
less than 10 t articulated	0.93	0.31	4.07	20.85	0.05	0.16	2.00	—	28.37
10 – 12 t articulated	0.96	0.32	4.23	21.67	0.05	0.17	2.09	—	29.49
13 – 14 t articulated	1.11	0.37	5.04	25.63	0.05	0.21	2.59	—	35.00
15 – 16 t articulated	1.11	0.37	5.08	25.87	0.05	0.21	2.61	—	35.30
17 – 18 t articulated	1.11	0.37	5.26	26.73	0.06	0.22	2.72	—	36.47
greater than 18 t articulated	1.29	0.43	6.37	32.44	0.08	0.28	3.43	—	44.32

3.2 RAIL TRANSPORT

Unlike road transport, the U.K. rail system is an integrated network so that goods fed into the rail system may follow a variety of routes to a given destination since the aim is to move maximum loads. Consequently the most useful data relate to the operation of the whole network apportioned between the volume of different types of traffic using the network [23, 24]. Detailed calculations of rail transport energies have been given in [1] and lead to the two sets of energy requirements given in Tables 3.2 and 3.3.

Table 3.2

Total energy required by general rail freight per tonne mile

Fuel type	Fuel production and delivery energy/MJ	Energy content of fuel /MJ	Feedstock energy /MJ	Total energy /MJ
Electricity	0.78	0.26	nil	1.04
Oil fuels	0.07	0.39	nil	0.46
Other fuels	nil	nil	nil	nil
Totals/MJ	0.85	0.65	nil	1.50

Table 3.3

Total energy required by circuit working freight trains per tonne mile

Fuel type	Fuel production and delivery energy/MJ	Energy content of fuel /MJ	Feedstock energy /MJ	Total energy /MJ
Electricity	0.12	0.04	nil	0.16
Oil fuels	0.16	0.77	nil	0.93
Other fuels	nil	nil	nil	nil
Totals/MJ	0.28	0.81	nil	1.09

3.3 SEA TRANSPORT

The energy associated with sea transport has also been discussed in detail elsewhere [1, 25] and this work uses the average requirements shown in Table 3.4.

Table 3.4
Total average energy required per tonne-mile for sea transport

Fuel type	Fuel production and delivery energy/MJ	Energy content of fuel /MJ	Feedstock energy /MJ	Total energy /MJ
Electricity	nil	nil	nil	nil
Oil fuels	0.03	0.14	nil	0.17
Other fuels	nil	nil	nil	nil
Totals/MJ	0.03	0.14	nil	0.17

CHAPTER 4

Beverage Systems

4.1 THE GENERAL BEVERAGE SYSTEM

The aim of this work is the evaluation of the energy and raw materials requirements of an overall system such as that shown in Fig. 4.1. It is however useful
to examine the nature of this system because it gives rise to some potentially
confusing aspects when the normalising procedure is not properly understood.

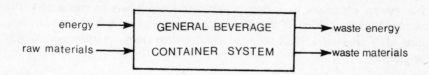

Fig. 4.1 – The general system describing the overall resource requirements of any
beverage container system.

The most important feature of the system shown in Fig. 4.1 is that there
is no readily identifiable normalising parameter since all ouputs are waste materials.
Consequently some alternative method must be devised for choosing a suitable
normalising parameter and two possible alternatives are shown in Fig. 4.2.

In the first method, (Fig. 4.2(a)), the **beverage** input is included and the
system is normalised with respect to this input. This is justified since the principal aim of the beverage container system is to assist the flow of this beverage
from the producer, who is external to the system of Fig. 4.1, to the consumer,
who is a sub-system within the overall system defined by Fig. 4.1. Hence normalisation of the resource requirements of the general beverage container system
with respect to this flow provides a measure of the resource efficiency of the
system. If an energy E is required to pack volume V of beverage, then the
normalised system energy E_s will be:

$$E_s = E/V .$$

(4.1)

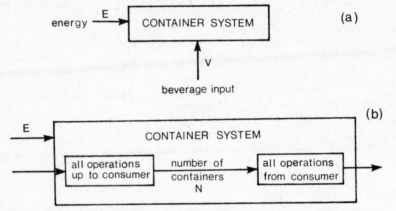

Fig. 4.2 — Alternative methods for normalising the energy requirements of the general beverage container system. See text for detailed explanation.

The second method of normalising the system energy of Fig. 4.1 is to divide the operation into two sub-systems as shown in Fig. 4.2(b). The first includes all operations up to the point at which the consumer takes possession of the container and the second all operations from consumer to disposal. These two sub-systems are linked by a flow of containers which may be measured as the number of units passing. If therefore a total energy input E is responsible for a flow of N containers, then the normalised system energy requirement, E_s', will be given by:

$$E_s' = E/N .$$ (4.2)

In general, E_s and E_s' will be different but they are related by the capacity of the container, v. Hence

$$V = N \cdot v$$ (4.3)

so that $$E_s' = v \cdot E_s .$$ (4.4)

It is of no significance whether system energies are calculated using volume of beverage delivered or number of containers delivered since the two are interchangeable. It is however important to note that when the energy requirements of two systems are compared, the comparison is only valid if the same normalising parameter is used for both systems.

4.2 PRACTICAL CONTAINER SYSTEMS FOR BEVERAGES

To analyse the energy requirements of systems such as that in Fig. 4.1 they must be divided into sub-systems for which operating data are available, as shown in Fig. 4.3 for the glass bottle system. Similarly, Figs. 4.4 and 4.5 define the corresponding systems for metal cans and PET bottles. These systems are discussed later in more detail.

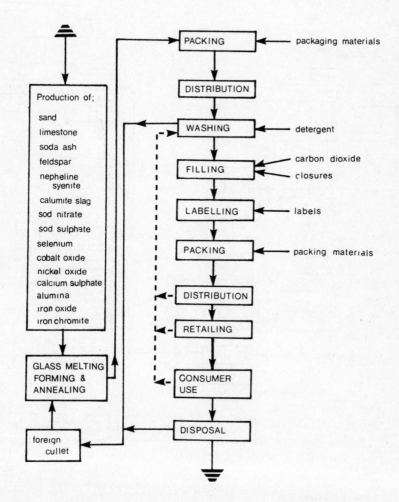

Fig. 4.3 – Schematic flow diagram of the materials flows within the glass bottle beverage system.

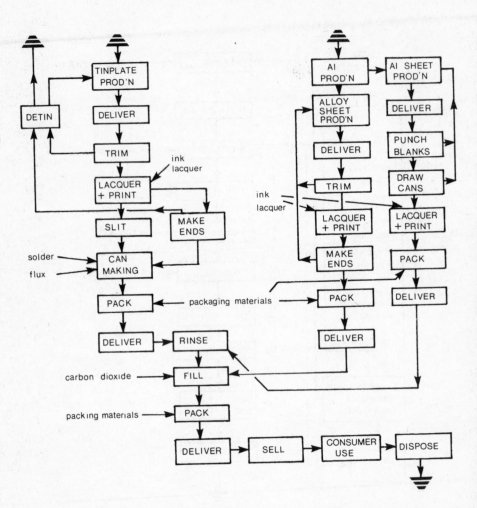

Fig. 4.4 – Schematic flow diagram of the operations needed to produce and use metal cans in beverage packaging.

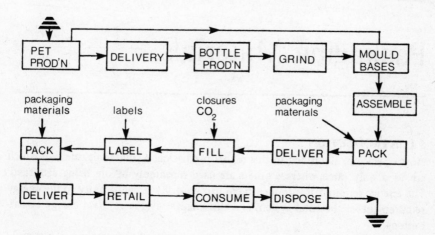

Fig. 4.5 – Schematic flow diagram of the operations needed to produce and use the PET bottle. This system assumes bases made from PET.

CHAPTER 5

Packaging Materials

5.1 INTRODUCTION

Most operations described in this report use packaging materials, some of which are used only once whereas others are used repeatedly before being discarded. The energy required to produce them can contribute significantly to the energy requirement of the total system and this chapter evaluates these energy contributions.

The major problem is that almost every commodity is packed in a slightly different way and it is impossible to take account of every variation. Products have therefore been chosen which are thought to be typical of the range in use. For example, manufacturers use pallets made to their own specifications so that overall pallet masses may lie in the range 10 kg to 40 kg. Most manufacturers however use pallets with overall masses close to 25 kg and this is chosen as typical.

A second problem arises in the use of returnable components of packaging since it is then necessary to estimate the number of trips made before they are 'lost' from the system. Few users maintain detailed records of the numbers of trips made by returnable packaging materials and often only guesses are available. The magnitude of any errors introduced as a result of these uncertainties becomes clear when the energy associated with the various packaging materials has been ascertained.

5.2 ENERGY REQUIREMENTS OF PACKAGING MATERIALS

This section presents data for the gross energy required to produce packaging materials of various types from raw materials in the ground. To illustrate the method used, detailed calculations have been given for the production of pallets. The data for the other materials are then given in summary form with notes explaining the basis of the calculations.

5.2.1 Pallets

A variety of types are in use and one typical specification is a wooden pallet of total finished mass 25.40 kg containing 17.91 kg soft wood, 6.33 kg hardwood and 1.16 kg of steel nails and wire. The energy required to construct such a pallet is shown in Table 5.1 and is based on the information outlined below.

Table 5.1

Calculation of the gross energy required to produce one pallet of total mass 25.40 kg of specification as detailed in the text

Operation	Electricity/MJ		Oil fuels/MJ			Other fuels/MJ			Total energy /MJ
	Fuel production and delivery	Energy content of fuel	Fuel production and delivery	Energy content of fuel	Feedstock energy	Fuel production and delivery	Energy content of fuel	Feedstock energy	
Production of 18.81 kg of softwood	nil	nil	25.58	124.90	nil	nil	nil	nil	150.48
Sea transport of softwood	nil	nil	0.56	2.63	nil	nil	nil	nil	3.19
Road transport of softwood	0.18	0.06	0.90	4.58	0.01	0.04	0.48	nil	6.25
Production of 6.65 kg of hardwood	nil	nil	9.04	44.16	nil	nil	nil	nil	53.20
Wood feedstock energy	nil	nil	nil	nil	nil	nil	nil	437.91	437.91
Production of steel products	7.47	2.49	0.81	3.82	nil	1.64	19.96	nil	36.19
Delivery of steel products	0.02	neg	0.07	0.38	neg	0.01	0.04	nil	0.52
Sawing and make-up	314.22	104.74	nil	nil	nil	nil	nil	nil	418.96
Total energy to produce one pallet	321.89	107.29	36.96	180.47	0.01	1.69	20.48	437.91	1106.70

The energy used in lumber operations is very variable and the published literature gives a range of energies from 3.33 MJ/kg to 12.59 MJ/kg for the production of raw wood from standing timber [54–56]. An average value of 8 MJ/kg has been chosen as reasonable and this is assumed to be all oil fuels. There is typically a 5% loss of wood during manufacture. Hence 18.81 kg of softwood must be supplied for the production of a single pallet.

The softwood most widely used in pallet making is Portugese pine imported by sea an average distance of 1000 miles. Hence the energy required to transport 18.81 kg of pine can be derived from Table 3.4. Within the U.K. the wood is transported by road. Assuming an average round trip delivery distance of 150 miles on fully loaded 20 tonne vehicles, the energy associated with this transport may be calculated from Table 3.1.

The hardwood used in pallet making is usually home grown. Again assuming that the provision of this wood requires an energy of 8 MJ/kg and that there is again a 5% loss during manufacture, the energy required to supply the wood can be readily calculated. Assuming an average round trip delivery distance of 50 miles for delivery using fully loaded 20 tonne lorries, the delivery energy may be obtained from Table 3.1.

The fuel content (feedstock energy) of the wood depends upon the moisture content and ranges from 19.75 MJ/kg for dry wood to less than 16 MJ/kg as the moisture content increases [38]. With a 12% moisture content, the fuel value is 17.2 MJ/kg and this is the value used here. Making a 5% allowance for losses during manufacture, the feedstock energy can be readily calculated from the mass of wood used.

A variety of steel products are used in pallet construction. The energy required to produce these products may be calculated using the data for steel production given in Chapter 8. For the delivery of steel products, a round trip delivery distance of 200 miles with full loads on 20 tonne lorries is assumed so that Table 3.1 can be used directly.

Published data [11] have been used for sawing and making up and the total energy associated with the production of one pallet is therefore the sum of all the above contributions as shown in Table 5.1.

Most companies regard pallets as returnable so the energy attributable to products carried on them depends upon the number of trips made by pallets before they 'disappear' from the system. From available information it is impossible to determine accurately the number of trips made by a pallet because pallets 'vanish' for a variety of reasons. For example, some firms use pallets but do not buy them; others repair broken pallets and yet others scrap pallets as soon as they are defective in any way. In view of these variations in practice, an average overall pallet lifetime of 10 trips has been assumed.

5.2.2 Paper and board
The energy needed to produce finished paper and board can be separated into

the energy to manufacture pulp and the energy to convert the pulp into paper
or board. Fuel requirements vary depending upon the type and quality of
product, the type of raw materials used as well as on the type and size of mill.
Integrated pulp and paper mills are the least energy intensive because part of
the steam required in paper-making can be produced from the waste products
of the pulping process. Much of the lower quality paper and board used in
general packaging, is made overseas in such plants and imported. This achieves
energy savings of the order of 10 to 20%.

Rather than attempt a detailed breakdown of the paper industry by paper
type, an average energy, thought to be typical of current practice, has been
evaluated. It is recognised that the energy used in paper making operations
depends upon the *area* of the paper produced rather than the *mass*. By choosing
typical packaging papers, the average has been related to mass of product.

The total energy required to produce paper can be evaluated by considering
in turn the operations involved from standing timber to finished paper. The
provision of timber has been based on the value of 8 MJ/kg quoted earlier.

Pulp may be produced either mechanically or chemically; the mechanical
method is the more energy intensive. Typical figures based on mill practice
have been published [40, 41] and both give total energy requirements for pulp
production lying in the range 29 to 31 MJ/kg. The value used here is 28.59
MJ/kg.

As already mentioned, the production energy of paper from pulp is sensitive
to the type of finished product. We have therefore obtained separate average
figures for the energy to manufacture.

   (a) a base paper as used for simple wrapping operations,
   (b) a coated paper as used for printed packaging materials,
   (c) solid board as used in the construction of cartons.

In addition to the energy directly used to carry out the processing, feedstock
energy is also involved because wood feedstock is a fuel. It may be argued
that wood feedstock energy should be excluded since it does not constitute a
recognised fuel in the industrialised world and because it is a renewable resource.
However, in integrated plants, the use of waste products from pulping as fuel
in paper making suggests that this argument is not valid.

Paper and board production both incur 2% fibre loss and an allowance
must be made for this in the calculations.

It is assumed that all wood and paper products are imported into the U.K.
by sea with an average delivery distance of 1000 miles. The energies required
to produce paper, board and some of the intermediate products are shown in
Table 5.2.

Table 5.2

Energy required to produce selected packaging materials and some intermediate products. For details of the assumptions made in the calculations, see text. Values include delivery inside the U.K.

Product and input	Electricity/MJ		Oil fuels/MJ			Other fuels/MJ			Total energy /MJ
	Fuel production and delivery	Energy content of fuel	Fuel production and delivery	Energy content of fuel	Feedstock energy	Fuel production and delivery	Energy content of fuel	Feedstock energy	
1 kg wood from standing timber	–	–	1.36	6.64	–	–	–	17.20	25.20
1 kg pulp from cut timber	9.18	3.06	–	–	–	0.05	16.30	–	28.59
1 kg base paper from standing paper	19.23	6.41	4.03	19.67	–	0.05	16.63	17.89	83.91
1 kg coated paper from standing timber	23.22	7.74	5.85	28.57	–	0.05	16.63	17.89	99.95
1 kg board from standing timber	28.05	9.35	5.25	25.63	–	0.05	16.63	17.89	102.85
1 kg naphtha from crude oil	–	–	1.30	6.35	45.00	–	–	–	52.65
1 kg ethylene from naphtha	0.66	0.22	3.75	18.31	–	–	–	–	22.94
1 kg polyethylene (L.D.) from ethylene	7.56	2.52	1.48	9.45	–	–	–	–	21.01
1 kg polyethylene (L.D.) from crude oil	8.28	2.76	7.08	36.82	49.95	–	–	–	104.89
1 kg polyethylene (H.D.) from crude oil	8.79	2.93	7.28	37.91	50.85	–	–	–	107.76
1 kg polypropylene from crude oil	10.02	3.34	7.72	40.39	52.65	–	–	–	114.12
1 kg polyethylene film from oil	48.33	16.11	12.07	61.26	50.95	–	–	–	188.72
1000 hole web of Hicone carrier from crude oil	31.23	10.41	7.38	38.27	47.04	–	–	–	134.33

5.2.3 Polyethylene

The production of polyethylene resin may be regarded as a three stage operation; (a) all operations from crude oil in the ground up to the production of naphtha by fractionation, (b) cracking of naphtha to produce ethylene and (c) the polymerisation of ethylene to produce polyethylene. Data from different plants show variations and although the values presented here are averages they are thought to be a reasonable description of current practice.

The calorific value of crude oil has been assumed to be 45 MJ/kg and the production efficiency of crude oil fractions from crude oil in the ground is 83% (measured as energy inputs and outputs – see Chapter 2). Hence the energy required to produce naphtha will be shown in Table 5.2.

Naphtha is cracked to ethylene by heating for a short time to a high temperature and then quenching. The output mixture of gases, one of which is ethylene, are then separated. All products are used within the petrochemical complex so the energy consumed by the cracking plant may be apportioned amongst them all. Table 5.2 shows the energy associated with the production of ethylene from naphtha by cracking.

Ethylene polymerisation is carried out by two distinct processes. One operates at high pressure and gives a branched, low density polymer which is widely used as packaging film. The other operates at low pressure and gives a high density product due to the greatly reduced side branches in the polymer molecules. For high density polyethylene, the feedstock requirements are some 1.8% higher than for low density polyethylene and the conversion energy is 6.2% higher [39]. Typical energy requirements for the production of both high and low density polyethylene are given in Table 5.2.

5.2.4 Polypropylene

Propylene is a co-product of ethylene cracking and the route to polypropylene follows much the same processing as high density polyethylene. The major difference is that the feedstock energy requirement for polypropylene production is 5.4% higher than for polyethylene production and the conversion energy is some 22% higher. Table 5.2 gives the overall energy requirement.

5.2.5 Polyethylene film

The commonest use of low density polyethylene in packaging is as film. The film producing process is very efficient in terms of materials conversion efficiency and total losses are of the order of 2% or less of the input polymer. Typical energy requirements for the production of film are given in Table 5.2.

5.2.6 Sacks

Low density polyethylene and paper sacks are both widely used in packaging. In general, the energy required to manufacture a sack from sheet material is negligible compared with the energy to produce the sheet in the first instance.

Therefore the energy associated with the production of sacks from raw materials in the ground may be calculated as the mass of the sack multiplied by the energy required to produce paper or polyethylene film from raw materials in the ground.

5.2.7 Hicone carrier

Hicone[†] carrier is a low density polyethylene web made from sheet in which circular holes have been punched and is widely used to hold filled metal cans in packs of 4 or 6. It is produced as a continuous roll, two holes wide, and is manufactured by extruding a continuous polymer sheet and stamping out the required design. Virtually all of the material stamped out of the sheet is recycled so that a 1000 hole web requires typically 0.941 kg of polymer resin. The production energy requirement is shown in Table 5.2.

†Hicone is the registered trade mark of I.T.W. Ltd.

CHAPTER 6

Glass Making Materials and Related Products

6.1 INTRODUCTION

Commercial silica glass is a complex mixture of minerals fused together at a temperature of about 1100 K. Over 97% of the mass of the glass is made up of the oxides of silicon, sodium and calcium with the remainder consisting of a number of minor additives used to impart colour to the glass or to improve the processing characteristics of the liquid mixture.

The basic raw materials are sand (SiO_2), limestone ($CaCO_3$) and sodium carbonate (Na_2CO_3 referred to in the glass industry as soda ash). In commercial glass making the furnace feed usually contains approximately 20% preformed glass (cullet) to improve the heat conducting properties of the powder during the initial stages of melting. Currently most of this cullet is derived from in-house waste with only small quantities being brought in from external sources (usually referred to as foreign cullet).

At the furnace temperature, both sodium carbonate and calcium carbonate decompose to the oxides with the evolution of carbon dioxide gas. As a consequence there is a net loss of mass of the input materials; for soda ash the loss is 42% and for limestone, 44%. In addition, there may be a loss of up to 5% in the mass of sand used because of the presence of residual moisture. The gases evolved during these processes are useful however since their movement stirs the contents of the furnace to give a homogeneous melt.

In addition to the major glass making ingredients mentioned above, a number of other materials are added in very small quantities to impart colour, improve melting characteristics or remove fine gas bubbles (refining agents). Since these minor glass making ingredients are present at very low concentrations significant errors in the calculated energy requirements of their production can be tolerated without affecting the total energy of glass making. For this reason, notional values for their energy of production will often suffice.

6.2 SAND

Silica sand, the principal component of commercial glass, occurs naturally in a relatively pure form in many locations in the U.K. The suitability of a specific sand deposit for glass manufacture depends, to a great extent, upon the concentration of impurities, particularly iron, which imparts undesirable colour especially to white flint glass. Suitable deposits are quarried by conventional techniques and subsequent treatment is directly related to the level of impurities present in the raw sand and the level acceptable to the glass maker. In general, all sands are usually subjected to a water wash to remove soluble impurities. A small quantity of very high purity sand is produced by hot acid treatment but this is used only where very high quality white flint glass production is demanded.

After treatment, wet sand is normally filtered to reduce the water content to approximately 5% by mass and this is adequate for most glass making purposes since residual moisture is readily removed in the early stages of melting. Some glass making sand is dried completely by heating, but the process is energy intensive and expensive. Moreover, if heat-dried sand is stored for prolonged periods, it tends to reabsorb moisture from the atmosphere. It is clear from discussions with glass makers that opinion seems to be divided over the need to use heat-dried sand.

The production of sand from four different locations has been examined. Each operation produces a wet sand (5% moisture) and a dry sand (heat-dried). The production energy requirements are given in Table 6.1. Note that the production of heat dried sand requires an additional 29% to 38% energy compared with that for wet sand production.

6.3 LIMESTONE

Limestone deposits occur widely in England and Wales but those used for glass making are situated in Derbyshire and North Wales. Most limestone is quarried but a small proportion is mined, with the mined rock being the more pure. The quarried or mined stone is subjected to crushing, grinding and drying. In some instances comminution and drying are carried out simultaneously in a fluidised bed. The resultant product is graded by screening and all of the output from a quarry or mine usually finds a use, ranging from argicultural applications (fines) through to road building (waste rock and oversize limestone). Typical values for the energy required to produce limestone are given in Table 6.2, with the higher value relating to mined rock.

6.4 SODA ASH

All soda ash used in the U.K. is synthesised from limestone and common salt. This situation differs from that in the U.S.A. where a significant proportion is

Table 6.1

Energy required to produce 1 kg of sand from four U.K. quarries. Wet sand refers to that containing 5% moisture. Dry sand refers to heat dried sand. The data for Quarry 4 is thought to be the most typical of general practice.

Product and source	Electricity/MJ		Oil fuels/MJ			Other fuels/MJ			Total energy /MJ
	Fuel production and delivery	Energy content of fuel	Fuel production and delivery	Energy content of fuel	Feedstock energy	Fuel production and delivery	Energy content of fuel	Feedstock energy	
Dry sand from									
Quarry 1	1.08	0.36	0.09	0.44	–	–	–	–	1.97
Quarry 2	0.27	0.09	0.03	0.17	–	–	–	–	0.56
Quarry 3	0.09	0.03	0.04	0.20	–	–	–	–	0.36
Quarry 4	0.15	0.05	0.07	0.32	–	–	–	–	0.59
Wet sand from									
Quarry 1	0.81	0.27	0.07	0.33	–	–	–	–	1.48
Quarry 2	0.21	0.07	0.03	0.13	–	–	–	–	0.44
Quarry 3	0.06	0.02	0.03	0.15	–	–	–	–	0.26
Quarry 4	0.12	0.04	0.05	0.23	–	–	–	–	0.44

Table 6.2

Energy required to produce 1 kg of glass making materials and related products from raw materials in the ground

Product	Electricity/MJ		Oil fuels/MJ			Other fuels/MJ			Total energy /MJ
	Fuel production and delivery	Energy content of fuel	Fuel production and delivery	Energy content of fuel	Feedstock energy	Fuel production and delivery	Energy content of fuel	Feedstock energy	
Dried limestone (mined)	0.15	0.05	0.07	0.35	–	–	–	–	0.62
Dried limestone (quarried)	0.12	0.04	0.04	0.20	–	–	–	–	0.40
Soda ash (synthetic)	0.23	0.08	1.60	8.20	–	0.26	3.86	–	14.23
Feldspar or Nepheline syenite	1.17	0.39	0.16	0.78	–	–	–	–	2.50
Calumite Brand Slag	0.42	0.14	0.16	0.77	–	–	–	–	1.49
Sodium nitrate	–	–	0.32	1.63	–	–	–	–	1.95
Sodium sulphate	0.81	0.27	0.61	2.99	–	–	–	–	4.68
Calcium sulphate	0.54	0.18	neg	0.01	–	–	–	–	0.72
Selenium	106.80	35.60	49.82	238.73	–	2.79	33.91	–	467.65
Iron chromite	1.17	0.39	0.16	0.78	–	–	–	–	2.50
Cobalt oxide/nickel oxide	0.90	0.30	0.75	3.65	–	–	–	–	5.60
Foreign cullet	0.02	0.01	0.14	0.74	neg	neg	0.05	–	0.96
Lime (calcium oxide)	0.54	0.18	0.10	0.53	–	0.37	4.46	–	6.18

derived from the naturally occurring sodium sesquicarbonate. The overall reaction for synthetic soda ash production may be written:

$$2NaCl + CaCO_3 = Na_2CO_3 + CaCl_2 ,$$

and typical process requirements are 1.6 tonne of sodium chloride and 1.3 tonne of limestone per tonne of sodium carbonate produced. For sodium chloride supplied as a saturated brine solution, this implies a brine requirement of 0.7 gallons per kilogramme of final product.

The manufacture of synthetic sodium carbonate is carried out in a plant in which high pressure steam is used to generate the electrical requirements of the plant as well as supply the process steam requirement. No electricity therefore appears in the energy table describing that plant.

The total energy required to produce soda ash from raw materials in the ground is shown in Table 6.2.

6.5 FELDSPAR AND NEPHELINE SYENITE

Feldspar is a naturally occurring mineral consisting chiefly of the alumino-silicates of sodium and potassium. The principal source is pegmatite ores in Finland where it occurs in combination with other minerals. A typical composition of such an ore is 50% feldspar, 10% mica, 20% quartz and 20% other minerals. Of these, mica, feldspar and quartz are saleable. The processing of the quarried rock is therefore a sequence of comminution and separation stages. If the known fuel requirements of each stage are partitioned between the masses of the saleable products through that stage and the energy associated with the very small amounts of chemicals that are used is ignored, the overall energy required to produce 1 kg of feldspar is as shown in Table 6.2. This includes an allowance for the energy required to supply water.

Nepheline syenite is a similar mineral quarried in Norway and the expected production energy requirement is similar to that for feldspar.

6.6 CALUMITE BRAND SLAG

Calumite Brand Slag[†] is widely used in the glass industry where it is frequently referred to simply as 'calumite'. It is derived from selected blast furnace slags by a patented process. The product is a siliceous material, which, it is claimed, improves the refining of the glass, increases the melting rate and reduces the overall glass furnace fuel consumption. The detailed process is not known but a production energy has been calculated from the overall fuel consumption of the plant to give the value shown in Table 6.2.

†Calumite is a registered trade mark of the Calumite Corporation of America.

6.7 SODIUM NITRATE

Crude sodium nitrate occurs in vast deposits in Chile and is extracted by the Guggenheim process [53], a large scale leaching and recrystallisation operation. Leaching is carried out at a temperature of 313 K and recrystallisation at a temperature of 278 K. The waste heat from the refrigeration process and in the exhaust gases from the electricity generators is used to heat the leaching water. Typically the yield is 27.5 tonne of nitrate for a fuel input of 1 tonne of oil fuel (assumed diesel). The energy requirements of the process are therefore as shown in Table 6.2.

6.8 SODIUM SULPHATE

Commercial sodium sulphate or salt cake is produced by reacting sodium chloride with sulphuric acid in a furnace. The resultant solid product is a saleable powder. The energy required to produce sodium sulphate (Table 6.2) has been calculated from the manufacturer's fuel consumption.

6.9 CALCIUM SULPHATE

Calcium sulphate occurs naturally as gypsum in a sufficiently pure form to be used directly. Selected mineral deposits are quarried, crushed, graded and bagged. From data supplied by two major producers, the average energy requirement shown in Table 6.2 has been calculated.

6.10 SELENIUM

Selenium is used in the manufacture of white flint glass where it acts as a decolorant. It is extracted from the anode slimes left behind after copper refining so that the total energy associated with the production of selenium from ore in the ground is the sum of the energies (a) to produce electrolytic copper and (b) to extract the selenium from the anode slimes. Data for this processing has been taken from [43] and is shown in Table 6.2. All selenium used in glass manufacture in the U.K. is imported — usually from Canada.

6.11 IRON CHROMITE

Iron chromite occurs naturally in a crude form and may also be synthesised from metallic iron. We have assumed production from the natural mineral so that the major processing is mineral treatment such as that described earlier for feldspar. The production energy is therefore as shown in Table 6.2.

6.12 COBALT OXIDE AND NICKEL OXIDE

Both cobalt oxide and nickel oxide are used as colorants in glass making. The simplest and least energy intensive sources of these products is the roasting of sulphide ores. We have therefore assumed mineral processing of the mined rock followed by roasting and using published data [4], the total energy requirement may be estimated as shown in Table 6.2.

6.13 CULLET

Approximately 20% of the furnace feed in all glass factories is cullet which improves the melting characteristics of the components. It has been estimated that approximately 80% of all cullet used is derived from waste glass generated within the factory [27]. This is usually referred to as **in-house cullet**. No energy has been attributed to the production and handling of in-house cullet since any fuels used in these operations will be included in the overall fuel requirements of the glass factory.

In addition to in-house cullet, many glass manufacturers take in small amounts of cullet from outside suppliers. This is known as **foreign cullet** and is derived from a number of sources. Of these, one of the most favoured is customers who handle substantial numbers of containers such as dairies and other bottlers. The cullet is usually well characterised by colour and composition although there is a reluctance on the part of many bottlers to store substantial quantities of broken glass on their premises until it can be removed. Moreover, there is also a reluctance on the part of glass firms to use container delivery lorries to transport broken glass back to the glass factory because of the potential hazards. As a consequence separate transport is usually hired to move the cullet. Cullet derived from customers is usually organised on a local basis and no general pattern seems to be established.

Cullet is also derived from cullet merchants. Few cullet merchants however regard the supply of cullet to U.K. glass factories as particularly economic and tend to treat it as a side-line accompanying other more profitable ventures such as scrap metal collection. Such merchants will usually only collect from sources where reasonable quantities of easily characterised cullet are likely to accumulate. As a result, most cullet is derived from dairies, soft drinks bottlers, breweries and glaziers in approximately equal amounts. Collection is usually achieved by leaving skips at the premises and collecting them at regular intervals. The average load collected is typically of the order of 5 tonne.

No cullet merchant appears to carry out any significant processing of the collected cullet apart from improving the compaction before onward despatch to the glass factories. It is claimed that loads of approximately 20 tonne are despatched to the glass factories. However, when the total mass of cullet moved is divided by the number of deliveries made, the average load is approximately 10 tonne.

Calculation of the energy associated with the recovery and delivery of foreign cullet is difficult because of the very fragmented nature of the business. An average fuel requirement for the collection of cullet is thought to be 1.32 gallons of diesel. Delivery of cullet is normally in 20 tonne lorries. Assuming an average load of 10 tonne over an average distance of 100 miles with empty return loads, an overall energy requirement for foreign cullet collection and delivery is as shown in Table 6.2.

6.14 LIME

Lime, calcium oxide, is manufactured by calcining limestone to drive off the carbon dioxide according to the equation

$$CaCO_3 = CaO + CO_2 .$$ (6.1)

Approximately 1.8 kg of limestone are required for each kilogramme of lime produced. The energy required to produce lime has been calculated from the limestone mining data of Table 6.2 and calcining data of [30]. The result is shown in Table 6.2.

CHAPTER 7

Aluminium and Aluminium Alloys

7.1 INTRODUCTION

Aluminium is used in the U.K. for the production of beverage containers and closures. It was first introduced in the form of easy-open ends on tinplate cans but two-piece all-aluminium cans are now entering the market in increasing numbers. The proportion of aluminium cans in circulation is likely to grow with the commissioning of the 60,000 tonne per year sheet rolling facility by Alcoa at Swansea, which is intended to produce can-stock for continental Europe as well as the U.K.

Aluminium is produced from bauxite mineral which is first purified to high grade alumina (Al_2O_3) in the Bayer Process. Because of the high energies involved, the whole production sequence is considered in some detail in this report. Aluminium is not used in the pure state in the production of beverage containers but is instead alloyed with magnesium and manganese. The effect of these alloying additions is therefore also considered.

7.2 BAUXITE

Bauxite is the source of primary aluminium metal and some 80 to 90% of world production of bauxite is by open pit mining [28]. Bauxites contain mainly hydrated alumina together with variable amounts of the oxides or iron, silicon and titanium as well as traces of lesser impurities. They are regarded as the end products of the slow weathering of rocks, taking place over millions of years in areas of heavy rainfall.

The products of this weathering are mainly surface deposits with little or no overburden and are eminently suitable for open cast quarrying. There are however some locations, notably in France, where conventional deep mining occurs. At present, most of the bauxite requirements of the western world are drawn from Australia. France, Ghana, Greece, Guyana, Jamaica, Malaya, Surinam and the U.S.A. Because deposits are usually remote from aluminium smelters, the crude ore is commonly treated at the mine to reduce the impurity level. It is usually crushed and given a water wash to separate physically some of the silica. The process may be followed by drying to reduce the shipped mass. Excess silica in bauxite reduces the recovery of alumina and increases consumption of caustic soda in the Bayer process, the next stage of treatment.

Bauxites are classified into two main groups according to whether their alumina content exists in the trihydrate form $Al_2O_3.3H_2O$ (gibbsite) or in the monohydrate form $Al_2O_3.H_2O$ (boehmite). Individual deposits may contain only one of these minerals or they may be a mixture of both. Typical analyses of good quality bauxites are given in Table 7.1.

Table 7.1
Chemical analysis of two typical bauxite ores. All values in mass %

Mineral type	Al_2O_3	SiO_2	Fe_2O_3	TiO_2	H_2O (combined)
Trihydrate	52	1.5	17	2	27.5
Monohydrate	53	7	25	3	12

© *Crown Copyright 1981*

Data for the energy required in bauxite mining and treatment is limited and published results give a spread of values ranging from 0.3 MJ/kg to 1.23 MJ/kg. Data from [29] have been chosen since they lie in the middle of the range and are well documented. These are shown in Table 7.2. See also references [4, 21, 29-31]. U.K. smelters import alumina rather than bauxite since there is no alumina production for aluminium smelters in the U.K.

Table 7.2
Energy required to produce 1 kg of bauxite from ore in the ground

Fuel type	Fuel production and delivery energy/MJ	Energy content of fuel /MJ	Feedstock energy /MJ	Total energy /MJ
Electricity	0.09	0.03	nil	0.12
Oil fuels	0.01	0.07	nil	0.08
Other fuels	0.04	0.46	nil	0.50
Totals/MJ	0.14	0.56	nil	0.70

© *Crown Copyright 1981*

7.3 ALUMINA

Alumina is produced from bauxite by the Bayer process (Fig. 7.1). The only alumina produced in the U.K. is for specialist chemical purposes. The object of the Bayer process is the separation of virtually pure anhydrous alumina

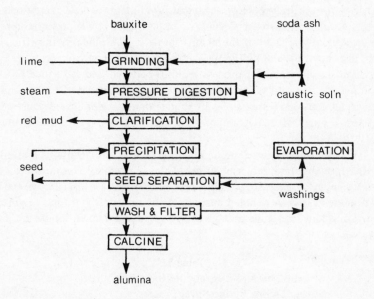

Fig. 7.1 – Schematic flow diagram showing the operations in the Bayer process for producing alumina from bauxite.

from bauxite. The ore is crushed and milled in the presence of sodium aluminate liquor to produce a mill slurry which is pumped into a series of steam heated autoclaves containing caustic soda. Here the chemical reaction is:

$$Al_2O_3.3H_2O + 2NaOH = 2NaAlO_2 + 4H_2O . \qquad (7.1)$$

Bauxites of the higher hydrate type are more readily soluble in hot caustic soda solutions than those of the monohydrate type and can be extracted effectively in low strength caustic solutions at boiling point. Monohydrate ores, however, require temperatures between 450 K and 520 K. Digestion conditions have to be adjusted to suit the type of ore.

Following digestion, reacted slurry is blown into a series of 'flash' vessels where steam is recovered and used for heating the caustic liquor. The solution is treated to remove dissolved silica and insoluble residues which contain most of the iron oxide and titanium oxide. After filtering, pure alumina trihydrate is precipitated by the reverse of reaction (7.1), with the precipitation being seeded by the addition of crystalline particles of the hydrate in solution. The product is filtered and the alumina hydrate calcined at 1500-1800 K to yield pure alumina.

The process consumes thermal fuels for steam raising and calcining and electricity for machinery. However, most Bayer plants recycle energy within the plant and use surplus steam to generate at least part of the power requirement. The energy consumption of the plant is in part determined by the efficiency of waste heat recovery and by the type of ore being processed. To produce 1 tonne of anhydrous alumina, the process requires between 2 and 3 tonne of bauxite according to type and grade and requires the replacement of caustic soda at a rate varying between 30 kg and 130 kg per tonne of alumina. Because of these variables, a range of energy requirements have been reported in the literature [4, 21, 29-34].

Jamaican bauxite, used predominantly in the U.K., contains a mixture of alumina monohydrate and trihydrate. A typical chemical composition is Al_2O_3 (49.0%), Fe_2O_3 (20.0%), SiO_2 (2.5%), TiO_2 (2.5%) ignition loss (25.0%) and minor constituents (1.0%). A detailed analysis of the materials and fuels requirements for treating Jamaican mineral is given in [34] and the main inputs are listed in Table 7.3.

Table 7.3
Materials and fuel requirements to produce 1 kg
of alumina by the Bayer Process

Component	Requirement	
Soda ash (Na_2CO_3)	0.067	kg
Lime (CaO)	0.066	kg
Steam for digestion, etc.	3.53	kg
Electricity	0.0698	kg
Natural gas for calcination	4.8	KJ

© *Crown Copyright 1981*

The total energy required to produce alumina and deliver it to the U.K. may be calculated from Tables 6.2, 7.2 and 7.3. In general, 2.3 kg of bauxite are required to produce 1 kg of alumina and the requirements of lime and soda ash (used to produce the caustic) are as in Table 7.3. Delivery is assumed by sea over a distance of 3000 miles for which the energy can be calculated from Table 3.4. Thus overall energy requirements are as shown in Table 7.4.

Table 7.4
Total energy required to produce and deliver 1 kg of alumina to the U.K.

Operation	Electricity/MJ		Oil fuels/MJ			Other fuels/MJ			Total energy
	Fuel production and delivery	Energy content of fuel	Fuel production and delivery	Energy content of fuel	Feedstock energy	Fuel production and delivery	Energy content of fuel	Feedstock energy	
Bauxite mining	0.21	0.07	0.02	0.16	–	0.09	1.06	–	1.61
Provision of lime	0.04	0.01	0.01	0.03	–	0.02	0.29	–	0.40
Provision of soda ash	0.02	0.01	0.11	0.55	–	0.02	0.26	–	0.97
Bayer fuel requirements	0.75	0.25	2.82	13.84	–	0.39	4.80	–	22.85
Alumina transport to U.K.	–	–	0.09	0.42	–	–	–	–	0.51
Total energy to produce alumina	1.02	0.34	3.05	15.00	–	0.52	6.41	–	26.34

© *Crown Copyright 1981*

7.4 PRIMARY ALUMINIUM PRODUCTION IN THE U.K.

Aluminium is produced by the electrolysis of alumina dissolved in cryolite (Na_3AlF_6). Alumina itself melts at 2293 K to give a non-conducting liquid which cannot therefore be electrolysed. However, at about 1300 K, molten cryolite will dissolve up to 15% alumina by mass to give a conducting electrolyte which can be electrolysed to give aluminium as one of the products. The process is known as the Hall-Heroult process.

The electrolytic cell consists of a steel box lined with carbon, in which steel conductor bars are embedded to form the cathode. High purity carbon anodes are suspended in the molten electrolyte. When a current is passed, aluminium is deposited to form a pool on the floor of the cell and this acts as the true cathode surface; at the same time, oxygen is liberated at the anodes where it reacts with the carbon of the anodes to form carbon dioxide and so consumes the anodes. The overall cell reaction is therefore:

$$2Al_2O_3 + 3C = 4Al + 3CO_2 .\tag{7.2}$$

A smelter with an output of 120,000 tonne per annum requires approximately 320 cells and is consequently highly capital intensive.

Modern aluminium smelters require a total of from 15.5 to 17.5 kWh of electricity for the production of 1 kg of aluminium and these very large power requirements have resulted in the location of smelters close to sources of cheap electrical power. Hydroelectricity is the best source of cheap power so that much of the world's aluminium production is centred in those areas where such power is available, for example Canada and Norway. In the U.K., hydroelectricity is not readily available and of the total production capacity of some 366,000 tonne, only 10% is produced with this source of electricity. Table 7.5 lists the U.K. smelters and shows their approximate capacities and sources of electrical power.

Table 7.5
U.K. aluminium smelters

Location	Annual capacity in tonne	Power Source	Percentage of total U.K. capacity
Lynemouth	120,000	Thermal generation from coal: on-site power station.	32.79
Anglesey	105,000	National Grid	28.69
Invergordon	102,000	National Grid	27.87
Kinlocheven	10,000	Hydroelectricity	2.73
Lochaber	29,000	Hydroelectricity	7.92
Totals	366,000		100.00

© Crown Copyright 1981

Because of the different power sources, the conversion efficiency of primary fuels to electricity must take account of the differing fuel production energies. An average fuel production energy for electricity used in primary aluminium smelting in the U.K. can be calculated using an equation of the form:

$$\eta_{average} = f_L \cdot \eta_L + f_A \cdot \eta_A + f_I \cdot \eta_I + f_K \cdot \eta_K + f_{Lo} \cdot \eta_{Lo} \qquad (7.3)$$

where f is the fraction of the U.K. aluminium production carried out at the smelter and η is the fuel production efficiency of the electricity consumed at the smelter. The subscripts L, A, I, K, and Lo refer to Lynemouth, Anglesey, Invergordon, Kinlochleven and Lochaber respectively.

Equation 7.3 can be used to calculate the present overall efficiency and will take into account any changes in the electricity supply to a particular smelter in the future. For example, should the Anglesey smelter draw its power from the nearby nuclear power station, η_A, will change. In this work, the f values in (7.3) are based on the stated capacities in Table 7.5 and the values assumed for for production efficiencies are shown in Table 7.6.

Table 7.6

Assumed fuel production efficiencies for electricity supplied to U.K. aluminium smelters

Plant	Primary fuel production efficiency	Primary fuel delivery efficiency	Generation efficiency	Electricity delivery efficiency	Overall efficiency
Lynemouth	0.95	1.0	0.35	1.0	0.33
Anglesey	NATIONAL GRID VALUES				0.25
Invergordon	NATIONAL GRID VALUES				0.25
Kinlochleven	–	–	0.85	1.0	0.85
Lochaber	–	–	0.85	1.0	0.85

© *Crown Copyright 1981*

Substituting these values into equation 7.3 and using the values of f from Table 7.5 gives

$$\eta_{average} = 0.3409 \text{ or } 34.09\% \qquad (7.4)$$

which is significantly higher than the average grid efficiency.

Much of the data used here for aluminium production and fabrication is based on information supplied by the Aluminium Federation who carry out detailed annual analyses *by fuel type* of energy consumption in the U.K. aluminium industry. In its own statistics, the Aluminium Federation uses a value of 30.77% as its grid efficiency in calculations relating to smelters drawing power from the national grid; this figure excludes the production and delivery of primary fuel and assumes distribution losses of 2.5%. However, in their 1976 survey [35], their average efficiency is 34% compared with the value of 34.09% calculated above. This difference is insignificant.

7.5 REQUIREMENTS OF U.K. ALUMINIUM SMELTERS

From the available industry data [35, 36], the materials and fuels requirements of the U.K. aluminium smelters are as shown in Table 7.7. The electricity requirement covers power losses in the rectifiers, power supplies to the cells and auxiliary power required in fume control, lighting, etc. Cell power requirements are approximately 92% of the total electricity consumption of the smelter. Natural gas and LPG are used for electrode baking. Petrol and diesel consumption relates only to in-plant use and does not include distribution.

Table 7.7

Typical materials and fuels requirements of aluminium smelters per 1 kg of hot metal produced (1976 data)

Input	Unit	Requirement
Materials		
Alumina	kg	1.92
Aluminium fluoride	kg	0.04
Cryolite	kg	0.03
Calcined petroleum coke	kg	0.463
Electrode pitch	kg	0.136
Fuels		
Electricity	kWh	17.41
Natural gas	Nm^3	0.027
LPG	kg	0.08
Heavy fuel oil	litre	0.0012
Light fuel oil	litre	0.0015
Gas oil	litre	0.0012
Lubricating oil	litre	0.0003
Petrol	litre	0.00034
Diesel	litre	0.0027

© *Crown Copyright 1981*

To evaluate the energy associated with the production of primary aluminium the energies required to produce the materials listed in Table 7.7 must be evaluated.

7.6 ALUMINIUM FLUORIDE

Aluminium fluoride is manufactured by a two stage process, First fluorspar is reacted with sulphuric acid to produce hydrogen fluoride according to the reaction:

$$CaF_2 + H_2SO_4 = CaSO_4 + 2HF \ . \tag{7.5}$$

The hydrogen fluoride is then reacted with alumina to give the desired aluminium fluoride according to the reaction:

$$Al_2O_3 + 6HF = 2AlF_3 + 3H_2O \ . \tag{7.6}$$

From the stoichiometry of (7.5), 1.95 kg of fluorspar are needed to yield 1 kg of hydrogen fluoride. Fluorspar is mined and treated in a manner similar to feldspar, so the energy associated with the provision of fluorspar is assumed to be the same as that given in Table 6.2 for feldspar. The energy associated with the production of sulphuric acid is assumed to be zero since it is produced in large quantities as a by-product of many chemical processes. Hence the energy associated with the production of hydrogen fluoride is as shown in Table 7.8.

From the stoichiometry of (7.6), 0.61 kg of alumina and 0.71 kg of hydrogen fluoride are needed to produce 1 kg of aluminium fluoride. The production energies of alumina and hydrogen fluoride are given in Tables 7.4 and 7.8. Hence the total energy required to produce aluminium fluoride may be calculated as shown in Table 7.8.

7.7 PETROLEUM COKE

Petroleum coke is produced by the complete dehydrogenation of oil products such as naphtha. The process essentially requires that the hydrocarbon be heated to a high temperature in the absence of air when the reaction proceeds as:

$$C_nH_m = nC + \tfrac{1}{2}mH_2 \ .$$

For alkane feedstocks, $m = 2n + 2$ in the above equation and the hydrogen content of the feedstock represents approximately 20% of the input mass. Hence the production of 1 kg of coke requires approximately 1.2 kg of petroleum feedstock. The production process may therefore be regarded as a two stage process consisting of (a) the production of the hydrocarbon feedstock from crude oil in the ground and (b) the dehydrogenation of this feedstock. The overall energy requirements from crude oil are given in Table 7.8.

7.8 PITCH

Pitch is used as the electrode sealant in the electrolytic cell. It is derived from the heavy residue remaining after oil fractionation. An overall energy production efficiency of 83% has therefore been used in accordance with Table 2.2 to give the value shown in Table 7.8.

Table 7.8

Energy associated with the production of 1 kg of chemicals and intermediates used in the production of aluminium. See text for details of production methods and assumptions made in the calculations

Operation	Electricity/MJ		Oil fuels/MJ			Other fuels/MJ			Total energy /MJ
	Fuel production and delivery	Energy content of fuel	Fuel production and delivery	Energy content of fuel	Feedstock energy	Fuel production and delivery	Energy content of fuel	Feedstock energy	
HF from raw materials	2.28	0.76	0.29	1.52	–	–	–	–	4.85
AlF$_3$ from raw materials	2.24	0.75	2.07	10.23	–	0.32	3.91	–	19.52
Petroleum coke from crude oil	0.90	0.30	6.04	29.50	54.00	–	–	–	90.74
Pitch from crude oil	–	–	1.30	6.35	45.00	–	–	–	52.65
NaOH from brine	18.36	6.12	–	–	–	–	–	–	24.48
Cl$_2$ from brine	18.36	6.12	–	–	–	–	–	–	24.48
Brine pumping per kg caustic produced	0.21	0.07	–	–	–	–	–	–	0.28
Brine heating per kg caustic produced	–	–	0.66	3.45	–	–	–	–	4.11
NaOH from brine in ground	18.57	6.19	0.66	3.45	–	–	–	–	28.87
Cryolite from raw materials in the ground	8.52	2.84	1.15	5.63	–	0.01	0.14	–	18.29

7.9 SODIUM HYDROXIDE

Sodium hydroxide is produced as a co-product with chlorine in the electrolysis of brine. The following data are based on the commercial process using mercury cells but it should be noted that the current trend is towards the use of diaphragm cells for which the energy consumption is expected to be slightly lower. In the process to which these data relate [26], the by-product hydrogen is burnt in the plant electricity generator. Thus the overall electrolysis energy is reduced by an amount equal to the energy recovered from this burnt hydrogen. This reduced electrolysis energy is apportioned equally between chlorine and caustic soda to give a direct electrical consumption per kilogramme of caustic soda of 6.12 MJ. This electrical energy is supplied by an in-plant generation facility which has a conversion efficiency of 30%; transmission losses are assumed to be negligible. Thus the indirect energy associated with the electrolysis is 20.40 MJ and the total energy due to electrolysis is shown in Table 7.8.

The brine feed to the cell is obtained by pumping water into underground salt deposits when the salt dissolves. Assuming a saturated brine solution containing 0.36 kg/litre and that this level is depleted by 15% during electrolysis, the production of 1 kg of caustic soda requires an input of about 7 gallons of brine for which about 6 gallons of water must be delivered to the well head. The energy associated with these operations is shown in Table 7.8.

Electrolytic chlorine cells operate at elevated temperatures to reduce electrical losses. Some of the heating is supplied by the unavoidable electrical losses within the cells but the remainder must be supplied from external sources. Assuming that the cells operate at a temperature of 350 K, that the brine output from the well is at a temperature of 280 K and that half of the heating is supplied from a boiler with an efficiency of 65%, the energy required per kilogramme of caustic or chlorine can be readily calculated to give the values shown in Table 7.8.

The energy required to produce chemicals needed to purify the brine and the losses of electrode materials during electrolysis are assumed to be negligible. The total energy required to produce caustic soda can therefore be calculated from the above data to give the result shown in Table 7.8.

7.10 CRYOLITE

Cryolite, the solvent for alumina in the Hall-Heroult cell, is a naturally occurring mineral but is in short supply. Consequently most of the cryolite used in the aluminium industry is synthetic. It is commonly manufactured from caustic soda, hydrogen fluoride and Bayer liquor, which contains sodium aluminate, by the reaction

$$NaAlO_2 + 2NaOH + 6HF = Na_3AlF_6 + 4H_2O .$$

From the stoichiometry of this reaction, it is clear that 1 kg of cryolite requires 0.39 kg sodium aluminate, 0.38 kg caustic soda and 0.57 kg hydrogen fluoride.

The energy associated with the production of sodium aluminate is that for the Bayer process less the energy required for the final calcining stage which produces alumina. The values presented for the Bayer process in Section 7.3 are for a final product of 1 kg alumina and hence refer to the production of 1.5625 kg of sodium aluminate. Values for the energies to produce hydrogen fluoride and caustic soda have already been calculated in Table 7.8. Hence the energy required to manufacture cryolite may be calculated and the result is also given in Table 7.8.

7.11 TOTAL ENERGY REQUIRED TO PRODUCE PRIMARY ALUMINIUM

The total energy associated with the production of aluminium hot metal from raw materials in the ground can now be calculated from Tables 7.4, 7.7 and 7.8 and this calculation is shown in Table 7.9.

7.12 ALUMINIUM INGOT CASTING

Aluminium ingots are produced from hot metal in two forms; slab for subsequent rolling to sheet and round stock for extrusion. Hot metal is syphoned from the reduction cells and kept in gas or oil fired holding furnaces where any alloying additions are made prior to casting. The fuel requirements per kilogramme of ingot cast are typically as given in Table 7.10 [35]. Using the data of Table 2.2 for all fuels other than electricity, the energy required to cast 1 kg of finished ingot from hot metal is shown as entry (a) in Table 7.11. As before electricity has been calculated using the production efficiency of 34.09% since the casting plants derive their electricity from the same source as the smelters.

Table 7.10

Fuel requirements for ingots cast in U.K. smelters in 1976

Fuel	Total consumption for all production		Consumption per kilogramme of ingot produced	
Electricity	20.9	GWh	0.061	kWh
Natural gas	7,063,100	Nm³	0.021	Nm³
LPG	20,427	t	0.060	kg
Medium fuel oil	133,000	l	0.0004	l
Light fuel oil	1,474,000	l	0.0043	l
Gas oil	538,000	l	0.0016	l
Petrol	43,000	l	0.00013	l
Diesel	123,000	l	0.000038	l

Table 7.9
Total energy required to produce 1 kg of aluminium hot metal from raw materials in the ground

Contributor	Electricity/MJ		Oil fuels/MJ			Other fuels/MJ			Total energy /MJ
	Fuel production and delivery	Energy content of fuel	Fuel production and delivery	Energy content of fuel	Feedstock energy	Fuel production and delivery	Energy content of fuel	Feedstock enegy	
Alumina production	1.96	0.65	5.86	28.80	–	1.00	12.31	–	50.58
Aluminium fluoride production	0.09	0.03	0.08	0.41	–	0.01	0.16	–	0.78
Cryolite production	0.26	0.09	0.03	0.17	–	neg	neg	–	0.55
Petroleum coke production	0.42	0.14	2.80	13.66	25.00	–	–	–	42.02
Pitch production	–	–	0.18	0.86	6.12	–	–	–	7.16
Smelter fuels	121.19	62.68	0.77	4.29	–	0.14	1.00	–	190.07
Production of aluminium hot metal from raw materials in the ground	123.92	63.59	9.72	48.19	31.12	1.15	13.47	–	291.16

Table 7.11

Summary of energy requirements for operations carried out on aluminium during all post-smelter stages. Unless otherwise stated, the outputs are produced from raw materials in the ground

Operation	Electricity/MJ		Oil fuels/MJ			Other fuels/MJ			Total energy /MJ
	Fuel production and delivery	Energy content of fuel	Fuel production and delivery	Energy content of fuel	Feedstock energy	Fuel production and delivery	Energy content of fuel	Feedstock energy	
(a) 1 kg finished Al ingot from hot metal	0.43	0.22	0.58	3.27	–	0.06	0.78	–	5.34
(b) 1 kg finished Al ingot	127.87	65.62	10.58	52.83	32.00	1.24	14.63	–	304.77
(c) 1 kg magnesium metal	154.17	51.39	0.73	3.61	5.63	14.99	182.28	–	412.80
(d) 1 kg manganese metal	107.04	35.68	–	–	–	–	–	–	142.72
(e) 1 kg can-stock ingot	128.06	65.16	10.36	51.74	31.34	1.38	16.33	–	304.37
(f) 1 kg end-stock ingot	129.27	64.97	10.13	50.62	30.73	1.88	22.35	–	309.95
(g) 1 kg remelt ingot from primary ingot (UK)	1.53	0.51	1.14	5.51	–	0.15	1.86	–	10.70
(h) 1 kg remelt ingot from primary ingot (US)	0.57	0.19	1.17	5.60	–	–	–	–	7.53
(j) 1 kg hot rolled sheet from remelt ingot	3.36	1.12	0.20	1.00	–	0.29	3.57	–	9.54
(k) 1 kg cold rolled sheet from hot rolled coil	6.06	2.02	0.58	2.90	–	0.14	1.75	–	13.45
(l) 1 kg cold rolled sheet from remelt ingot	10.86	3.62	0.87	4.33	–	0.56	6.86	–	27.10
(m) 1 kg cold rolled sheet from ingot (US data)	9.30	3.10	–	–	–	0.77	9.33	–	22.50
(n) 1 kg can-stock	143.04	71.04	12.67	62.92	27.36	2.21	26.29	–	345.53
(p) 1 kg end-stock	144.31	70.84	12.42	61.73	26.84	2.71	32.66	–	351.51
(q) 1 kg tab-coil	142.90	71.54	12.90	64.06	27.93	2.06	24.58	–	345.97

The melting and casting operation typically results in some 2.5% melt loss of input materials due to oxidation. In addition, approximately 10% of the metal cast is subsequently trimmed from the ingot and recirculated through the melting furnace. A materials balance for the holding furnace/casting operation is shown in Fig. 7.2 for the production of 100 kg of good ingot. Table 7.11

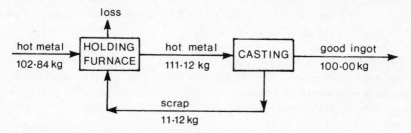

Fig. 7.2 – Materials balance for the holding furnace and casting operations for aluminium.

therefore takes account of the need to process 1.1396 kg of hot metal to produce 1 kg of finished ingot but to obtain the energy associated with the production of finished ingot from bauxite in the ground, Table 7.9 must be increased by a factor of 1.0284 and added to entry (a) in Table 7.11; this gives entry (b) in Table 7.11.

7.13 ALUMINIUM ALLOYS

Aluminium for can manufacture is alloyed with varying amounts of magnesium and manganese depending upon the application. Typical additive concentrations are given in Table 7.12.

Table 7.12

Typical compositions of aluminium alloys used
for the production of can- and end-stock

	Mass % present in the alloy	
Element	Can stock Alloy 3004	End stock Alloy 5182
Aluminium	97.2 – 98.2	94.5 – 95.8
Magnesium	0.8 – 1.3	4.0 – 5.0
Manganese	1.0 – 1.5	0.2 – 0.5

7.13.1 Magnesium production

Most of the magnesium used is extracted from seawater by the Dow electrolytic process. Seawater is filtered and mixed with a slurry of calcium hydroxide obtained by mixing lime with water. The magnesium in solution precipitates as magnesium hydroxide in large thickeners or settling tanks, is separated by filtration and neutralised with hydrochloric acid. The solution is evaporated in direct fired evaporators and driers to form 'wet' magnesium chloride granules which are charged to the electrolytic cells.

Magnesium chloride is decomposed to magnesium at the cathode of a cell operating at a potential of 6 V and a current of 30,000 A. Graphite electrodes suspended in the fused chloride bath act as anodes. Chlorine liberated at the anode is sent to the hydrochloric acid plant and recycled to the neutralisation stage. The requirements for the production of 1 kg of magnesium are given in Table 7.13 [43].

Table 7.13

Fuel and materials requirements for the production of 1 kg of magnesium metal from seawater

Input	Requirement
Seawater pumping	0.3973 kWh (electricity)
Lime	1.75 kg
Mixing energy	0.0167 kWh (electricity)
Evaporation	159.3 MJ (natural gas)
Electrolysis	13.766 kWh (electricity)
Electrode consumption	0.1 kg
HCl recovery	14.81 MJ (natural gas)

© Crown Copyright 1981

The energy associated with the production of lime has been taken from Table 6.2 although in many plants, it is obtained from sea shells rather than limestone rock. High purity carbon electrodes for use in the electrolytic reaction cells are manufactured from petroleum coke and pitch. The green mixture of the two constituents is formed into shape in hydraulic presses or vibratory formers then cured at a temperature of approximately 1500 K in gas fired furnaces. Volatile constituents are driven off as the mass carbonises. The requirements for the production of 1 kg of electrodes are typically 0.86 kg of petroleum coke, 0.22 kg of pitch and 0.096 Nm^3 of natural gas. Hence, using the data of Tables 2.2, 6.2, 7.9 and 7.13, the energy required to produce magnesium metal is as shown by entry (c) in Table 7.11.

7.13.2 Manganese
In contrast to magnesium, the energy associated with the production of manganese is poorly documented. From published data [42] an approximate value can be calculated as shown by entry (d) of Table 7.11. For present purposes however, this approximate value is thought to be adequate.

7.13.3 Aluminium can alloys
The energy associated with the production of aluminium alloy ingot for can- and end-stock may now be calculated using the data of Tables 7.9 and 7.11. These have been calculated using the mid-range concentrations of magnesium and manganese indicated by Table 7.12 and are shown as entries (e) and (f) of Table 7.11.

7.14 PRODUCTION OF SHEET FOR CAN BODIES AND CAN ENDS
Fig. 7.3 shows the sequence of operations used to produce sheet from primary ingot [44]. 1.78 kg of metal must be melted and processed to produce 1 kg of can- or end-stock sheet. However, by recirculating in-house scrap, a primary ingot mass of only 1.04 kg is required to produce 1 kg of sheet.

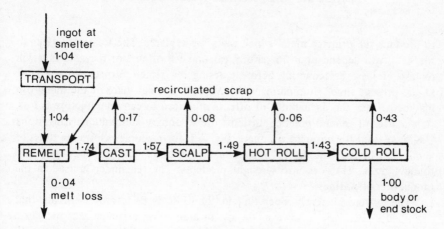

Fig. 7.3 – Materials balance for aluminium sheet production from primary ingot.

In practice, the fabrication remelt facility may include process scrap from can manufacturers as well as any other available clean scrap in addition to in-house scrap and primary ingot. For example, Fig. 7.4 shows the materials flows in a U.S. can-stock production facility [44]. For the system of Fig. 7.4, scrap replaces primary ingot so that the system energy requirement is lower than that for the system described in Fig. 7.3. This is because the energy associated with scrap treatment is always very much less than the energy required

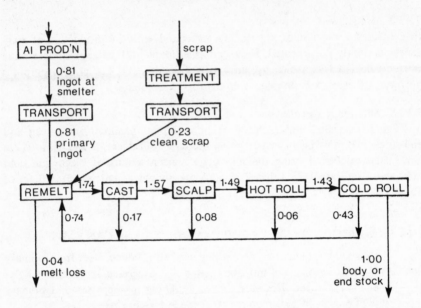

Fig. 7.4 – Materials balance for aluminium sheet production from primary ingot and scrap.

to produce the primary metal which the scrap replaces. The treatment given to the scrap will depend upon its nature; can-makers' offcuts or recycled cans will require at least delacquering before entering the remelt furnace. In the U.K. at the present time, aluminium cans have only a small share of the metal can market so that the availability of offcuts produced in can manufacture is low. As a matter of general practice, little of the secondary aluminium arising in the U.K. is re-used for wrought products. Fig. 7.5, for example, shows the flows of aluminium in the U.K. in 1975. Out of a total of 179,000 tonne, only 7,000 tonne (about 4%) goes into wrought products. The remainder is used in the production of castings.

The system currently operating in the U.K. is therefore essentially that described in Fig. 7.3. If however the use of aluminium beverage cans increases, systems such as that shown in Fig. 7.4 will become feasible. Furthermore, with can recycling schemes, the ratio of scrap to primary metal in the feed to the remelt furnace may grow beyond that described by Fig. 7.4.

The operations leading to sheet production can be conveniently split into two components; the remelting stage which produces ingot suitable for subsequent working and the rolling stage (both hot and cold) which produces sheet.

At the time this report was being prepared, Alcoa had just commissioned a 60,000 tonne per year sheet production facility at Swansea to provide sheet for aluminium can manufacturers in the U.K. and Europe. However this plant cannot be used to give representative data for can-stock production since it is

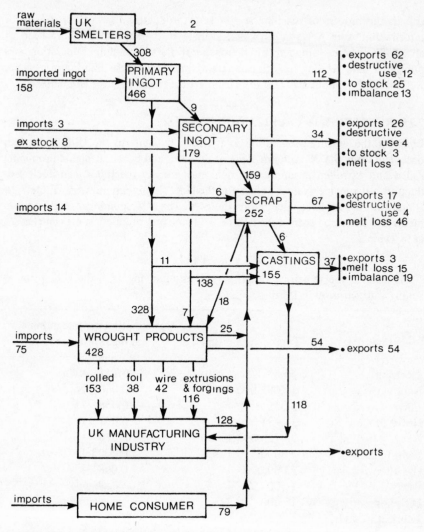

Fig. 7.5 – Flow of aluminium in the U.K. in 1975. All values are in thousands of tonnes.

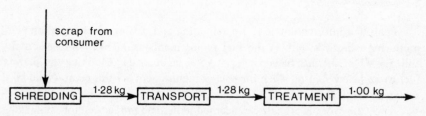

Fig. 7.6 – Typical flow of aluminium can scrap in the U.S.A.

still in the process of reaching steady state production. Two other sources of information were however available; average U.K. data for the fabrication of all product types supplied by members of the Aluminium Federation and published data relating to U.S. can-making practices where aluminium cans are well established. Both sets of information are presented here for comparison.

7.15 REMELTING IN FABRICATION PLANTS

Data for the remelting operation at fabrication plants in the U.K. have been obtained from the Aluminium Federation [35] and cover all plants producing rolled and extruded products. As such they are not specific to can-stock production but provide average values for the U.K. aluminium industry. Table 7.14 shows the fuel requirements of this industry. The corresponding energy requirements are shown as entry (g) in Table 7.11; all fuel production efficiencies are as in Table 2.2.

Table 7.14

Fuels used in the remelting of primary aluminium ingots in the U.K. prior to rolling and extrusion in 1976

Fuel type	Total consumption		Consumption per kilogramme of ingot produced	
Electricity	50.55	GWh	0.141	kWh
Natural gas	18,091,000	Nm^3	0.050	Nm^3
LPG	3,041	t	0.008	kg
Heavy fuel oil	34,952,000	l	0.097	l
Medium fuel oil	7,079,000	l	0.020	l
Light fuel oil	77,000	l	0.0002	l
Gas oil	2,909,000	l	0.008	l
Petrol	1,400	l	negligible	
Diesel oil	13,600	l	negligible	
Lubricating and hydraulic oils	147,700	l	0.0004	l

Published information [44] for remelting and casting in a fabrication plant producing can-stock, quotes the fuel requirements as 0.054 kWh of electricity and 5.6 MJ of thermal fuels per kg of good ingot made. Assuming the thermal fuel to be heavy fuel oil, then the energy requirement for this plant would be as shown by entry (h) in table 7.11.

Since this work is primarily concerned with sheet production for can-making, the value given as entry (h) in Table 7.11 has been used for the remelt energy rather than the average U.K. value (entry (g) in Table 7.11). It is however useful to be able to compare the energy for the production with the general average energy requirements of the industry.

7.16 ROLLING INGOT AND SHEET

There is little reliable published information for the energy associated with the rolling of metal ingot to sheet and no U.K. data at all for the energy to manufacture can-stock from primary ingot. Information from the Aluminium Federation again represents an average value for the production of all products and does not specifically refer to can-stock.

7.16.1 Hot rolling

The fuel requirement for hot rolling is shown in Table 7.15 and relates to practices in the whole of the U.K. aluminium industry which made some 333,978 tonne of hot rolled products in 1976. The corresponding energy requirement is therefore as shown by entry (j) in Table 7.11 using the fuel production efficiencies of Table 2.2.

Table 7.15

Fuel requirements for the production of hot-rolled products from ingot in the U.K. in 1976

Fuel type	Total consumption		Consumption per kilogramme of hot-rolled product	
Electricity	102.93	GWh	0.31	kWh
Natural gas	32,014,000	Nm^3	0.096	Nm^3
LPG	1,624	t	0.0049	kg
Heavy fuel oil	2,163,000	l	0.0065	l
Gas oil	3,601,000	l	0.0107	l
Diesel oil	7,000	l	negligible	
Lubricating and hydraulic oils	704,000	l	0.0021	l

© *Crown Copyright 1981*

7.16.2 Cold rolling

The fuel requirements for cold rolling are shown in Table 7.16 and again relate to practice in the whole of the U.K. aluminium industry which produced some 263,288 tonne in 1976. Using the fuel production efficiencies of Table 2.2, the corresponding energy requirements for cold rolling are as shown by entry (k) in Table 7.11.

Table 7.16

Fuel requirements for the production of 1 kg of coiled rolled products from hot rolled material in 1976

Fuel type	Total consumption	Fuel consumption per kilogramme of rolled product	
Electricity	152.32 GWh	0.56	kWh
Natural gas	12271×10^3 Nm^3	0.047	Nm^3
LPG	2369 tonne	0.0090	kg
Heavy fuel oil	10219×10^3 l	0.039	l
Medium fuel oil	95×10^3 l		
Gas oil	3462×10^3	0.013	l
Diesel	203×10^3 l	0.00077	l
Lubricating oil	2125×10^3 l	0.0081	l

© Crown Copyright 1981

Figure 7.3 shows that 1.43 kg of hot rolled product must be fed to the cold rolling stage to yield 1 kg of finished product. Hence an average energy requirement for the combined hot and cold rolling stages can be calculated using the appropriate multiplier to give entry (l) in Table 7.11.

Published data [44] do not separate hot and cold rolling stages but quote a fuel energy requirement of 0.86 kWh of electricity and a thermal fuel requirement of 9.33 MJ per kg of finished can stock produced from remelt ingot. If the thermal fuel is regarded as natural gas, then using the fuel production efficiencies of Table 2.2, the energy associated with can-stock production is as shown by entry (m) in Table 7.11. Comparison of entries (l) and (m) in Table 7.11 shows that the energy requirements are similar. Again the U.S. values (entry (m)) have been used in this work since it is thought that they are more representative of the energy that will probably be achieved when can-stock is produced in the U.K.

7.17 TOTAL ENERGY REQUIRED TO PRODUCE CAN- AND END-STOCK

Regardless of whether the metal input to the fabrication plant is all primary aluminium or some mixture of primary aluminium and clean scrap, the energy for sheet manufacture from metal at the plant will be given by the remelt energy plus the fabrication energy. Fig. 7.3 shows that 1.75 kg of remelt ingot is required to produce 1 kg or final sheet.

The total energy required to produce 1 kg of can-stock from ore in the ground must take account of the production and delivery of 1.04 kg of primary can-stock alloy (Fig. 7.3). The transport energy has been estimated by assuming that the fabrication plant is 200 miles from the smelter and that delivery is by articulated vehicle of 20 tonne payload with a performance as shown by Table 3.1. The total energy required to produce 1 kg of can-stock sheet from ore in the ground is therefore as shown by entry (n) in Table 7.11.

Making similar assumptions, the total energy required to produce 1 kg of end-stock from raw materials from ore in the ground can be calculated to give the value shown by entry (p) in Table 7.11.

A small amount of pure aluminium sheet is used in the preparation of the ring pull tabs on easy-open ends for cans and the same procedure as outlined above leads to a total production energy from raw materials in the ground as shown by entry (q) in Table 7.11.

7.18 THE EFFECT OF RECYCLING SECONDARY ALUMINIUM

In view of the energy intensive nature of the processes for primary aluminium production, it is to be expected that reclamation and recycling of aluminium scrap will significantly affect the total energy required in the production of can- and end-stock. In the U.K. where the all-aluminium can is a relative new-comer to the scene, such recycling schemes are little practised apart from the recovery of can making scrap. This is in sharp contrast to the U.S.A. where all-aluminium cans have a significant share of the market. The effect of recycling is considered in Chapter 11.

CHAPTER 8

Iron, Steel and Tinplate

8.1 INTRODUCTION

Iron and steel products form by far the largest production of metal, in tonnage terms, in the U.K. The energy required to produce steel is of direct interest here because tinplate cans are manufactured from cold-rolled steel sheet which has been electrolytically plated with tin. Crown closures are also made from tinplate. In addition, steel products find use in other operations as for example in the steel strapping used for bulk packaging.

The production of steel from iron ore involves a number of processes. Iron ore is mined and shipped to the iron works where it may be fed directly to the blast furnace or it may undergo a preliminary treatment known as sintering in which it is burned with limestone and coke to form a porous mass (sinter) that is more easily reduced to iron in the blast furnace.

The blast furnace itself is a large countercurrent shaft furnace. Ore or sinter, coke and limestone, together with several minor inputs are charged to the top of the furnace. The mass is heated by the combustion of coke and oil in a jet of preheated air (the blast) which is injected through holes known as tuyères in the side of the furnace. The ore is reduced by the carbon in the coke and by the carbon monoxide produced by partial combustion of the coke. Some of the carbon monoxide produced, remains unreacted and is taken off from the top of the furnace as a mixture with carbon dioxide and nitrogen. This mixture, called blast furnace gas, has a significant calorific value and is usually burnt to recover the fuel content. The heat so produced may be used within the ironworks for heating the incoming blast to the furnace and for generation of electricity on site. Occasionally it is piped to outside factories which use it as fuel (as for example in some glass factories).

The main output from the blast furnace is pig iron, known also as hot metal, because in an integrated works, where steel making plant is associated with the iron making operation, the molten iron is normally charged to the steelmaking furnace while still hot so as to retain the sensible heat. Iron and slag, which contains the unwanged gangue materials associated with the iron ore, are tapped from the base of the blast furnace.

The hot metal is charged to a steel making furnace where it reacts with elemental oxygen or iron oxide to remove excess carbon picked up in the blast furnace. Other impurities are also removed at this stage, again, largely by oxidation.

In the U.K. there are now essentially only two types of steel making furnace in common use. That with the largest output is the basic oxygen furnace (BOF) sometimes referred to as the basic oxygen converter (BOC). Iron is charged to a pear shaped converter and oxygen is blown in. The metal, oxygen and slag form an emulsion in which the iron is rapidly refined to steel. This furnace is only used in integrated works since it requires a supply of hot metal. Scrap is also charged to the BOF to control temperature because the removal of impurities by oxidation is strongly exothermic and the heat evolved greatly exceeds the thermal losses from the furnace. The process is therefore essentially autogenous with a small amount of electricity being required to rotate and tilt the furnace.

The second type of furnace in widespread use is the electric arc furnace (EAF). The metal charge is heated by means of an electric arc struck between graphite electrodes and the charge. The EAF is most commonly employed with 100% scrap charges although it is extremely flexible and will accomodate a wide range of inputs. It is not used in tinplate production in the U.K.

In the U.K. iron is made from both indigenous and imported ore. Although both are extremely varied in composition and mineral type, they differ mainly in the iron content and the nature of the associated gangue. U.K. ores exhibit iron contents typically in the range 23 to 25% and the gangue is a mixture of siliceous and limey material giving a basicity which is nearly ideal for blast furnace operations. Imported ore is generally of a siliceous nature but has an iron content of about 60%. It therefore requires less energy to extract the iron but on the other hand requires more limestone to control the basicity. Transport distances will also be quite different being typically 50 miles for U.K. ores and up to 2400 miles for imported material [48]. In practice, the blast furnace charge is usually a blend of home-produced ore and imported ores.

Steel for tinplate manufacture is made in the BOF and is usually conventionally cast in capped ingots. It is therefore produced in integrated iron and steel works which include all operations from ore treatment (sintering) through to the production of hot rolled coil — see Fig. 8.1. In 1977, tinplate was manufactured in the U.K. from hot rolled coil at three plants in South Wales. These plants were supplied in turn from two integrated steelworks. The analysis in the following sections is based on data supplied by the British Steel Corporation and is representative of practices at the above locations.

Throughout the analysis, production processes are treated as unit operations, but the inputs and outputs from the operations up to the manufacture of hot rolled coil have been obtained from flows within the integrated works. Within the integrated works, 36% of the electricity requirement is generated on-site

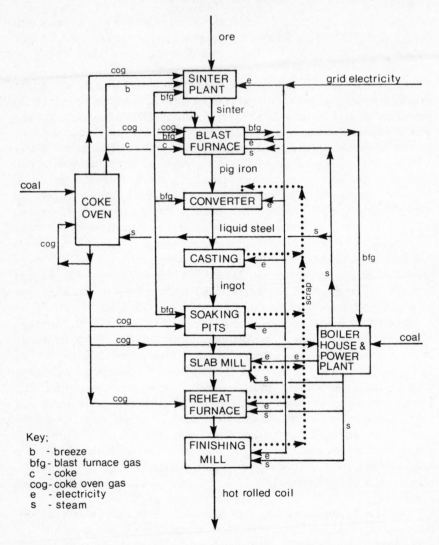

Fig. 8.1 – Selected materials and energy flows in an integrated iron and steel works.

and the primary fuels needed for this generation are included in the primary fuel inputs to the site. Hence the grid value of electricity production efficiency has been applied to only 64% of the total electricity requirement of the site. Published data for iron and steel production may be found in [45–47], [49] and [50].

Table 8.1

Energy requirements of some operations in iron and steel production. See text for full explanation

Operation	Electricity/MJ		Oil fuels/MJ			Other fuels/MJ			Total energy /MJ
	Fuel production and delivery	Energy content of fuel	Fuel production and delivery	Energy content of fuel	Feedstock energy	Fuel production and delivery	Energy content of fuel	Feedstock energy	
(a) 1 kg iron ore from ore in the ground	0.42	0.14	0.03	0.14	–	0.03	0.42	–	1.18
(b) Delivery of 1 kg of U.K. produced ore	0.01	neg	0.01	0.04	–	–	–	–	0.06
(c) Delivery of 1 kg of overseas produced ore	–	–	0.07	0.32	–	–	–	–	0.39
(d) 1 kg of sinter from ore in the ground	0.63	0.24	0.11	0.50	–	0.34	2.75	–	4.57
(e) Fuel energy for 1 kg pig iron with no credit for BF gas	0.12	0.06	0.46	2.21	–	2.30	16.58	–	21.73
(f) Fuel energy for 1 kg pig iron with 4.2 MJ credit for BF gas	0.12	0.06	0.46	2.21	–	2.30	12.38	–	17.53
(g) Provision of materials to blast furnace per 1 kg iron	1.09	0.41	0.19	0.89	–	0.55	4.45	–	7.58
(h) 1 kg pig iron from raw materials in the ground	1.21	0.47	0.65	3.10	–	2.85	16.83	–	25.11
(j) 1 kg liquid steel from ore in the ground (BF/BOF)	1.37	0.59	0.51	2.45	–	2.23	13.25	–	20.40
(k) 1 kg hot rolled slab from liquid steel	0.18	0.10	–	–	–	–	1.27	–	1.55
(l) 1 kg hot rolled strip from slab	0.64	0.33	0.56	2.67	–	–	0.13	–	4.33
(m) 1 kg hot rolled strip from ore in the ground	2.49	1.15	1.18	5.64	–	2.70	17.51	–	30.67

8.2 IRON ORE MINING AND DELIVERY

Iron ore is largely mined in open pit operations. A number of values for mining energy have been reported in the literature from 0.41 MJ/kg to 5.80 MJ/kg [7, 21, 29, 30, 52, 57]. The data of [30] are the best documented and are based on the energy to mine ore when 3.5 kg of waste are removed per kg of ore mined. The process requirements are typically (for 1 kg of iron ore), 0.396 ft^3 of natural gas, 0.000726 gallons of oil fuels, 0.0308 kWh of electricity and 3.21 gallons of water. The corresponding energy requirements are shown as entry (a) in Table 8.1.

Iron ore is supplied to the U.K. iron and steel industry both from indigenous sources and from Canada, Norway, Liberia, Mauritania, Sweden, Venezuala, Brazil and the U.S.S.R.

The average delivery distance by rail for home produced ore is 50 miles [40]. In contrast most overseas ores have an average delivery distance of 2300 miles by ship. The average energy required to deliver home produced ore may therefore be derived from Table 3.2 and is shown as entry (b) in Table 8.1. The delivery energy of overseas ore is calculated from Table 3.4 and is shown as entry (c) in Table 8.1.

8.3 ORE PREPARATION AND SINTER PLANT

Typical inputs to the sinter plant are given in Table 8.2 and the energy requirement for sinter production from ore in the ground forms row (d) of Table 8.1. This entry has been calculated from the data of section 8.2 assuming that all ore is derived from overseas sources. This is acceptable since indigenous ores supply a decreasing proportion of U.K. demand. The energy to mine and supply 1 kg of ore to the sinter plant is then the sum of entries (a) and (c) in Table 8.1.

Table 8.2
Materials and fuels requirements to produce 1 kg of sinter

Input	Requirement
Ore fines and screenings	1.0322 kg
Other iron (flue dust, etc)	0.0481 kg
Losses of iron	0.3013 kg
Flux (predominantly limestone)	0.1411 kg
Coke breeze	0.0790 kg
Electricity	0.025 kWh
Steam	0.007 MJ
Blast furnace gas	0.298 MJ

© Crown Copyright 1981

No fuel production energy has been charged for the blast furnace gas or for the provision of thermal energy as steam since these services are both produced by the combustion of fuels elsewhere in the system. Their production energy is therefore automatically charged to the appropriate unit operation.

8.4 THE IRON BLAST FURNACE

Typical fuels and materials inputs required in the production of 1 kg of pig iron in works which supply hot rolled coil to U.K. tinplate plant, are given in Table 8.3. All ore used is imported.

Table 8.3

Fuel and raw materials requirements for the production of 1 kg of pig iron from imported ore in works supplying the tinplate mills in the U.K.

Input	Requirement	
Electricity	0.017	kWh
Steam	1.70	MJ
Blast furnace gas	2.60	MJ
Fuel oil	0.0519	kg
Coke	0.5856	kg
Manganese	0.0199	kg
Sinter	1.5838	kg
Ore and Pellets	0.2214	kg
Other iron	0.0855	kg

© *Crown Copyright 1981*

In addition to the output of 1 kg of pig iron, the blast furnace also produces 6.8 MJ of blast furnace gas, 2.6 MJ of which is used for air preheating, leaving a net export of 4.2 MJ of gas per kg pig iron. Using the data of Table 8.3, the fuel energy required to produce pig iron may be calculated to give the values shown as row (e) in Table 8.1. However, all blast furnace gas is used within an integrated works. The blast furnace may be credited therefore with the production of this gas and each subsequent user operation individually charged when blast furnace gas is used. Row (f) in Table 8.1 shows the energy to produce pig iron when this gas credit is made.

The energy associated with the provision of materials for the blast furnace is shown in row (g). Hence the total energy required to produce 1 kg of pig iron from raw materials in the ground is the sum of rows (f) and (g) and this is shown as row (h) in Table 8.1.

8.5 STEELMAKING

All steel used in the manufacture of tinplate in the U.K. is made in the basic oxygen furnace for which the inputs are as shown in Table 8.4. The energy of scrap provision is assumed to be zero, as is the energy associated with the provision of cold iron since 0.082 kg of cold iron is produced for every 1 kg of pig iron in the blast furnace. The energy required to produce 1 kg of steel is therefore as shown in Table 8.1, row (j)

Table 8.4

Fuel and materials requirements of the basic oxygen
furnace for the production of 1 kg of liquid steel

Input	Requirement
Hot metal	0.7831 kg
Scrap	0.2796 kg
Cold iron	0.0257 kg
Fluxes (mostly lime)	0.0837 kg
Other additives (not specified)	0.0126 kg
Electricity	0.055 MJ
Steam	0.010 MJ
Blast furnace gas	0.060 MJ
Oxygen (electricity requirement)	0.158 MJ
Nitrogen (electricity requirement)	0.003 MJ

© *Crown Copyright 1981*

8.6 PRODUCTION OF HOT ROLLED COIL

At present, all tinplate steels are produced in the U.K. through the conventional ingot casting route. The production sequence for hot-rolled coil from liquid steel is therefore as shown in Fig. 8.2. The materials flows are also shown on

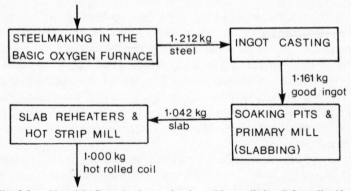

Fig. 8.2 – Materials flows in the production of hot rolled coil from liquid steel within an integrated works.

this diagram. Table 8.5 summarises the fuel requirements for the operations involved in the production of hot rolled coil in the integrated steelworks.

Table 8.5

Fuel requirements for the manufacture of hot rolled coil from liquid steel

Operation	Requirement
Soaking pits and primary mill	Electricity: 0.095 MJ/kg slab
	Steam : 0.016 MJ/kg slab
	Gas : 1.249 MJ/kg slab
Slab reheaters and hot strip mill	Electricity: 0.332 MJ/kg coil
	Steam : 0.13 MJ/kg coil
	Gas or oil : 2.667 MJ/kg coil

© *Crown Copyright 1981*

The energies per unit output for each of these operations are given in rows (k) and (l) of Table 8.1. Gas usage in the soaking pits is assumed to be blast furnace gas and no production energy is therefore charged. Oil is assumed to be the fuel burned in the slab reheating furnace. In each case no production energy is charged for steam use and 36% of the electricity is generated on site with pass-out steam.

The total energy associated with the production of 1 kg of hot rolled strip from ore in the ground may therefore be determined from rows (j), (k) and (l) using the multipliers shown in Fig. 8.2. Note that in this analysis, the casting operation has no energy associated with it. The total energy for the production of hot rolled coil in an integrated works in the U.K. from ore in the ground is therefore as shown in row (m) of Table 8.1.

8.7 TINPLATE PRODUCTION

Tinplate is produced from hot rolled coil at a site which is separate from the steel making facility. All fuels used at the tinplate works are therefore bought in and must be charged as in Table 2.2. Each tinplate works produces both single- and double-reduced plate as well as tin-free steel and a typical product route for such a plant is shown in Fig. 8.3. The percentages give an approximate indication of the proportion of the total production which passes through each operation.

The fuels consumed by the unit operations involved in tinplate and tin-free steel manufacture are given in Table 8.6. The fuel energy requirements for each of these unit operations are given in Table 8.7. Multi-stack and single-stack annealing are both batch processes and an average value has been assumed for batch annealing.

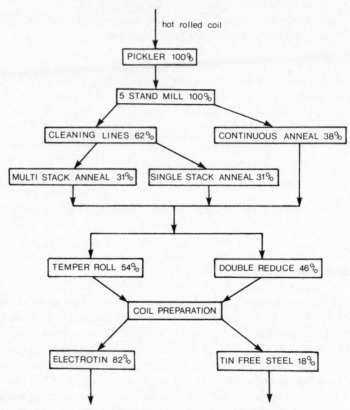

Fig. 8.3 – A typical product route for a tinplate manufacturing plant.

Table 8.6

Fuels consumed by unit operations in the production of tinplate and tin-free steel

Process	Requirement (MJ/kg of product)			
	Electricity	Steam	Natural gas	Yield
Pickling (in H_2SO_4)	0.024	0.321	nil	0.9435
5 Stand rolling mill	0.432	0.055	nil	0.9976
Electrolytic cleaning	0.048	0.624	nil	0.9875
Multi-stack annealing	0.054	nil	0.897	1.0000
Single-stack annealing	0.068	nil	0.634	1.0000
Continuous annealing	0.151	0.590	1.182	0.9805
Temper mill	0.085	0.056	nil	0.9940
Double reduction mill	0.180	0.388	nil	0.9825
Coil preparation	0.024	nil	nil	0.9373
Tin-free steel coating	0.011	0.057	0.006	0.9062
Electrotinning	0.017	0.029	nil	0.9005

If both batch annealing processes are combined, then Fig. 8.3 can be used to identify the routes from hot-rolled coil by which coil suitable either for electro-tinning or for chromating to produce tin-free steel may be produced. These routes together with the materials flows (derived from the yields given in Table 8.6) are shown in Fig. 8.4. The fuel energy requirement for each of these routes may now be calculated by multiplying the energies of the relevant unit operations shown in Table 8.7 by the appropriate mass throughputs. These energy requirements are given in Table 8.8. The totals are sufficiently close to allow an average value to be used to represent the energy of coil manufacture and this average is also shown in Table 8.8.

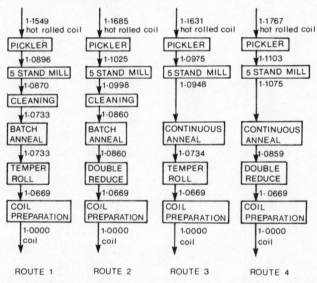

Fig. 8.4 — Possible routes from hot rolled coil to coil suitable for electrotinning or chromating to give tin-free steel.

The fuel energy requirement for the production of 1 kg of tinplate or tin-free steel may be calculated by noting from the yields in Table 8.6 that 1.295 kg of hot rolled coil is needed to produce 1.1105 kg of coil which in turn yields 1 kg of tinplate. Also, 1.287 kg of hot rolled coil produces 1.1035 kg coil which yields 1 kg of tin-free steel. The input values for hot rolled coil have been obtained by assuming that on average, 1.1658 kg of hot rolled coil is required to manufacture 1 kg of coil suitable for making tinplate or tin-free steel; that is the average of the input values to the four systems shown in Fig. 8.4.

The fuel energy requirement for the manufacture of tinplate from hot rolled coil is therefore given by:

$$(1.1105 \times \text{Table 8.8. (average)}) + (\text{Entry (k) in Table 8.7})$$

This is shown as row (l) in Table 8.7.

Table 8.7

Energy requirements of some unit processes in tinplate production in the U.K.

Operation	Electricity/MJ		Oil fuels/MJ			Other fuels/MJ			Total energy /MJ
	Fuel production and delivery	Energy content of fuel	Fuel production and delivery	Energy content of fuel	Feedstock energy	Fuel production and delivery	Energy content of fuel	Feedstock energy	
(a) 1 kg of pickled steel from hot rolled coil	0.072	0.024	0.067	0.321	–	–	–	–	0.484
(b) 1 kg cold rolled sheet from pickled steel	1.296	0.432	0.011	0.055	–	–	–	–	1.794
(c) 1 kg electrolytically clean sheet from cold rolled steel	0.144	0.048	0.130	0.624	–	–	–	–	0.946
(d) 1 kg batch annealed steel from electrolytically clean sheet	0.183	0.061	–	–	–	0.063	0.765	–	1.072
(e) 1 kg continuously annealed steel from cold rolled steel	0.453	0.151	0.123	0.590	–	0.097	1.182	–	2.596
(f) 1 kg temper milled sheet from annealed sheet	0.255	0.085	0.012	0.056	–	–	–	–	0.408
(g) 1 kg double reduced sheet from annealed sheet	0.540	0.180	0.081	0.388	–	–	–	–	1.189
(h) 1 kg prepared steel coil from temper milled or double reduced sheet	0.072	0.024	–	–	–	–	–	–	0.096
(j) 1 kg tin-free steel from temper milled or double reduced coil	0.033	0.011	0.012	0.057	–	neg	0.006	–	0.119
(k) 1 kg electrotinned sheet from temper milled or double reduced coil	0.051	0.017	0.006	0.029	–	–	–	–	0.103
(l) Fuel energy requirement for 1 kg									

	Electricity/MJ		Oil fuels/MJ			Other fuels/MJ			Total energy/MJ
	Fuel production and delivery	Energy content of fuel	Fuel production and delivery	Energy content of fuel	Feedstock energy	Fuel production and delivery	Energy content of fuel	Feedstock energy	
rolled coil	2.70	0.90	0.31	1.51	—	0.10	1.17	—	6.69
(n) Energy to provide ancilliary services (fuels + materials) to tinplate plant per 1 kg tinplate	0.23	0.08	0.18	0.89	—	0.04	0.50	—	1.92
(p) 1 kg of tin from ore in the ground	18.00	6.00	7.00	37.00	—	3.78	46.00	—	117.78
(q) 1 kg tinplate sheet from hot rolled coil	3.18	1.07	0.57	2.79	—	0.18	2.19	—	9.98
(r) 1 kg tinplate sheet from ore in the ground	6.40	2.56	2.10	10.09	—	3.68	24.87	—	49.70
(s) 1 kg tin-free steel from hot rolled coil	2.93	0.98	0.49	2.40	—	0.14	1.67	—	8.61
(t) 1 kg tin-free steel sheet from ore in the ground	6.13	2.46	2.01	9.66	—	3.61	24.21	—	48.08
(u) 1 kg of steel strapping from ore in the ground	4.05	1.68	1.44	6.92	—	2.87	18.60	—	35.56

Table 8.8

Fuel energy required to produce 1 kg of sheet from hot rolled coil using the four routes shown in Fig. 8.5

Route	Electricity/MJ		Oil fuels/MJ			Other fuels/MJ			Total energy/MJ
	Fuel production and delivery	Energy content of fuel	Fuel production and delivery	Energy content of fuel	Feedstock energy	Fuel production and delivery	Energy content of fuel	Feedstock energy	
1	2.18	0.73	0.24	1.14	—	0.07	0.82	—	5.18
2	2.51	0.84	0.31	1.51	—	0.07	0.83	—	6.07
3	2.33	0.78	0.23	1.11	—	0.10	1.27	—	5.82
4	2.66	0.89	0.31	1.47	—	0.11	1.28	—	6.72
Average	2.42	0.81	0.27	1.31	—	0.09	1.05	—	5.95

Similarly the fuel energy requirement for the production of tin-free steel from hot rolled coil may be calculated as:

(1.1035 × Table 8.8 (average)) + (Entry (j) in Table 8.7)

This is shown by row (m) in Table 8.7.

In addition to the energy and materials directly consumed by the process, account must also be taken of ancillary materials and fuels consumed by the plant. These requirements are given in Table 8.9. Approximately 0.00575 kg of rolling oil and 1.7 gallons of process water are also used per kg of tinplate or tin-free steel produced. The energy associated with the ancillary fuels (Table 8.9) and the provision of raw materials can be calculated to give row (n) in Table 8.7. Finally to calculate the total energy required to produce tinplate, a value is needed for the production of tin from raw materials in the ground.

Table 8.9
Other fuels consumed in the tinplate works

Process	Requirement (MJ/kg tinplate or tin-free steel)		
	Electricity	Steam	Natural gas
Acid recovery	0.006	0.009	0.166
Effluent plant	0.026	0.015	0.009
Rolling oil recovery	0.009	0.132	nil
Hydrogen-nitrogen atmosphere for annealing	0.012	0.036	0.328
Space heating	0.011	0.448	nil

© Crown Copyright 1981

8.8 TIN PRODUCTION

There is little quantitative, published information for the production of tin, a point noted by Wright [51]. A notional energy requirement for the production of this metal has therefore been calculated using data available for similar production techniques applied elsewhere. These calculations possess no great accuracy but do allow an estimate of the significance of this energy contribution in tinplate production.

Tin occurs as the oxide, cassiterite, or as complex sulphides. Ore grades vary from 0.4% to 0.01% measured as the total tin content. Although the oxide is usually uncontaminated, it is widely dispersed. In contrast, sulphides tend to be more concentrated but are usually contaminated with other metals. The oxide therefore requires extensive mineral processing and the sulphide needs extensive preliminary chemical and thermal treatment. The production processes have therefore been divided into four main groups; mining, mineral processing, roasting and smelting.

From ore with an average tin content of 0.2%, the production of 1 kg of tin requires the mining of 500 kg of ore. This is enriched to about 80% concentration during mineral processing and roasting operations. The data for mining and mineral processing have been obtained from a number of British and European operations, which, although not used for tin specifically, are used for minerals having similar properties. The roasting process has been assumed to require about 10% of the energy used in smelting and the smelting energy has been taken from [51] for British tin smelting practice. The total energy required for the production of tin has therefore been calculated as that shown by entry (p) in Table 8.7.

Calculation of the energy requirement for tin use in tinplate mills is complicated by the fact that the thickness of the tin coatings applied varies with the final end use of the tinplate. In this analysis, a coating rate of 0.0112 kg m^{-2} has been assumed although this probably overestimates slightly the use of tin for beverage containers. For a constant thickness of tin coating, the energy due to tin relative to that due to steel will vary with steel thickness. Within the systems examined here, three different tinplate thicknesses are involved; for can bodies the thickness is 0.1676 mm, for can ends it is 0.2997 mm and for crown closures it is 0.26 mm. An average plate thickness for all applications of 0.25 mm has therefore been assumed.

Writing A as the area of steel sheet that uses 1 kg of tin, then from the above data

$$A = \frac{1}{2 \times 0.0112} = 44.64 \text{ m}^2$$

With an average steel density of 7860 kg m^{-3}, this corresponds to a steel mass of 87.72. Hence 1 kg of steel will require 0.0114 kg of tin to convert it to tinplate.

8.9 TOTAL ENERGY FOR TINPLATE PRODUCTION

The total energy required to produce tinplate from hot rolled coil at the tinplate work is now given by

(Row (l), Table 8.7) + (Row (n), Table 8.7) + 0.0114 (Row (p), Table 8.7)

This is shown as row (q) in Table 8.7.

The energy required to produce tinplate sheet from ore in the ground is obtained by recognising that 1.295 kg of hot rolled coil is needed to produce 1 kg of tinplate. The required energy is therefore calculated as

(1.295 × Row (m), Table 8.1) + (Row (q), Table 8.7)

This yields row (r) in Table 8.7 so that the gross energy requirement for the production of tinplate is 49.70 MJ/kg.

8.10 TOTAL ENERGY FOR TIN-FREE STEEL PRODUCTION

Tin-free steel is protected by a thin chromate layer. The production of this coating requires only 0.3 kg chromic oxide per tonne of tin-free steel so that the energy associated with its production can be safely ignored.

The total energy required to produce tin-free steel from hot rolled coil at the finishing works is therefore

$$(\text{Row (m), Table 8.7}) + (\text{Row (n), Table 8.7})$$

This result appears as row (s) in Table 8.7. Since 1.287 kg of hot rolled coil is required for the production of 1 kg of tin-free steel, the total energy for tin-free steel production from ore in the ground is

$$(1.287 \times \text{Row (m), Table 8.1}) + (\text{Row (s), Table 8.7}) \ .$$

This appears as row (t) in Table 8.7 and shows that the gross energy requirement for tin-free steel production is 48.08 MJ/kg.

8.11 STEEL STRAPPING FOR PACKAGING

Steel strapping for packaging is essentially cold rolled mild steel strip and is similar to the product from the 5-stand mill in the tinplate works. If it is assumed that this is the method of production, then the materials flow required to produce steel strapping will be an input of 1.065 kg of hot rolled coil leading to a production of 1.002 kg of pickled coil which would give 1.00 kg of cold rolled coil. The energy required to accomplish this sequence is therefore as shown by row (u) of Table 8.7 and takes into account rolling oil use, acid recovery and space heating requirements.

CHAPTER 9

Miscellaneous Materials and Services

9.1 INTRODUCTION

In addition to those inputs to systems discussed in Chapters 5 to 8, there are a number of others which are common to many industrial processes and these are considered in this Chapter. The overall energies required for their production from raw materials in the ground are summarised in Table 9.1 and the notes given below explain the basis of the calculations.

9.2 WATER

Almost all industrial processes use water derived from the public main, from private wells or from rivers and canals. Public water supplies require energy in pumping and in producing the additives needed to counteract bacterialogical contamination. One of the most widely used energy requirements for water production is that due to Smith [2]. Assuming that all energy used in water production is electricity, the Smith value has been used here, after correcting to the fuel production efficiencies of Table 2.2.

When water is derived from sources other than the public main, it is often extremely difficult to separate electricity used in water pumping operations from other uses, because it is often included as a small addition to the total factory energy and is seldom monitored. However, in those instances where electricity used for these purposes is monitored, the associated energy is found to be very close to that shown in Table 9.1 for mains water production. The value shown in Table 9.1 has therefore been used irrespective of the source of the water.

9.3 STEAM

Most plants use steam for a variety of purposes and, in general, the fuel required to generate steam forms part of the total fuel consumption of the factory.

Table 9.1

Energy required to produce miscellaneous materials from raw materials in the ground. All values per 1 kg of product unless otherwise stated. The notes in the text provide the basis of the calculation

Product	Electricity/MJ		Oil fuels/MJ			Other fuels/MJ			Total energy /MJ
	Fuel production and delivery	Energy content of fuel	Fuel production and delivery	Energy content of fuel	Feedstock energy	Fuel production and delivery	Energy content of fuel	Feedstock energy	
Water (1 gallon)	0.03	0.01	–	–	–	–	–	–	0.04
Steam	–	–	0.80	3.92	–	–	–	–	4.72
Carbon dioxide	2.43	0.81	–	–	–	–	–	–	3.24
Lead	6.00	2.00	1.37	3.53	–	4.80	35.90	–	53.60
Crystal flux	7.65	2.55	1.99	8.08	6.00	1.48	17.97	–	45.72
Can soldering flux	3.57	1.19	4.59	22.34	30.50	neg	0.02	–	62.21
Can sealing compound	8.52	2.84	12.14	60.30	96.70	–	–	–	180.50
Crown sealing compound	18.78	6.26	9.74	49.36	31.99	–	–	–	116.13
Screw sealing compound	13.11	4.37	6.78	34.36	22.27	–	–	–	80.89
Sodium nitrite	0.40	0.13	1.25	6.41	–	0.51	6.40	5.85	20.95
Benzene, toluene, xylene	3.46	1.15	6.26	30.58	45.00	–	–	–	86.45
Methanol	0.66	0.22	2.41	11.76	28.14	–	–	–	43.19
Lacquer	18.46	6.15	20.38	100.58	151.54	6.23	73.38	–	376.72
Varnish	15.43	5.13	17.59	85.83	99.13	3.52	41.49	–	268.12
Printing ink	2.06	0.69	4.34	21.17	36.57	–	–	–	64.83
Epoxy adhesive	2.49	0.83	6.94	33.86	45.00	–	–	–	89.12

However, there are instances in which large sites generate steam in a central boiler-house and distribute it around the plant. Whilst in principle it should be possible to calculate the proportion of the boiler-house energy attributable to each of the satellite plants, there are many instances where no satisfactory apportionment is possible using the available data. In addition, much published information frequently refers to a steam requirement without indicating the energy required to produce it. A general value has therefore been calculated for the energy required to produce intermediate pressure steam, assuming that oil fuels are used in the boiler and taking account of typical losses in the distribution system. This average energy requirement is shown in Table 9.1.

9.4 CARBON DIOXIDE

Carbon dioxide is derived from two main sources; from the combustion of coal, coke and oil and from by-product streams as for example in fermentation and petrochemical operations. The costs of carbon fuels mean that most carbon dioxide used is derived from by-product streams so that the energy requirements for the production of carbon dioxide for the beverage industry are primarily those for running compression and liquefaction equipment. Typical average requirements derived from a number of sources are shown in Table 9.1.

9.5 LEAD

Limited information exists for the manufacture of lead [11, 21] but no detailed breakdown of energy use by fuel type is given. Since significant quantities of lead are produced simultaneously with zinc by a blast furnace process, the same fuel breakdown is assumed for lead [58]. This is shown in Table 9.1. The value quoted here is probably an overestimate.

9.6 SOLDER

Three-piece tinplate cans have a body seam secured by solder. For beer and food cans, the solder composition is 2% tin and 98% lead, whereas for soft drinks cans, pure tin solder (100% tin) is used to prevent lead pickup by the contents. In practice, lead/tin solder is not taken into the can factory. Pure lead is used and the tin content arises from the pickup of tin by the solder bath during can production. Indeed one of the problems in can making is ensuring that the tin level does not rise above the 2% value. For lead/tin solder, the energy requirement for pure lead has been used (Table 9.1) and for pure tin solder the value for tin has been used (Table 8.7).

9.7 SOLDER FLUXES

Two types of solder flux are used in can making; crystal flux which is floated in the solder bath and can-soldering flux which is applied to the can surface immediately prior to solder application. In both cases the quantities used per can produced are small so that notional values for the production energy of these fluxes will suffice.

9.7.1 Crystal flux
Crystal flux is predominantly a mixture of the chlorides of zinc and ammonia. For the purposes of the present work, an average composition has been assumed:

$$1.1 \, Zn: \, 2 \, NH_3: \, 1 \, Cl \; .$$

Hence the production of 1 kg of flux requires approximate inputs of 0.5 kg of zinc, 0.23 kg of ammonia and 0.25 kg of chlorine. The mixture is usually made as a mixture of the appropriate chlorides. From manufacturers' data the energy used in the production of crystal flux from raw materials is as shown in Table 9.1. In can making, crystal flux is typically added at the rate of 4 kg per bath and this would normally be sufficient to last an eight hour shift.

9.7.2 Can-soldering fluxes
A variety of can-soldering fluxes are in use. These are predominantly organic mixtures such as isopropanol with glycerine and hydrochloric acid or isopropanol with sebacic acid. Because of the wide variety of formulations in use, energy requirements have been calculated for a number of different compositions and the value shown in Table 9.1. is the average of these results.

9.8 SEALING COMPOUND (LINING COMPOUND)
Sealing compounds are applied to can ends and to all types of bottle closures to ensure a hygienic, gas-tight seal after filling. Sealing compounds are essentially suspensions of polymers in a solvent (water or hexane), together with smaller quantities of plasticisers, antioxidants and inert fillers. This work is primarily concerned with three different types of sealing compound, namely:

(a) can sealing compound which is a suspension of either SBR or natural rubber in hexane or water,
(b) crown cork sealing compound which is typically a suspension of PVC in plasticiser usually of the phthalate type, and
(c) sealing compound for screw closures which is essentially similar to (b) but, due to the different application, the basic properties are modified, usually by increasing the amount of plasticiser.

From an examination of a number of formulations, the production energies have been calculated for each of these three types and the values shown in Table 9.1 are the average of these results. Note that the normalisation parameter for these compounds is the residual mass of the compound after evaporation of the solvent, even though the energy associated with production of the solvent has been included in the calculations.

9.9 SODIUM NITRITE

Sodium nitrite, used in detinning operations, is manufactured by absorbing the oxidation products of ammonia in sodium carbonate (soda ash). The reaction sequence is:

Ammonia synthesis: Natural gas + air + steam = NH_3 + by-products .

$$(9.1)$$

Ammonia oxidation: $4NH_3 + 5O_2 = 4NO + 6H_2O$. $\qquad$ (9.2)

Absorption: $Na_2CO_3 + 2NO + \frac{1}{2}O_2 = 2NaNO_2 + CO_2$. $\qquad$ (9.3)

Typically a yield of 1 kg of sodium nitrite requires an input of 0.768 kg of sodium carbonate and 0.247 kg of ammonia. The energy associated with ammonia synthesis has been taken from [53]. Reaction (9.2) is self-sustaining and the requirements of the absorption reaction (9.3) can be calculated from the demand for sodium carbonate (See Table 6.2), to give the overall energy requirement shown in Table 9.1.

9.10 BENZENE, TOLUENE, XYLENE

Benzene, toluene and xylene are widely used as solvents in the printing industry and in the production of lacquers and varnishes. They are manufactured as co-products from naphtha by reforming reactions. The values shown in Table 9.1 have been derived by partitioning total plant energy requirements on a mass basis between the three products.

9.11 METHANOL

Methanol is used as a solvent in printing and in coating lines in can making. The most common production route is the oxidation of natural gas or cracked oil products, and the energy associated with this production sequence is shown in Table 9.1.

9.12 LACQUERS AND VARNISHES

Lacquers and varnishes are complex mixtures of organic compounds in solvents. A variety of formulations are used and from inspection of a number of such commercial specifications, the energy requirement has been calculated for the production of typical examples. These are:

Lacquer: 13% epoxy; 13% vinyl; 37% xylene/toluene; 37% methanol.
Varnish: 46% alkyd; 54% xylene.

The production energy requirement from raw materials in the ground is shown in Table 9.1. Note that the normalising parameter is the residual mass, even though the energy of the solvents has been included in the calculations.

9.13 PRINTING INKS

Printing inks consist of fine dispersions of pigments or dyes in solvent. The mass of the residual pigment or dye after evaporation of the solvent is small compared with the mass of the solvent. Here we have assumed that the ink can reasonably be regarded solely as solvent and typically this would be toluene or methanol. Table 9.1 has been calculated assuming a 50/50 mix of these two solvents.

9.14 EPOXY BASED RESINS

Epoxy resins are used as adhesives and as a component in many lacquers. They are a family of compounds based on the oxides of ethylene and propylene. The common processing route from crude oil is via naphtha and ethylene (or propylene) followed by oxidation. Thereafter the precise processing route depends upon the final formulation of the resin. The value shown in Table 9.1 is the average of a number of different formulations.

CHAPTER 10

Glass Container Production

10.1 INTRODUCTION

In 1977, the glass container industry produced over one thousand million glass bottles of various types for the packaging of beer, cider and carbonated soft drinks. A detailed breakdown of this production is given in Table 10.1 From these data it can be seen that carbonated soft drinks are responsible for approximately 76% of the number of containers sold with the remaining 24% being used for beer and cider. Approximately 32% of all containers were returnable. However, the use of returnable bottles is much more widespread in the packaging of beer and cider where they form approximately 52% of the market compared with only 25% for carbonated soft drinks.

Within this market, the number of different types of container available is very large. Table 10.2 for example shows a selection of the containers marketed in 1977 for the non-returnable carbonated soft drinks market. Similar ranges exist for returnable bottles and for containers produced specifically for beer and cider. The existence of this wide range of bottle masses means that it is difficult, if not impossible, to describe a 'typical' bottle for a given capacity. For example, a 1 litre non-returnable bottle for carbonated soft drinks may have a mass anywhere in the range 546 to 740 g. A graphical method that can be used to give an average bottle mass is demonstrated by Fig. 10.1, which shows a plot of the data in Table 10.2. The mean line gives the required relationship between mass and volume. Similar graphs can be prepared for the other types of containers.

10.2 SOURCES AND TREATMENT OF DATA

Information concerning the inputs of raw materials and fuels and the output of finished saleable glass containers has been obtained directly from U.K. glass manufacturers who were asked to supply data relating to the calendar year 1977. However, in many instances, the most accurate and readily available information related to the company's accounting period, which only occasionally coincided

Table 10.1

Production of glass containers in the U.K. in 1977 for beer, cider and carbonated soft drinks. Values are in millions of units

Container size (ml)	Beer		Cider		Carbonated soft drinks		Totals
	Returnable	Non-returnable	Returnable	Non-returnable	Returnable	Non-returnable	
up to 142	–	–	–	–	36.90	4.33	41.23
143 – 199	–	–	–	–	36.48	80.59	117.07
up to 199	14.17	14.99	24.89	nil	–	–	54.05
200 – 256	0.66	6.49	nil	0.08	2.21	284.28	293.72
257 – 284	29.77	42.64	7.67	0.09	1.11	22.46	103.74
285 – 379	21.36	6.45	nil	0.07	nil	17.52	45.40
380 – 568	25.83	8.90	0.04	nil	0.02	111.27	146.06
569 – 758	0.63	0.02	nil	1.86	64.23	12.49	79.23
759 – 1136	1.08	6.98	10.53	39.31	66.52	90.97	215.39
over 1136	1.33	–	0.18	0.01	0.03	nil	1.55
Totals	94.83	86.47	43.31	41.42	207.50	623.91	1097.44

Table 10.2

Mass-volume relationships for some non-returnable glass bottles produced in 1977 for carbonated soft drinks.

Volume/ml	Mass/g	Volume/ml	Mass/g	Volume/ml	Mass/g
129	162	284	230	760	518
181	160	284	252	977	618
190	223	293	249	977	630
193	159	293	252	977	640
193	281	288	247	1000	546
199	173	334	380	1000	553
250	176	341	196	1000	567
250	187	356	227	1000	600
250	188	500	347	1000	616
250	191	509	349	1000	625
250	193	509	352	1000	670
250	195	515	312	1000	694
250	204	592	444	1000	740
250	208	600	440	1014	465
250	216	700	489	1018	670
259	219	738	468	1136	703
263	198	750	496	1136	710
264	181	753	533	1176	740

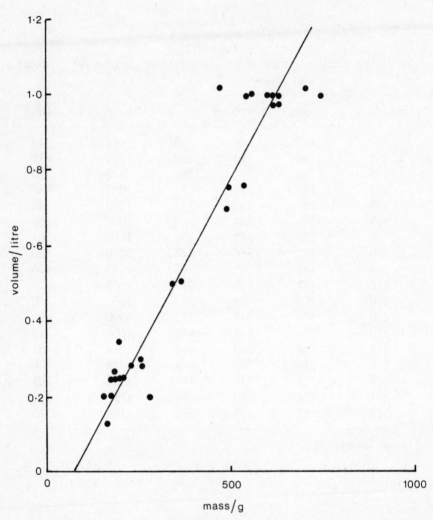

Fig. 10.1 – Mass-volume relationship for non-returnable bottles for carbonated soft drinks.

with the calendar year, so that data was chosen which most closely approximated to the calendar year. This means that the information presented here covers the period from August 1976 to March 1978.

The total number of containers produced by the companies who supplied information represents some 91% of the total production indicated in section 10.1. The following calculations are therefore thought to be a reasonable representation of glass container manufacture in the U.K. in 1977.

All of the glass companies were concerned that information concerning their operations should not be identified and so each factory has been assigned an arbitrary reference letter. This letter possesses no physical significance but the same letter is used to identify the same factory throughout.

Raw materials and fuel requirements of glass factories are considered under a number of headings, namely:

(a) raw materials requirements,
(b) energy required to produce raw materials,
(c) energy required to deliver raw materials,
(d) energy required to operate the glass making factory, and
(e) total energy required to produce glass containers, which is the sum of the energies calculated in (b), (c) and (d).

Glass factory requirements can also be considered in a number of other ways, depending upon the amount of detail sought. There are essentially four levels of detail that may be considered, namely;

(i) Total U.K. production data
(ii) Production data for each factory,
(iii) Production data for each furnace and associated forming lines, and
(iv) Production detail within any production line.

The present work concentrates on (i) and (ii) since these can be calculated from information that is relatively easily assembled by glass factories in an accurate form. Item (iii) has been considered only in relation to determining the influence of glass colour on production energy requirement and item (iv) has not been examined at all.

10.3 TOTAL REQUIREMENTS OF RAW MATERIALS IN THE U.K. IN 1977

The total consumption of raw materials in glass making in the U.K. in 1977 has been calculated by aggregating the information supplied for the individual factories. This is shown in Table 10.3, together with the total output of saleable glass containers in the same period. No distinction has been made between glass of different colours and no allowance has been made for the fact that the factories included within these totals produced containers other than beer, cider and carbonated soft drinks bottles.

Table 10.3

Total materials requirements of all glass container factories producing bottles for beer, cider and carbonated soft drinks in 1977 and which have supplied information in this study

Item	Total consumption	Requirements per tonne of saleable glass produced	
Inputs			
Sand	1,364,082 t	0.7572	t
Limestone (including dolomite)	312,078 t	0.1732	t
Soda ash	407,055 t	0.2260	t
Nepheline syenite	20,401 t	0.0113	t
Feldspar	26,034 t	0.0145	t
Alumina	19,450 t	0.0108	t
Calumite Brand Slag	42,490 t	0.0236	t
Foreign cullet	11,861 t	0.0066	t
Sodium sulphate	4,205 t	0.0023	t
Calcium sulphate	11,859 t	0.0066	t
Selenium	18 t	0.00001	t
Cobalt oxide	0.7 t	0.0000003	t
Iron chromite	1,406 t	0.0008	t
Sodium nitrate	69 t	0.00004	t
Water	929,013,222 gall	516	gall
Saleable output			
Glass as containers	1,801,481 t	1.0000	t

Note that the total mass of the input materials is 2,221,008.7 tonne. The major cause of mass loss during glass making (here 19%) is evolution of gases during decomposition of some inputs.

© *Crown Copyright 1981*

10.4 ENERGY REQUIRED TO PRODUCE RAW MATERIALS FOR TOTAL U.K. PRODUCTION

The energy associated with the production of raw materials requirements (Table 10.3) is somewhat more difficult to calculate with any accuracy because of the varying sources of supply from factory to factory. Sand for example may attract a production energy anywhere in the range 0.26 MJ/kg to 1.97 MJ/kg (See Table 6.1) and similar variations are encountered in many of the other materials. Nevertheless an approximate overall energy requirement has been calculated as shown in Table 10.4.

Table 10.4

Energy required to produce raw materials for an output of 1 tonne of saleable glass in the U.K. in 1977

Input	Electricity/MJ		Oil fuels/MJ			Other fuels/MJ			Total energy /MJ
	Fuel production and delivery energy	Energy content of fuel	Fuel production and delivery energy	Energy content of fuel	Feedstock energy	Fuel production and delivery energy	Energy content of fuel	Feedstock energy	
Sand	90.86	30.29	37.86	174.16	nil	nil	nil	nil	333.17
Limestone	20.78	6.93	6.93	34.64	nil	nil	nil	nil	69.28
Soda ash	51.98	18.08	361.60	1853.20	nil	58.76	872.36	nil	3215.98
Nepheline syenite	13.22	4.41	1.81	8.81	nil	nil	nil	nil	28.25
Feldspar	16.97	5.66	2.32	11.31	nil	nil	nil	nil	36.26
Alumina	11.02	3.67	32.94	162.00	nil	5.62	69.23	nil	284.48
Calumite slag	9.91	3.30	3.78	18.17	nil	nil	nil	nil	35.16
Foreign cullet	0.13	0.07	0.92	4.88	nil	neg	0.33	nil	6.33
Sodium sulphate	1.86	0.62	1.40	6.88	nil	nil	nil	nil	10.76
Calcium sulphate	3.56	1.19	0.01	0.07	nil	nil	nil	nil	4.83
Selenium	1.07	0.36	0.50	2.39	nil	0.03	0.34	nil	4.69
Cobalt oxide	neg	neg	neg	neg	nil	nil	nil	nil	neg
Iron chromite	neg	neg	neg	neg	nil	nil	nil	nil	neg
Sodium nitrate	nil	nil	neg	neg	nil	nil	nil	nil	neg
Water	15.48	5.16	nil	nil	nil	nil	nil	nil	20.64
Totals	236.84	79.74	450.07	2276.51	nil	64.41	942.26	nil	4049.83

© Crown Copyright 1981

10.5 DELIVERY OF RAW MATERIALS FOR TOTAL U.K. PRODUCTION

Each of the factories provided information on the source of their raw materials inputs, the mode of transport used in their delivery and the typical average loading of the vehicles. Imported materials, which were typically minor additives such as selenium, were transported to the U.K. by ship with internal delivery by road. Bulk raw materials such as sand, limestone and soda ash, produced within the U.K., are generally transported by rail or 20 tonne payload lorries. Minor inputs are invariably transported by road and are often packed in 50 kg sacks. These minor ingredients are frequently delivered as part loads.

For simplicity, transport elements have been divided into four groups:

(a) ships for importing raw materials,
(b) rail for bulk deliveries using a roundabout system whereby full loads are carried in one direction and empty loads in the reverse direction,
(c) bulk road deliveries of large tonnage materials inputs in 20 tonne payload lorries with a full load in one direction only,
(d) delivery of minor inputs on 10 tonne payload lorries operating in a circuit with a number of dropping points. The basis of the calculations for these lorries is full loads in one direction and empty loads in the reverse, with distances calculated as the most direct route.

Table 10.5 summarises the transport requirements of all the factories and Table 10.6 shows the corresponding energy requirements normalised with respect to the production of 1 kg of saleable glass output as containers based on Tables 3.1 to 3.4.

Table 10.5

Transport requirements for the delivery of raw materials to glass factories in the U.K. in 1977.

Transport mode	Total U.K. glass factory requirement	Requirement per tonne of saleable glass
20 tonne payload road vehicles (vehicle-miles)	5,778,364	3.208
10 tonne payload road vehicles (vehicle-miles)	42,126	0.023
Rail (tonne-miles)	99,368,986	55.160
Sea (tonne-miles)[†]	99,227,369	55.081

† Including barge transport.

Table 10.6

Energy associated with the delivery of raw materials to glass factories in the U.K. in 1977 to produce 1 tonne of saleable glass as containers

Fuel type	Fuel production and delivery energy/MJ	Energy content of fuel /MJ	Feedstock energy /MJ	Total energy /MJ
Electricity	11.03	3.60	nil	14.63
Oil fuels	31.00	154.73	0.26	185.99
Other fuels	1.81	11.05	nil	12.86
Totals/MJ	43.84	169.38	0.26	213.48

© *Crown Copyright 1981*

10.6 TOTAL FUEL CONSUMPTION BY GLASS FACTORIES

The total fuel consumption of U.K. glass container factories in 1977 is given in Table 10.7. Normalising with respect to unit mass output of saleable glass and using Table 2.2, the energy requirement corresponding to these fuels will be shown as Table 10.8. Note that these calculations include the consumption of lubricating oil and greases.

Table 10.7

Total fuel consumption by all glass factories producing beer, cider and carbonated soft drinks bottles in the U.K. in 1977.

Fuel	Total consumption	Average consumption per tonne of saleable glass
Electricity/kWh	612,523,459	340.011
Natural gas/therms	77,391,227	42.960
Heavy fuel oil/gall	56,876,868	31.572
Medium fuel oil/gall	244,971	0.136
Gas oil/gall	750,705	0.417
LPG/kg	7,001,452	3.886
Kerosine/gall	6,090	0.003
Diesel/gall	37,417	0.021
Petrol/gall	3,075	0.002
Lubricating oil/gall	161,213	0.089
Grease/kg	53,324	0.030
Manufactured gas/therms	7,649,366	4.246
Coke/kg	67,000	0.037

© *Crown Copyright 1981*

Table 10.8

Average fuel energy requirement for all glass factories in the U.K. in 1977 which produced bottles for beer, cider and carbonated soft drinks. Values are expressed per tonne of saleable glass produced

Fuel type	Fuel production and delivery energy/MJ	Energy content of fuel /MJ	Feedstock energy /MJ	Total energy /MJ
Electricity	3672.12	1224.04	nil	4896.16
Oil fuels	1288.72	6195.16	nil	7483.88
Other fuels	547.59	4976.62	nil	5524.21
Totals/MJ	5508.43	12395.82	nil	17904.25

© Crown Copyright 1981

10.7 TOTAL AVERAGE ENERGY REQUIRED TO PRODUCE GLASS CONTAINERS

The total average energy required to produce glass containers in the U.K. in 1977 is the sum of Tables 10.4, 10.6 and 10.8 and is shown in Table 10.9. It is important to note that neither the total energy requirement for the production of saleable containers (22167.56 MJ/t in Table 10.9) nor the energy used by the glass factory (17904.25 MJ/t in Table 10.9) is the energy required to melt glass which is only one factor in the above calculations. Many criticisms of published data argue that energy requirements of the type shown by Table 10.9 cannot be correct because they disagree with known melting energies. Such arguments reveal a misunderstanding of the total system energy requirement because Table 10.9 refers to the system which takes as inputs raw materials in the ground and yields saleable glass containers as output.

10.8 RAW MATERIALS CONSUMPTION IN INDIVIDUAL GLASS FACTORIES

Whilst Table 10.9 is a useful guide to the average performance of U.K. glass factories in 1977, it gives no indication of the range of energies associated with the different contributing factories. The calculations have therefore been repeated for each factory in turn to establish the significance of this variation.

The raw materials requirements of a factory have been calculated as the total input to the factory divided by the total output of saleable glass containers. These requirements are shown in Table 10.10 where the letters A, B, C, etc. refer to individual factories. As before, no distinction has been made between the different colours of glass produced.

Table 10.9

Total average energy required to produce 1 tonne of saleable glass containers in the U.K. in 1977

Operation	Electricity/MJ		Oil fuels/MJ			Other fuels/MJ			Total energy /MJ
	Fuel production and delivery energy	Energy content of fuel	Fuel production and delivery energy	Energy content of fuel	Feedstock energy	Fuel production and delivery energy	Energy content of fuel	Feedstock energy	
Production of raw materials	236.84	79.74	450.07	2276.51	nil	64.41	942.26	nil	4049.83
Delivery of raw materials	11.03	3.60	31.00	154.73	0.26	1.81	11.05	nil	213.48
Fuel use in glass factories	3672.12	1224.04	1288.72	6195.16	nil	547.59	4976.62	nil	17904.25
Totals/MJ	3919.99	1307.38	1769.79	8626.40	0.26	613.81	5929.93	nil	22167.56

Table 10.10

Raw materials consumption per tonne of saleable glass containers produced by individual glass factories in 1977

Factory	A	B	C	D	E	F	G	H	J	K	L	M	N	P	Q	R
Inputs																
Sand/kg	747.1	768.6	845.1	730.0	809.5	778.2	724.1	782.3	818.1	779.0	755.1	683.1	747.2	812.9	722.1	724.7
Limestone/kg	160.6	142.2	201.4	196.2	226.0	202.0	191.3	175.6	192.1	158.7	150.4	179.8	159.6	155.3	167.8	177.0
Soda ash/kg	230.7	225.7	231.8	216.3	218.3	225.1	229.9	207.5	238.1	249.2	243.9	207.9	221.5	228.5	225.8	237.2
Nepheline syenite/kg	–	–	–	51.0	–	38.9	36.8	16.7	37.6	19.9	–	–	7.2	–	17.0	–
Feldspar/kg	38.7	33.4	–	–	–	–	–	–	–	–	24.4	35.0	–	–	–	–
Alumina/kg	–	–	0.5	–	–	–	–	–	–	–	–	–	–	–	–	–
Calumite slag/kg	38.9	26.6	–	–	4.9	16.4	16.3	23.7	17.5	35.1	40.9	15.3	16.4	44.6	20.0	28.8
Foreign cullet/kg	–	25.9	2.7	2.4	–	–	–	15.4	–	–	35.3	–	0.4	–	4.4	–
Sodium sulphate/kg	1.3	3.5	0.8	–	9.0	6.6	–	9.4	–	–	–	–	–	–	–	–
Calcium sulphate/kg	10.4	9.5	3.4	7.6	–	–	6.9	1.7	8.4	11.4	6.4	6.2	4.5	7.6	9.0	9.9
Selenium/kg	0.02	0.02	0.01	0.01	–	0.006	0.008	0.01	0.004	0.02	0.01	0.009	0.002	0.011	0.005	0.017
Cobalt oxide/kg	neg	–	0.001	0.001	–	0.001	–	–	–	0.001	0.0003	0.0009	0.0006	0.0007	0.0002	0.0005
Iron chromite/kg	–	4.9	–	–	2.3	–	1.5	1.1	–	–	–	–	3.03	–	1.17	–
Sodium nitrate/kg	–	0.6	0.02	–	–	–	–	–	–	–	–	–	–	–	–	–
Water/gall	393	154	446	614	281	584	1001	881	867	659	1	156	1133	510	493	306

Table 10.11

Total energy required to produce 1 kg of saleable glass as containers in the U.K. factories in 1977

Factory	Item	Electricity/MJ		Oil fuels/MJ			Other fuels/MJ			Total energy /MJ
		Fuel production and delivery energy	Energy content of fuel	Fuel production and delivery energy	Fuel content of fuel	Feedstock energy	Fuel production and delivery energy	Energy content of fuel	Feedstock energy	
A	Materials	0.24	0.08	0.43	2.16	nil	0.06	0.89	nil	3.86
	Transport	0.02	0.01	0.04	0.17	neg	neg	0.01	nil	0.25
	Fuels	3.50	1.17	1.53	7.34	nil	0.07	0.88	nil	14.49
	Totals/MJ	3.76	1.26	2.00	9.67	neg	0.13	1.78	nil	18.60
B	Materials	0.23	0.08	0.42	2.13	nil	0.06	0.87	nil	3.79
	Transport	0.06	0.02	0.12	0.58	neg	neg	0.02	nil	0.80
	Fuels	3.59	1.20	1.33	6.39	nil	0.33	4.04	nil	16.88
	Totals/MJ	3.88	1.30	1.87	9.10	neg	0.39	4.93	nil	21.47
C	Materials	0.20	0.07	0.42	2.15	nil	0.06	0.90	nil	3.80
	Transport	0.01	neg	0.03	0.14	neg	neg	0.01	nil	0.19
	Fuels	4.91	1.64	2.68	12.84	nil	0.23	2.84	nil	25.14
	Totals/MJ	5.12	1.71	3.13	15.13	neg	0.29	3.75	nil	29.13
D	Materials	0.24	0.08	0.40	2.02	nil	0.06	0.84	nil	3.64
	Transport	0.01	neg	0.03	0.15	neg	neg	0.01	nil	0.20
	Fuels	4.25	1.42	0.61	2.91	nil	4.18	12.18	nil	25.55
	Totals/MJ	4.50	1.50	1.04	5.08	neg	4.24	13.03	nil	29.39
E	Materials	0.15	0.05	0.39	1.99	nil	0.06	0.84	nil	3.48
	Transport	0.01	neg	0.02	0.12	neg	neg	0.01	nil	0.16
	Fuels	3.82	1.27	1.39	6.66	nil	1.28	3.90	nil	18.32
	Totals/MJ	3.98	1.32	1.80	8.77	neg	1.34	4.75	nil	21.96

Table 10.11 — *continued*

Total energy required to produce 1 kg of saleable glass as containers in the U.K. factories in 1977

Factory	Item	Electricity/MJ		Oil fuels/MJ			Other fuels/MJ			Total energy /MJ
		Fuel production and delivery energy	Energy content of fuel	Fuel production and delivery energy	Fuel content of fuel	Feedstock energy	Fuel production and delivery energy	Energy content of fuel	Feedstock energy	
F	Materials	0.32	0.11	0.40	2.05	nil	0.06	0.87	nil	3.81
	Transport	0.02	0.01	0.06	0.30	neg	neg	0.03	neg	0.42
	Fuels	3.48	1.16	2.14	10.25	nil	0.16	1.97	nil	19.16
	Totals/MJ	3.82	1.28	2.60	12.60	neg	0.22	2.87	nil	23.39
G	Materials	0.25	0.08	0.42	2.13	nil	0.06	0.89	nil	3.83
	Transport	0.02	0.01	0.05	0.28	neg	neg	0.02	neg	0.38
	Fuels	3.46	1.15	2.16	10.31	nil	0.10	1.24	nil	18.42
	Totals/MJ	3.73	1.24	2.63	12.72	neg	0.16	2.15	nil	22.63
H	Materials	0.23	0.08	0.39	1.98	nil	0.05	0.80	nil	3.53
	Transport	0.01	neg	0.01	0.06	neg	neg	neg	neg	0.08
	Fuels	4.94	1.65	1.97	10.11	nil	0.03	0.32	nil	19.02
	Totals/MJ	5.18	1.73	2.37	12.15	neg	0.08	1.12	nil	22.63
J	Materials	0.26	0.09	0.44	2.22	nil	0.06	0.92	nil	3.99
	Transport	0.02	0.01	0.04	0.20	neg	neg	0.01	neg	0.28
	Fuels	4.38	1.46	2.54	12.16	nil	0.12	1.43	nil	22.09
	Totals/MJ	4.66	1.56	3.02	14.58	neg	0.18	2.36	nil	26.36
K	Materials	0.31	0.10	0.44	2.22	nil	0.06	0.96	nil	4.09
	Transport	0.03	0.01	0.15	0.74	neg	0.01	0.08	nil	1.02
	Fuels	4.97	1.66	2.17	10.46	nil	0.46	1.20	nil	20.92

									Totals
	Materials	...	...	...	...	...	...	...	...
	Transport	0.02	0.01	0.08	0.40	neg	0.01	0.04	nil	0.56
	Fuels	4.09	1.36	2.62	12.66	nil	neg	0.06	nil	20.79
	Totals/MJ	4.40	1.47	3.13	15.25	neg	0.07	1.04	nil	25.36
M	Materials	0.23	0.08	0.40	2.00	nil	0.05	0.80	nil	3.56
	Transport	0.01	neg	0.07	0.34	nil	0.01	0.03	nil	0.46
	Fuels	2.03	0.68	0.88	4.20	nil	0.47	5.75	nil	14.01
	Totals/MJ	2.27	0.76	1.35	6.54	neg	0.53	6.58	nil	18.03
N	Materials	0.22	0.07	0.40	2.05	nil	0.06	0.86	nil	3.66
	Transport	0.01	neg	0.02	0.08	neg	neg	0.01	nil	0.12
	Fuels	3.34	1.11	0.16	0.77	nil	0.91	11.12	nil	17.41
	Totals/MJ	3.57	1.18	0.58	2.90	neg	0.97	11.99	nil	21.19
P	Materials	0.46	0.16	0.43	2.18	nil	0.06	0.88	nil	4.17
	Transport	neg	neg	0.02	0.10	neg	neg	0.01	nil	0.13
	Fuels	3.72	1.24	0.14	0.66	nil	1.05	12.74	nil	19.55
	Totals/MJ	4.18	1.40	0.57	2.94	neg	1.11	13.63	nil	23.85
Q	Materials	0.21	0.07	0.44	2.05	nil	0.06	0.87	nil	3.70
	Transport	0.02	neg	0.07	0.37	neg	0.01	0.04	nil	0.51
	Fuels	3.57	1.19	1.38	7.32	nil	0.53	6.42	nil	20.41
	Totals/MJ	3.80	1.26	1.89	9.74	neg	0.60	7.33	nil	24.62
R	Materials	0.26	0.09	0.45	2.27	nil	0.06	0.92	nil	4.05
	Transport	0.02	0.01	0.03	0.15	neg	neg	neg	nil	0.21
	Fuels	3.09	1.03	0.01	0.06	nil	0.94	11.44	nil	16.57
	Totals/MJ	3.37	1.13	0.49	2.48	neg	1.00	12.36	nil	20.83

10.9 ENERGY REQUIRED TO PRODUCE AND DELIVER RAW MATERIALS INPUTS TO FACTORIES

The energy associated with the production of the raw materials inputs to individual glass factories may be calculated as before using the data in Tables 6.1, 6.2, 7.9 and 9.1. These are shown in the summary Table 10.11. As in Section 10.5, raw materials delivery transport has been divided into four components:

(a) sea,
(b) rail with empty return journeys,
(c) 20 tonne bulk road deliveries with empty return journeys, and
(d) 10 tonne circular deliveries with an average of half loading.

Table 10.12 shows, for each factory, the relative contributions of each of these components and Table 10.11 shows the corresponding total delivery energies based on this information and calculated using Tables 3.1 to 3.4.

Table 10.12

Transport contributions for the delivery of raw materials to glass factories in the U.K. in 1977. All values are expressed per tonne of saleable glass produced by the factory

Factory	Vehicle-miles in 20 tonne vehicles	Vehicle-miles in 10 tonne vehicles	Tonne-miles by rail	Tonne-miles by sea
A	1.452	0.0004	110.0	65.85
B	3.835	0.0135	468.8	73.64
C	1.108	0.0191	101.4	0.10
D	1.266	0.0001	87.6	102.03
E	1.749	0.0670	32.82	0.01
F	4.590	0.1300	43.22	77.83
G	3.955	negligible	63.22	59.58
H	0.084	0.0390	54.76	53.70
J	1.855	0.0300	115.43	56.43
K	13.306	negligible	nil	39.87
L	7.042	0.2200	nil	48.83
M	5.807	0.2280	nil	52.60
N	0.878	0.0120	28.58	49.52
P	1.751	0.0010	5.25	0.05
Q	6.572	0.0010	nil	53.58
R	0.433	0.0010	146.60	63.44

10.10 FUEL USE IN INDIVIDUAL GLASS FACTORIES

The fuels directly used per unit output of saleable glass containers may be readily calculated from the total data for the 12 month period and is shown in Table 10.13. Using the data of Table 2.2, these fuel requirements may be converted to energy requirements to give the values shown in Table 10.11. Published data relating to fuel use in glass factories may be found in [65-67].

10.11 TOTAL ENERGY REQUIRED TO PRODUCE GLASS CONTAINERS IN U.K. FACTORIES

The total energy required to produce unit output of saleable glass containers from raw materials in the ground may now be calculated as the sum of the three energy components (a) energy to produce raw materials, (b) energy to deliver the raw materials to the glass factory and (c) energy directly used within the glass factories. These total energies are shown for each glass factory in Table 10.11.

10.12 EFFECT OF GLASS COLOUR

Since glass bottles are produced in different colours, it is important to examine whether colour is a significant factor in determining energy requirement. The essential test of significance in this instance is the determination of whether or not the variation in production energies for the different colours is greater than the variation in the average factory energies shown in Table 10.11.

It is generally thought within the glass industry that the fuel required to operate a coloured glass furnace is no different from that required to operate a white flint glass furnace and so we may reasonably assume that the fuel requirements will be unaffected by the colour of the glass. Table 10.9 shows that this factor accounts for some 81% of the total average energy requirement to produce glass containers, with the remaining 19% being split as 18% for the production of materials inputs and 1% for their delivery. Inspection of the values for the individual factories (Table 10.11) shows no striking variations in this general conclusion.

Any variation in the total production energy must therefore arise from differences in the composition of the feed to the glass furnace. This factor has been examined by calculating in detail the energy requirements for four factories which make more than one colour of glass and Table 10.14 shows the results for one of them.

From Tables 10.14, it can be seen that the energy required to produce white flint glass is only 1% higher than the factory average whereas both green and amber glass are approximately 2% lower. The absolute values of these percentage changes will obviously vary with the relative tonnages of the different

Table 10.13

Fuel consumption of individual glass factories per tonne of saleable glass containers produced in 1977

Factory	A	B	C	D	E	F	G	H	J	K	L	M	N	P	Q	R
Electricity kWh	323.75	332.28	454.68	393.33	353.80	322.66	320.40	457.36	405.77	460.31	378.85	188.30	308.81	344.11	330.45	286.14
Gas—natural and manufactured/th	8.37	38.34	26.95	115.55	37.02	18.68	11.78	2.99	13.59	11.35	0.56	54.54	105.49	120.69	60.89	108.47
Heavy fuel oil/gall	38.78	33.61	68.28	14.18	35.52	54.57	54.66	45.75	64.97	53.01	61.58	21.97	3.56	2.53	–	–
Gas oil/gall	0.45	0.52	0.36	1.31	0.05	0.27	0.52	0.55	0.10	–	–	0.09	0.06	0.78	2.05	–
Kerosine/gall	0.007	0.015	–	–	–	–	–	–	–	–	0.01	–	–	–	–	0.005
Diesel/gall	–	–	–	–	–	–	–	–	–	–	0.25	0.01	0.01	–	–	0.12
Petrol/gall	–	0.012	0.0003	–	–	–	–	–	–	–	neg	–	–	–	–	–
Propane/kg	0.035	0.054	–	–	–	–	–	–	–	10.94	1.13	1.62	1.68	0.06	–	0.01
Butane/kg	–	–	–	–	–	–	–	29.5	–	–	21.25	–	–	–	140.44	–
Lubricating oil/gall	0.177	0.164	0.311	0.256	0.2	0.2	0.2	0.2	0.2	0.2	0.2	0.02	0.02	0.25	0.05	0.18
Grease/kg	0.004	0.002	0.003	0.005	0.003	0.003	0.003	0.003	0.003	0.003	0.003	0.09	0.06	0.11	0.06	0.08
Medium fuel oil/gall	–	–	–	–	–	–	–	–	–	–	–	–	–	–	1.55	–
Coke/kg	–	–	–	–	–	–	–	–	–	–	–	–	–	0.47	–	–

Table 10.14

Total energy required to produce 1 kg of containers of different colours in a glass factory using the different known compositions of furnace feeds but retaining the same direct factory energy consumption

Glass colour	Electricity/MJ		Oil fuels/MJ			Other fuels/MJ			Total energy /MJ
	Fuel production and delivery energy	Energy content of fuel	Fuel production and delivery energy	Energy content of fuel	Feedstock energy	Fuel production and delivery energy	Energy content of fuel	Feedstock energy	
White flint	3.88	1.30	1.89	9.20	nil	0.39	4.98	nil	21.64
Green	3.86	1.29	1.83	8.88	nil	0.38	4.83	nil	21.07
Amber	3.85	1.29	1.82	8.84	nil	0.38	4.83	nil	21.01
Factory average (all colours)	3.88	1.30	1.87	9.10	nil	0.39	4.93	nil	21.47

colours produced, but it is clear that a variation of the order of 3% in total average energy requirement for glass production encompasses the energies required to produce all colours of glass. Similar calculations for the other factories confirm this conclusion.

Table 10.15 shows the variation in the total average energy requirements of the different factories when compared with the 'national' average derived in Table 10.9. Excluding factories C and D, which produce large numbers of very small containers and so present a special case, the maximum variation in total average energy requirements is ± 19%, although the average variation, weighted by saleable output is closer to ± 9%. The magnitude of this variation

Table 10.15

Variations in the average energy to produce glass containers by the different glass factories expressed as a percentage of the average energy calculated in Table 10.9

Factory	Energy required to produce 1 kg of glass bottles /MJ	$\left(\dfrac{\text{Factory energy} - \text{Average energy}}{\text{Average energy}}\right)\%$
A	18.66	− 15
B	21.47	− 3
C	29.13	31[†]
D	29.39	32[†]
E	21.96	1
F	23.39	6
G	22.63	2
H	22.63	2
J	26.36	19
K	26.03	17
L	25.36	14
M	18.03	− 19
N	21.19	− 4
P	23.85	8
Q	24.62	11
R	20.83	− 6
Average	22.16	−

†These values are thought to result from a high level of production of very small non-beverage containers.

is significantly higher than the variation introduced by producing glass containers of different colours. The effect of glass colour has therefore been ignored.

10.13 ENERGY TO PRODUCE GLASS BOTTLES

The data presented here relates to the energy to produce unit mass of glass as
containers. Hence for any given bottle mass, the energy associated with its
production will be the product of the container mass and the energies given
in Table 10.9 (average) or Table 10.11 (specific factories).

CHAPTER 11

Production of Metal Cans

11.1 INTRODUCTION

The number of metal cans produced in the U.K. for beverage packing is shown in Table 11.1 with approximately equal numbers being used for beer and soft drinks packaging. However, the carbonated soft drinks market is dominated by the 12 fl. oz. size (86%), whereas the beer market is dominated by the 10 fl. oz. and 16 fl. oz. sizes (94%).

Table 11.1

Beverage can manufacture in the U.K. in 1976. All values are in millions of cans

Capacity fl. oz. or pints	Beer	Soft drinks	Total
6 fl. oz.	–	46 (3.0%)	46 (1.5%)
10 fl. oz.	720 (47.6%)	–	720 (23.8%)
12 fl. oz.[†]	78 (5.2%)	1452 (96.0%)	1530 (50.6%)
16 fl. oz.	697 (46.0%)	15 (1.0%)	712 (23.5%)
4 pint	8 (0.5%)	–	8 (0.3%)
5 pint	4 (0.3%)	–	4 (0.1%)
7 pint	6 (0.4%)	–	6 (0.2%)
Totals	1513 (100%)	1513 (100%)	3026 (100%)

†Inclusive of 330 ml size.

© *Crown Copyright 1981*

Beverage cans exist in a variety of designs and the following notes summarise the current position.

(a) Aluminium cans are available in sizes up to 16 fl. oz. (455 ml).

(b) All-aluminium cans are of two-piece design since the properties of aluminium readily permit drawing.

(c) Easy-open ends are made from aluminium alloy.

(d) Cans with capacities up to 16 fl. oz. (455 ml) are usually fitted with an easy-open end. Both ends of the larger can sizes are invariably made from solid tinplate or steel.

(e) Two-piece tinplate cans are available in sizes up to 12 fl. oz. (340 ml). Larger can sizes may become available but technical problems currently limit the size to which tinplate may be successfully drawn commercially for an acceptable reject rate.

(f) Tin-free steel is currently only used in three-piece cans. Two-piece cans made from tin-free steel may become commercially available in the future but current problems are associated with finding an effective, non-toxic and inexpensive lubricant to permit adequate drawing.

(g) Three-piece tin-free steel cans use an organic adhesive instead of solder to bond the can-body side seam.

(h) Lead-tin solder is used to join the bodies of three-piece tinplate beer cans but pure tin solder is used on soft drinks cans to eliminate any possibility of lead pickup by the beverage.

(i) Originally can ends were manufactured to the same diameter as the body but, for materials savings, the body is now necked in to accept a smaller end.

Typical finished masses of different cans are given in Table 11.2.

11.2 PRODUCTION SEQUENCES FOR METAL CANS

The output from the primary metal manufacturer is thin sheet in the form of coil. Subsequent stages of processing leading to saleable metal cans, whilst varying in detail for different can types may be broadly summarised by the operations shown in Fig. 11.1.

Tinplate or tin-free steel coil for three-piece can bodies is usually cut into rectangular sheets at the primary metal works. In contrast coil for the production of can ends is usually cut by the can manufacturer into sheets with serrated edges so that waste is minimised. (This operation is usually referred to as primary scroll shearing). The circular blanks which are subsequently drawn into bodies for two piece cans are usually pressed directly from the coil without prior cutting into sheets.

Table 11.2

Typical mass in grammes of the components in finished metal cans. Abbreviations used: 3TP = three-piece tinplate, 2TP = two-piece tinplate; 3TFS = three-piece tin-free steel; 2AL = two-piece aluminium

Component	6 fl. oz.	10 fl. oz.	12 fl. oz.				16 fl. oz.			4 pint	5 pint	7 pint
	3TP	3TP	3TP	3TFS	2TP	2AL	3TP	3TFS	2AL	3TP	3TP	3TP
Body												
Tinplate	20.53	25.77	30.34	–	31.71	–	39.25	–	–	162.95	204.31	239.08
Tin-free steel	–	–	–	30.10	–	–	–	37.50	–	–	–	–
Aluminium	–	–	–	–	–	15.83	–	–	17.60	–	–	–
End 1												
Tinplate	3.67	–	–	–	–	–	–	–	–	42.22	42.22	72.59
Aluminium	–	5.05	5.05	5.05	5.05	5.05	5.05	5.05	5.05	–	–	–
End 2												
Tinplate	7.35	10.55	10.55	–	–	–	10.55	–	–	42.22	42.22	72.59
Tin-free steel	–	–	–	10.20	–	–	–	10.20	–	–	–	–
Coatings												
Lacquer/varnish	0.48	0.48	0.53	0.68	0.62	0.31	0.64	0.82	0.35	1.51	1.76	2.16
Inks	0.15	0.07	0.08	0.08	0.08	0.06	0.11	0.11	0.08	0.60	0.75	0.90
Joints												
Solder	0.37(Sn)	0.64(Pb)	0.43(Sn)	–	–	–	–	–	–	–	–	1.10(Sn)
Adhesive	–	–	–	0.13	–	–	–	0.18	–	–	–	–
End seals												
End 1 compound	0.05	0.09	0.09	0.09	0.09	0.06	0.09	0.09	0.06	0.13	0.13	0.15
End 2 compound	0.05	0.07	0.07	0.07	–	–	0.07	0.07	–	0.13	0.13	0.15
Totals/g	32.65	42.72	47.14	46.40	37.55	21.31	56.67	54.02	23.14	249.76	291.52	388.72

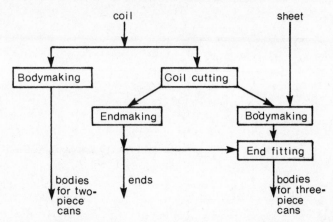

Fig. 11.1 – Schematic diagram of the principal operations used in can body and can end manufacture.

11.2.1 Production of plain ends

Starting with scrolled sheet (Fig. 11.2(a)), plain ends are made in tinplate or tin-free steel by the processing sequence shown in Fig. 11.3. The sheet is first lacquered and then cured in an oven. Coated sheets are rescrolled at right angles to the original scrolling to produce secondary scrolled strips (Fig. 11.2(b)) again to maximise the use of material. Circular blanks are then pressed from these strips and processed in two stages so that a flange or curl is formed around the edge (Figs. 11.2(c) and (d)). Lining compound is sprayed onto the inside of the curl while the end is rapidly rotated. This spreads the lining compound into a uniform coat (Fig. 11.2(e)). Finished ends are fed directly to a can making line or may be packed for despatch to customers.

11.2.2 Production of easy-open ends

Easy-open ends are made from aluminium alloy. The shell (that is the end minus the pull tab) follows the same processing sequence as for plain ends (Fig. 11.3) until after application of the lining compound. There is then an additional operation in which the pull tab is stamped from aluminium strip, the shell is scored and the pull-tab is fixed onto the shell to give the finished end (Fig. 11.4).

11.2.3 Production of bodies for three-piece cans

Bodies for three-piece cans are produced from rectangular sheets using the typical production sequence shown in Fig. 11.5. Sheets are lacquered and printed, with the exception of uncoated margins which will subsequently become the edges of the body blanks (to allow for soldering). The sheets are then dried in an oven. A number of passes along the printing lines may be needed depending upon the number of colours to be laid down. The print is protected by a final coat of varnish which may be applied as the last operation during the final pass along the printing line.

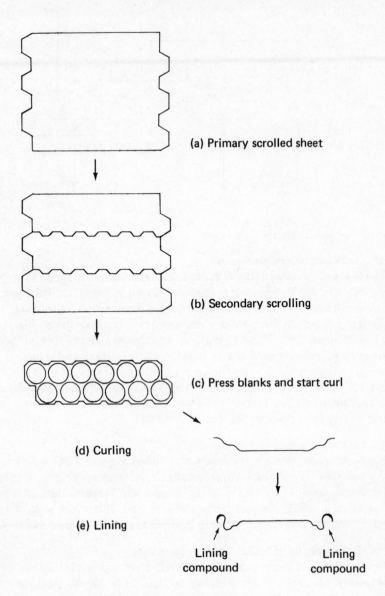

(a) Primary scrolled sheet

(b) Secondary scrolling

(c) Press blanks and start curl

(d) Curling

(e) Lining

Lining
compound

Lining
compound

Fig. 11.2 – Stages in the production of can ends.

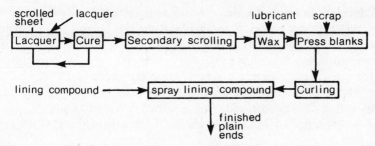

Fig. 11.3 – Sequence of operations needed to produce finished plain ends from tinplate or tin-free steel.

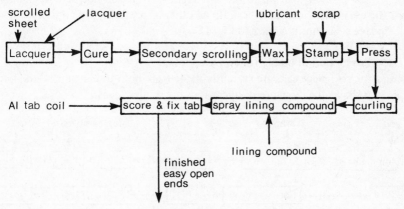

Fig. 11.4 – Sequence of operations leading to the production of finished easy-open ends from aluminium alloy sheet and aluminium tab coil.

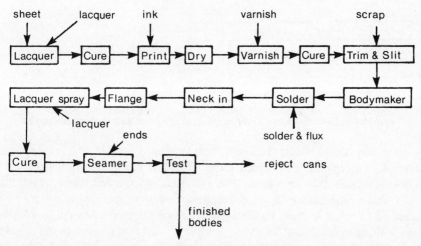

Fig. 11.5 – Operations involved in the production of bodies for three-piece cans.

Printed and varnished sheets are trimmed, slit and fed to the bodymaker where they are bent around a mandrel and the side seam is formed. Solder is applied to the joint and any excess is wiped off. The ends of the cylinder are then necked in and flanged to provide the seating for the ends. To complete the internal can coating, lacquer is sprayed onto the inside of the can and oven cured. The finished body cylinder is then closed at one end by fitting either a plain end or an easy-open end. This operation is known as seaming. Cans are then pressure tested for leaks before leaving the production lines for the packing area.

Bodies for three-piece cans made from tin-free steel use an organic adhesive instead of solder. The major advantage of this process is that it allows the body to be printed right up to the side seam since it will not be heat damaged as is the case with the soldered can. In all other respects, the production sequence for tin-free steel cans is similar to Fig. 11.5.

11.2.4 Production of bodies for two-piece cans
Bodies for two-piece cans are currently made from both tinplate and aluminium. Fig. 11.6 shows the sequence of operations for tinplate cans.

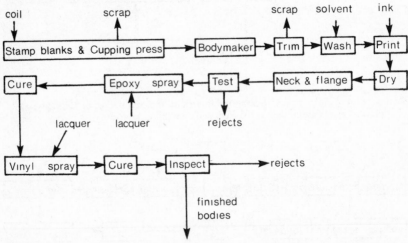

Fig. 11.6 – Operations used in the production of bodies for two-piece can bodies.

Circular blanks, stamped directly from lubricated tinplate or aluminium coil are fed to a cupping press which forms a shallow cup shape from the blank. The bodymaker then draws the cup and irons out the walls to produce the body shape. The cans walls are deliberately overdrawn so that the top may be trimmed to give a smooth, uniform rim. The body is then subjected to a vigorous washing to remove all traces of lubricant before being printed on the outside and cured in an oven. Tinplate bodies are then necked and flanged ready to receive the end and pressure tested to detect leaks. The untreated metal on the inside of the can-body is then sprayed with an epoxy coat which is oven cured. A further vinyl coat is applied over this, and the can is again passed through the curing oven.

In contrast, aluminium can bodies are necked, flanged and pressure tested after application of the internal lacquer system so that the lacquer acts as a lubricant during the subsequent necking and flanging operations. After inspection, the cans are sent to packing.

11.3 INFORMATION ON CAN MAKING

Beverage cans are available in a variety of sizes and specifications and whilst overall performance data for can making factories is obtainable, it is important to recognise that few factories make only one product. It is common for can-making factories to produce both beverage and non-beverage cans as well as ends for despatch to customers and for use in the production of bodies for three-piece cans. Moreover, when a company operates more than one factory, it is common for some operations such as coil cutting to be concentrated at certain sites only.

An added complication is that the industry is currently changing over from 211 beverage cans to 209/211[†] necked-in cans in the popular sizes as well as increasing the production of two-piece cans. This latter innovation has entailed the installation of can making lines based on new technology. In common with all new production lines, such installations seldom reach steady state operation until some months after commissioning and data collected during this shake-down period are seldom representative of the steady state performance. This problem was encountered in a number of the factories examined.

Consequently no single factory can be regarded as representative of the industry. Data has therefore been obtained from a number of factories and this information is presented in the following sections, which attempt to build up a composite but representative picture of can making practices in the U.K.

For simplicity, the energy associated with can production may be divided into that associated with all operations carried on within the can making factory and that associated with the production and delivery of materials consumed in can making. These two contributions are treated separately.

11.4 ENERGY USED IN COIL CUTTING

The first operation in can making is coil cutting and this is commonly concentrated in a specific area of the factory, which is sometimes monitored separately. A composite energy requirement for this operation has been calculated from three sets of data as shown in Table 11.3.

†The can diameter is 211 mm but the necked-in end diameter is 209 mm.

Table 11.3

Energy required by coil cutting to produce 1000 usable sheets of tinplate and aluminium from coil. See text for explanation of the basis of the calculations.

	Electricity/MJ		Oil fuels/MJ		
Source	Fuel production and delivery	Energy content of fuel	Fuel production and delivery	Energy content of fuel	Total energy /MJ
Tinplate					
Calculation 1	73.69	24.56	–	–	98.25
Calculation 2	–	–	35.70	171.88	207.58
Calculation 3	63.57	21.19	–	–	84.76
Representative average for cutting and heating	73.69	24.56	35.70	171.88	305.83
Aluminium					
Calculation 1	70.66	23.55	–	–	94.21
Calculation 2	60.97	20.32	–	–	81.29
Representative average for cutting and heating	70.66	23.55	35.70	171.88	301.79

Calculation 1

This calculation relates to a factory concerned with the conversion of tinplate and aluminium coil into primary scrolled sheet for the production of ends. Fig. 11.7(a) summarises the flow of materials through the factory.

Since the energy required to cut tinplate will be very little different from that used to cut aluminium, the total fuel consumption (E in Fig. 11.7(a)) may be partitioned between aluminium and tinplate cutting using the total number of sheets cut as the partitioning parameter. From Fig. 11.7(a),

$$\text{Electricity used to cut tinplate} = E . \frac{N_T}{N_T + N_A}$$

and

$$\text{Electricity used to cut aluminium} = E . \frac{N_A}{N_T + N_A}$$

Hence the system of Fig. 11.7(a) may be divided into the two systems shown in Fig. 11.7(b), one of which deals only with tinplate, the other deals only with aluminium.

For tinplate can manufacture, the desired product is prime plate. Any defective plate is kept separate from prime plate and is used fully within the company for other products. Hence from Fig. 11.7(b), the total output of usable tinplate sheet product is (947.6 + 3.8) = 951.4 sheets from every 1000 sheets of total output. Normalising with respect to sheets of usable product,

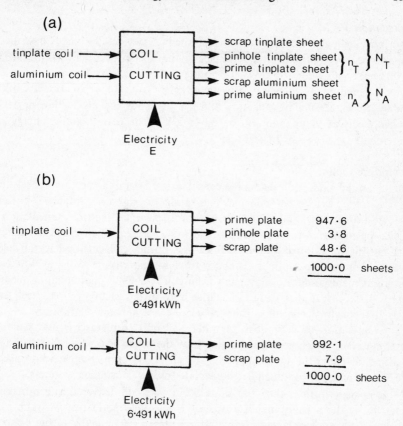

Fig. 11.7 – Materials flows in coil cutting. (a) Generalised system, (b) separated tinplate and aluminium cutting systems each with a total output of 1000 sheets based on data discussed in the text in a specific example.

the electricity requirement of the system is 6.823 kWh per 1000 usable sheets for which the corresponding energy requirement is as shown in Table 11.3.

Using an identical procedure to Fig. 11.7(b) for aluminium, the electricity requirement for the production of 1000 usable sheets of aluminium is 6.543 kWh and the corresponding energy requirement is as shown in Table 11.3. No data for space heating in the coil shop was available from this factory.

Calculation 2

In contrast to the factory described in Calculation 1, this coil shop was such that the electricity consumption in cutting could not be easily identified. However, it was possible to partition the site heating between the different operational areas on the basis of floor area. The heating oil attributable to the production of 1000 sheets of tinplate is 0.924 gallons of medium fuel oil which yields an energy requirement for space heating in the coil shop as shown in Table 11.3.

Calculation 3

As a check on the measured energy required to cut coil estimates have been made based on machine ratings. Typically the rating for a machine producing 1000 sheets per hour is 5.6 kW. Adopting the same levels of scrap production as in calculation 1, the electricity requirement for the production of 1000 usable sheets of tinplate will be 5.886 kWh and for the production of 1000 usable sheets of aluminium will be 5.645 kWh. The corresponding energy requirements are shown in Table 11.3.

Composite value

The measured values for electricity consumption in coil cutting are sufficiently close to those estimated from machine ratings to warrant confidence in their use. Adopting these, and taking the value for space heating from Calculation 2 yields the representative averages given in Table 11.3 for tinplate and aluminium coil cutting operations. The materials losses during coil cutting have all been estimated from Calculation 1. Thus for tinplate, there is a net loss of 4.68% of the input material during conversion from coil to sheet. For aluminium, the net loss is 0.79%. Note that these losses arise both in the cutting process itself and from defects already in the coil but which are not apparent until it is unreeled.

Examination of a number of other coil cutting plants suggests that the total materials rejection in coil cutting is of the order of 5%. We therefore feel justified in using the value of 4.68% for all plants even though, some plants are able to make use of much defective plate for less demanding products.

It is also worth noting that some 5% of the coil delivered to can making factories by the primary metals producer, is rejected upon arrival when obvious physical damage is detected. This coil is returned immediately to the primary producer without entering the can making factory. It is usually resold by the primary manufacturer for alternative applications. In the later analysis, this defective plate is treated as scrap which is recycled within the system after suitable treatment. Such a procedure leads to a slight overestimate of the system energy requirement for can manufacture but this is partly offset by the fact that the primary metals manufacturer may himself treat such rejected sheet before resale.

11.5 ENERGY USE IN PRINTING AND COATING

Sheet used in can manufacture is coated with a variety of lacquers, inks and varnishes depending upon subsequent use. For example, tinplate for end production is usually given a coat of clear lacquer to protect the sheet from attack by the contents of the can. Similarly, the surface of tinplate which will ultimately become the inside surface of a three-piece can is also lacquered. In contrast, the surface of tinplate which will eventually form the outside surface of a three-piece can is usually printed and varnished. The number of coats received by such decorated plate will obviously depend upon the design to be printed. It is therefore difficult to estimate an average value for the number of passes that a sheet must make through a coating or printing line. In general, all surface coatings are stoved after application to evaporate the solvent and to harden the coating. However, it is possible to apply a number of layers of print with only a single pass through the oven and the final varnish coat may well be applied as the last stage of the final pass along the printing line.

These operations may be separated into two groups:

11.5.1 Coating operations
In these the whole surface of the sheet is coated with a single material such as lacquer, base coat white or varnish. Also included in this group would be the lacquering of sheet for can ends (in which a series of circular patches only are coated) and those passes in which some print is applied but is immediately followed by a coat of varnish. In all cases the application of the coat is followed immediately by stoving.

11.5.2 Printing operations
In these, the sheet is decorated with a pattern using coloured inks. The pass along the printing line may lay down a number of colours but only one pass through the oven is needed to cure all the layers.

The throughput in printing and coating operations is usually monitored in terms of the number of sheet passes along the line. Thus a sheet which receives three layers of print followed by a single curing operation in the oven will represent a single sheet pass in the same way as the application of a single coat of lacquer followed by a pass through the oven. This parameter has therefore been used to partition the total energy consumption of a line between the sheets passing along it.

The following information is based on the operation of a print shop which is thought to be typical of those in the can making industry.

11.5.3 Energy used in coating lines
The total consumption of electricity for machinery and natural gas for ovens has been divided by the total number of sheet passes to give an average fuel requirement per 1000 sheet passes of 11.276 kWh of electricity and 2.8595 therms of natural gas. The corresponding energy requirement is therefore as shown by entry (a) in Table 11.4.

11.5.4 Energy used in printing lines
The total consumption of fuels has again been divided by the total number of sheet passes to give an average fuel consumption per 1000 sheet passes of 9.626 kWh of electricity and 2.6989 therms of natural gas, for which the energy requirement is as shown by entry (b) in Table 11.4.

Table 11.4

Energy required to coat and decorate sheet used in can making. See text for full explanation

Operation	Electricity/MJ		Other fuels/MJ		Total energy /MJ
	Fuel production and delivery	Energy content of fuel	Fuel production and delivery	Energy content of fuel	
(a) 1000 sheet passes for coating	121.78	40.59	24.79	301.51	488.67
(b) 1000 sheet passes for decorating	103.96	34.65	23.40	284.57	446.58
(c) All coating for 1000 sheets for can ends assuming no losses	243.56	81.18	49.58	603.02	977.34
(d) All coating and decorating for 1000 sheets assuming no losses	573.26	191.07	121.17	1473.67	2359.17
(e) All coating for 1000 good sheets for ends including materials losses	244.78	81.59	49.83	606.04	982.24
(f) All decorating of 1000 good sheets inclusive of materials losses	576.13	192.03	121.77	1481.04	2370.97

11.5.5 Energy to coat sheet for can end manufacture

Sheet for can ends is lacquered on one side and varnished on the other. It therefore requires two passes along a coating line and so the energy associated with these operations will be as shown by entry (c) in Table 11.4, that is twice entry (a).

11.5.6 Energy use in decorating sheet

Decorated sheet in can body production experiences a number of passes along printing and coating lines. By dividing the total number of printing passes recorded by the total number of sheets treated, it is found that a sheet typically passes twice along a printing line. The following overall sequence has therefore been chosen as typical of the production of decorated sheet:

    Pass 1: Lacquer coat on side 1
    Pass 2: Base coat on side 2
    Pass 3: First print on side 2
    Pass 4: Second print on side 2.
    Pass 5: Additional print plus varnish on side 2.

Hence the total energy associated with the production of such sheet will be given by

$$(3 \times \text{entry (a) in Table 11.4}) + (2 \times \text{entry (b) in Table 11.4})$$

This is shown as entry (d) in Table 11.4.

11.5.7 Materials losses during printing and coating

Information from a number of factories indicates that the total materials losses during passage of a sheet through printing and coating lines is typically 0.5%. These losses arise in part from defective plate reaching the beginning of the printing/coating line, in part from the malfunction of the lines themselves and, in the case of print shops which print sheet for outside customers, from plate which is returned by these customers. As a consequence, entries (c) and (d) in Table 11.4 must be increased to account for these materials losses; these revised values are shown as entries (e) and (f) in Table 11.4.

Although the defective production during printing and coating is of the order of 0.5%, it is not necessarily all scrapped. Reject printed tinplate may find use in other less cosmetic applications, although sheet cut for beverage cans will need recutting for these other applications thus generating additional scrap. Furthermore, this reuse scheme is only applicable to those companies who make alternative products or who can find a buyer for the decorated scrap. An actual average scrap level of 0.25% has therefore been assumed for tinplate. In general these alternative uses do not exist for aluminium and so a full 0.5% scrap level is assumed for this metal.

11.6 PRODUCTION OF TINPLATE AND TIN-FREE STEEL ENDS

The volumes of the more popular sizes of can are obtained by changing the length of the body cyclinder whilst leaving the end diameter constant. Thus 10, 12 and 16 fl. oz. cans all have the same diameter. Similarly, 4 and 5 pint cans also have the same end diameter. Therefore the number of different types of ends produced is less than the number of different sizes of can.

Table 11.5 gives the materials requirements for the different types of tin-plate and tin-free steel ends manufactured. The pressing of circular blanks from rectangular sheet produces considerable unavoidable loss of materials, although this is minimised by choosing the optimum layout on the sheet. The waste is also recycled.

The outline flowsheet for this production sequence is given in Fig. 11.8 and the corresponding materials flows are shown in Table 11.5. Note that two types of ends are produced for 4 and 5 pint cans and of these, the high pressure end which employs heavier plate represents approximately 10% of all production. The materials flows in Fig. 11.8 and Table 11.5 make allowance for this.

Data have been obtained from two factories producing ends for 10, 12 and 16 fl. oz. cans and in which electricity is either metered separately or can be readily separated from the consumption for other purposes. The fuel requirement per 1000 ends, calculated as an average of these two sets of data, weighted by the output from the two factories, leads to the energy requirement shown in Table 11.6.

Table 11.5

Materials requirements for tinplate and tin-free steel ends

Item	Nominal tinplate can capacity				10, 12 and 16 fl. oz. tin-free steel cans
	6 fl. oz.	10 fl. oz 12 fl. oz. 16 fl. oz.	4 pint 5 pint	7 pint	
Sheet area/m^2	0.6597	0.7637	0.6882	0.6446	
Sheet thickness/mm	0.25	0.27	0.30[†]	0.38	
Sheet mass/kg	1.295	1.618	1.622[†]	2.884	
Number of blanks per sheet	144	132	30	30	
Mass of blanks/kg	1.059	1.393	1.221[†]	2.178	
Inevitable materials loss as a percentage of input materials	18.2%	13.9%	24.7%	24.5%	
Materials flows — see Fig. 11.7					
A (sheets)	7.009	7.653	33.649	33.669	7.706
A (kg)	9.007	12.382	56.608	97.100	12.051
B (kg)	9.059	12.368	56.546	96.993	12.038
C (kg)	7.410	10.649	42.579	73.230	10.365
D (kg)	7.354	10.553	42.213	72.600	10.200
a (kg)	0.018	0.014	0.062	0.107	0.013
b (kg)	1.649	1.719	13.967	23.763	1.673
c (kg)	0.056	0.096	0.366	0.630	0.165

[†]Approximately 10% of ends of this size are made in a high pressure form which uses sheet of thickness 0.41 mm; the corresponding mass of the sheet is 2.225 kg and the mass of the blanks is 1.675 kg.

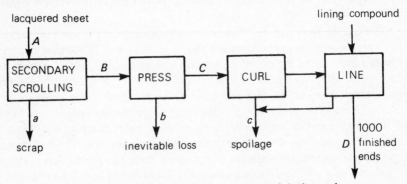

Fig. 11.8 – Materials flows in the production of tinplate ends.

Table 11.6

Average energy required to produce 1000 tinplate or tin-free steel ends from lacquered sheet

Fuel type	Fuel production and delivery energy/MJ	Energy content of fuel /MJ	Feedstock energy /MJ	Total energy /MJ
Electricity	34.52	11.51	nil	46.03
Oil fuels	2.68	14.86	nil	17.54
Other fuels	0.25	3.07	nil	3.32
Totals/MJ	37.45	29.44	nil	66.89

It is thought that there is little difference in the energy required to produce ends of different sizes and so the values calculated in Table 11.6 have been used for all sizes of tinplate ends.

11.7 PRODUCTION OF EASY-OPEN ENDS

Easy-open ends are fitted to cans with capacities in the range 6 to 16 fl. oz.

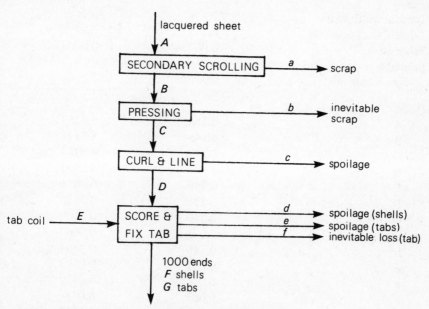

Fig. 11.9 – Materials flows in the production of easy-open ends. Flows refer to mass of aluminium only and are given in Table 11.7.

Table 11.7 shows the materials specification for the production of these ends and also gives the materials flows in the production sequence shown in Fig. 11.9. Production energy requirements (excluding the provision of materials) have been obtained from two factories and are shown in Table 11.8.

Table 11.7

Materials requirements for easy-open ends

Item	Can capacity	
	6 fl. oz.	10, 12 and 16 fl. oz.
Shell		
Sheet area/m^2	0.696	0.754
Sheet thickness/mm	0.32	0.33
Sheet mass/kg	0.5845	0.6602
Number of blanks per sheet	144	120
Mass of blanks/kg	0.4793	0.5691
Inevitable materials loss	18%	13.8%
Tab		
Tab coil width/mm	27	30
Tab coil thickness/mm	0.36	0.29
Tab mass/kg	0.00034	0.00031
Inevitable materials loss in producing tab	56%	56%
Process flows in kg (see Fig. 11.9)		
A	4.150	5.6253
B	4.145	5.6191
C	3.399	4.8437
D	3.362	4.7904
E	0.746	0.7076
F	3.328	4.7425
G	0.340	0.3100
a	0.005	0.0062
b	0.746	0.7754
c	0.037	0.0533
d	0.034	0.0479
e	0.007	0.0031
f	0.339	0.3945

Table 11.8

Total energy associated with the conversion of lacquered sheet and tab coil into 1000 easy-open ends

Fuel type	Fuel production and delivery energy/MJ	Energy content of fuel /MJ	Feedstock energy /MJ	Total energy /MJ
Electricity	34.78	11.59	nil	46.37
Oil fuels	nil	nil	nil	nil
Other fuels	0.60	7.35	nil	7.95
Totals/MJ	35.38	18.94	nil	54.32

© *Crown Copyright 1981*

11.8 PRODUCTION OF BODIES FOR THREE-PIECE TINPLATE CANS

11.8.1 Materials flows
Bodies for three-piece tinplate cans are made from rectangular decorated sheet. Table 11.9 gives the materials specification for typical products as well as the materials flows in the production flowsheet shown in Fig. 11.10.

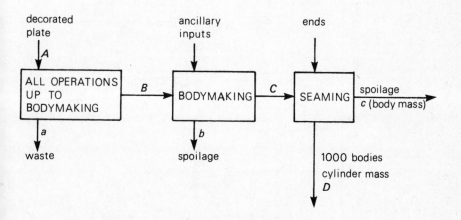

Fig. 11.10 – Mass flows in the production of 1000 bodies for three-piece tinplate cans. See Table 11.9 for numerical data.

Table 11.9

Materials requirements for the manufacture of bodies for three-piece tinplate cans

Item	Can capacity						
	6 fl. oz.	10 fl. oz.	12 fl. oz.	16 fl. oz.	4 pint	5 pint	7 pint
Specification							
Sheet area/m^2	0.7059	0.7088	0.7302	0.6755	0.3569	0.4573	0.5450
Sheet thickness/mm	0.15	0.15	0.15	0.15	0.26	0.26	0.26
Sheet mass/kg	0.8312	0.8346	0.8599	0.7954	1.3194	1.2420	0.9728
Number of bodies/sheet	40	32	28	20	8	6	4
Materials efficiency/%	98.8	98.8	98.8	98.7	98.8	98.7	98.3
Materials flows in production flowsheet (Fig. 11.10)							
A (kg)	21.171	26.853	31.241	40.457	168.905	211.996	247.304
A (sheets)	25.470	32.175	36.331	50.864	128.017	170.689	254.219
B (kg)	20.917	26.531	30.866	39.931	166.878	209.240	243.100
C (kg)	20.603	26.133	30.403	39.332	164.592	206.373	239.210
D (kg)	20.531	26.081	30.341	39.253	162.946	204.309	239.066
E (number)	1002	1002	1002	1002	1010	1010	1001
a (kg)	0.254	0.322	0.375	0.526	2.027	2.756	4.204
b (kg)	0.314	0.398	0.463	0.599	2.286	2.867	3.890
c (kg)	0.072	0.052	0.061	0.079	1.646	2.064	0.144

11.8.2 Energy requirement

Evaluation of the energy associated with the production of bodies for three-piece tinplate cans poses two main problems. First, factories frequently produce beverage cans alongside other types of can and the internal metering of fuel composition by different can lines seldom permits isolation of data relating specifically to beverage can production without mathematical partitioning. Secondly, different sizes of beverage cans are often made on the same line. For example, a line making 10 fl. oz. cans may, with minor adjustments, be used to produce 12 and 16 fl. oz. cans. Similarly, a 4 pint can making line may also be used to manufacture 5 pint cans. Even when the total consumption of individual can lines is monitored, this seldom extends to measuring the variation in consumption with the type of can being produced. Body making energies must then be deduced for example from the different rates of production of the different body sizes assuming that total line consumption of fuels per unit time is independent of the product being made.

By examining the production fuel requirements of a number of factories, the average production energy requirements are as shown in Table 11.10.

Table 11.10

Typical average energy required to produce 1000 three-piece tinplate cans of different sizes from decorated plate but excluding the energy used to produce the end

Nominal can size	Electricity/MJ		Other fuels/MJ		Total energy /MJ
	Fuel production and delivery	Energy content of fuel	Fuel production and delivery	Energy content of fuel	
6 fl. oz.	88.92	21.02	7.29	88.56	205.79
10 fl. oz.	98.80	23.35	8.10	98.40	228.65
12 fl. oz.	98.80	23.35	8.10	98.40	228.65
16 fl. oz.	112.92	26.69	9.25	112.46	261.32
4 pint	348.30	116.10	23.23	282.47	770.10
5 pint	426.82	142.27	28.46	346.16	943.71
7 pint	447.89	105.85	36.72	446.08	1036.54

11.9 PRODUCTION OF THREE-PIECE TIN-FREE STEEL CAN BODIES

The major difference between three-piece tinplate cans and three-piece tin-free steel cans is the type of side seam used. The soldered seam on tinplate cans is replaced by a simple overlap joint cemented with an epoxy based adhesive. This allows all round can decoration since the decorated plate near the side seam is not subjected to such high temperatures as in soldering. An additional advantage is that it eliminates the remote possibility of lead pick-up from the solder by the can contents.

In practice, the adhesive is extruded as a strip along one side of the body blank. The cylinder is folded over and the seam is hammered and rapidly cooled by refrigerated wheels to reduce the temperature of the adhesive so that the side seam develops sufficient mechanical strength to remain in position. The remainder of the production sequence is identical to that for the three-piece tinplate cans.

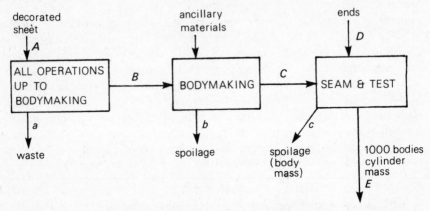

Fig. 11.11 – Materials flows in the production of 1000 bodies for three-piece cans made from tin-free steel. See Table 11.11 for numerical data.

Fig. 11.11 shows the flowsheet for the production of tin-free steel cans from decorated plate. Table 11.11 gives the materials flows and the associated energy requirements are given in Table 11.12.

Table 11.11

Materials flow for the production sequence shown in Fig. 11.11 for tin-free steel can bodies from decorated sheet

Flow (see Fig. 11.11)	Can size	
	12 fl. oz.	16 fl. oz.
A (kg)	31.471	39.207
B (kg)	31.093	38.737
C (kg)	30.160	37.575
D (kg)	30.100	37.500
E (number)	1002	1002
a (kg)	0.378	0.470
b (kg)	0.933	1.162
c (kg)	0.060	0.075

Table 11.12

Energy required to produce 1000 bodies in tin-free steel for 12 or 16 fl. oz.
beverage can bodies but excluding the manufacture of the end

Fuel type	Fuel production and delivery energy/MJ	Energy content of fuel /MJ	Feedstock energy /MJ	Total energy /MJ
Electricity	123.44	41.15	nil	164.59
Oil fuels	nil	nil	nil	nil
Other fuels	9.56	116.30	nil	125.86
Totals/MJ	133.00	157.45	nil	290.45

© Crown Copyright 1981

11.10 PRODUCTION OF TWO-PIECE TINPLATE CAN BODIES

Two-piece tinplate cans are an increasingly important feature of can making.
However, the relatively recent commissioning dates of much of the purpose
built plant makes it difficult to obtain practical data which can be regarded
as truly representative of this technology.

From data supplied, typical materials flows are shown in Fig. 11.12. The

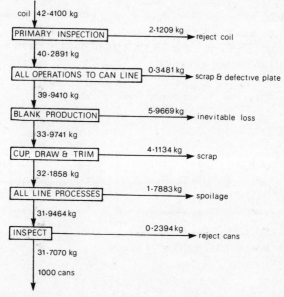

Fig. 11.12 – Materials flows in the production of bodies for 12 fl. oz. tinplate
two-piece cans.

mass of coil fed to the line per 1000 finished saleable cans (40.2891 kg) is based on direct factory data. However, some 5% of all delivered coil is rejected before being taken into the factory. The production energy requirements have been calculated as the average of a number of operations and are shown in Table 11.13.

Table 11.13

Average energy required to produce 1000 bodies for two-piece tinplate cans from coil at the can making factory

Fuel type	Fuel production and delivery energy/MJ	Energy content of fuel /MJ	Feedstock energy /MJ	Total energy /MJ
Electricity	385.53	128.51	nil	514.04
Oil fuels	30.69	147.77	nil	178.46
Other fuels	30.39	369.52	nil	399.91
Totals/MJ	446.61	645.80	nil	1092.41

© *Crown Copyright 1981*

11.11 PRODUCTION OF BODIES FOR TWO-PIECE ALUMINIUM CANS

Fig. 11.13 shows the flowsheet for the production of the bodies for two-piece aluminium cans and Table 11.14 gives the materials flows. The average production energy, shown in Table 11.15 is based on U.S. practice [44] corrected for U.K. fuel production efficiencies.

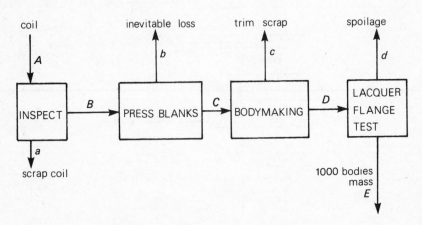

Fig. 11.13 — Materials flows in the production of 1000 bodies for aluminium two piece cans. See Table 11.14 for numerical data.

Table 11.14

Materials flows in kg for the production of 1000 two-piece aluminium cans using the system shown in Fig. 11.13

Materials flow (see Fig. 11.13)	Can size	
	12 fl. oz.	16 fl. oz.
A	22.162	24.513
B	21.054	23.287
C	18.138	20.062
D	16.660	18.427
E	15.830	17.506
a	1.108	1.226
b	2.916	3.225
c	1.478	1.635
d	0.830	0.921

Table 11.15

Energy required to produce 1000 two-piece aluminium can bodies from coil

Fuel type	Fuel production and delivery energy/MJ	Energy content of fuel /MJ	Feedstock energy /MJ	Total energy /MJ
Electricity	399.54	133.18	nil	532.72
Oil fuels	35.88	172.76	nil	208.64
Other fuels	36.90	448.75	nil	485.65
Totals/MJ	472.32	754.69	nil	1227.01

11.12 SCRAP RECYCLING IN THE METAL CAN INDUSTRY

The production of both can ends and can bodies involves the production of significant quantities of well characterised scrap metal. Such metal is fed back into the metal production system as a matter of course. The energy associated with the provision of metal to the can making industry must therefore take this scrap recycling into account.

11.12.1 Aluminium recycling

Of all the metals involved in can making, aluminium is the simplest to recycle. At present all aluminium scrap is transported to one site where it is melted and cast into remelt ingots which are then transported to the sheet fabrication works for the production of new rigid container sheet. The scrap treatment is therefore the sum of three components: (a) energy to transport scrap from can making factory to the melting plant, (b) energy to produce remelt ingot from can scrap and (c) energy required to transport remelt ingot from the scrap plant to the sheet production facility.

Can making scrap is baled in the can factory and transported on 20 tonne payload vehicles to the treatment works. The transport distance obviously depends upon the location of the can factory relative to the treatment plant, but a one-way distance of 100 miles is thought to be a reasonable average. Outward journeys are assumed to be fully loaded and return journeys, empty. The energy associated with this transport can be calculated therefore from Table 3.1.

Within the scrap treatment plant, body scrap must be distinguished from end and tab scrap because the different compositions of the alloys from which they are made give rise to different losses during melting. The most significant loss is due to the oxidation of magnesium. Thus the higher magnesium content of end stock results in a typical remelt loss for this alloy of 11% whereas that arising from remelting body stock is typically only 5.5%.

Within the U.K. there is presently a greater use of end stock compared with body stock because aluminium ends are also fitted to both tinplate and tin-free steel cans which dominate the U.K. market. A typical mix for scrap melting is 75% end stock and 25% body stock with average materials losses of the order of 6.88%. The melting energy has been calculated from data supplied by the operator of the melting process.

Remelt ingot is delivered by road to the aluminium fabrication works, a distance of 170 miles. Assuming full outward loads and empty return loads, the delivery energy can be calculated from Table 3.1.

At present ingot derived from scrap is remelted at the fabrication plant and cast into a form suitable for further processing. It replaces primary ingot and the materials lossess occurring during the second remelting will be as discussed in Chapter 7 for primary metal.

The total energy associated with this complete recycling sequence is as shown in Table 11.16. The production of 1 kg of good remelt ingot requires a scrap input of 1.101 kg. It should be noted that at present, approximately 10% of all can scrap is used in the production of alloys for general casting purposes. Subsequent calculations, make no allowance for this practice since can scrap also replaces primary aluminium in this system. The net energy saving is therefore the same.

Table 11.16

Energy required to process scrap aluminium to yield 1 kg of remelt ingot

Operation	Electricity/MJ			Oil fuels/MJ			Other fuels/MJ			Total energy /MJ
	Fuel production and delivery	Energy content of fuel	Feedstock energy	Fuel production and delivery	Energy content of fuel	Feedstock energy	Fuel production and delivery	Energy content of fuel	Feedstock energy	
Delivery of scrap	0.01	neg		0.06	0.31	neg	neg	0.03	–	0.41
Scrap melting	0.39	0.13		–	–	–	1.24	15.13	–	16.89
Delivery of ingot to fabrication	0.02	0.01		0.09	0.48	neg	0.01	0.05	–	0.66
Second melting	0.38	0.13		–	–	–	0.81	9.89	–	11.21
Ingot casting	0.66	0.22		0.58	3.27	–	0.06	0.78	–	5.57
Totals/MJ	1.46	0.49		0.73	4.06	neg	2.12	25.88	–	34.74

11.12.2 Tinplate recycling

Tinplate scrap is sent to one of two plants for detinning. As with aluminium scrap treatment, the energy associated with the treatment of tinplate scrap is the sum of three components; (a) the delivery of scrap from can maker to treatment plant, (b) the treatment of scrap and (c) the delivery of detinned scrap to the point of use.

For can making scrap the average one-way delivery distance is approximately 150 miles. A typical lorry load is 15 tonnes using 20 tonne skips. Hence the transport energy can be calculated from Table 3.1 to give the values shown as entry (a) in Table 11.17.

At the treatment plant, scrap is shredded and the tin coating removed chemically leaving a high grade steel scrap which is washed and baled. The tin bearing solution is electrolysed to recover tin metal which is then melted and cast into ingots. The tin stripping agent is an aqueous solution of sodium nitrite and sodium hydroxide. The hydroxide is however recovered and recycled so that consumption is negligible. Typical recovery rates per tonne of scrap treated are 980 kg of steel and 5.2 kg of ingot tin. Some lead is also recovered but this has been ignored in the analysis. It is important to note that the mass of tin recovered is variable since it depends upon the amount of tin present on tinplate (which is itself variable) as well as on the presence of tin in any solder in the scrap.

This total treatment system may be regarded in two ways. First it may be regarded as a system whose function is to treat input tinplate scrap to yield detinned steel. The fuel requirements per tonne of input scrap are 75.06 kWh of electricity, 1.48 therm of natural gas and 8.81 gallons of medium fuel oil and the corresponding energy requirements form row (b) in Table 11.17. This entry includes the energy required to produce the sodium nitrite consumed (Table 9.1).

Detinned steel is returned to the steelworks, an average one way distance of 25 miles. The transport energy may again be calculated from Table 3.1 to yield row (c) in Table 11.17.

The total energy required to detin 1 kg of scrap and deliver the products is therefore the sum of the above three components and is shown as row (d) in Table 11.17. This approach to the analysis loads all of the processing energy onto the steel product and ignores the production of tin.

An alternative way of viewing the detinning process is to regard it as a method of separating high grade tin and steel from otherwise low value scrap. If the fuel requirements of the shredding and stripping operations are partitioned between steel and tin on the basis of the relative mass throughputs and the electrolysis energy is attributed solely to the tin, then the fuel requirements for the production of 1 kg of tin from tinplate scrap are 6.63 kWh of electricity, 0.13 therm of natural gas and 0.65 gallons of medium fuel oil, and those for the production of 1 kg of baled high grade steel scrap from scrap tinplate are 0.04

Table 11.17

Energies associated with detinning tinplate scrap

Operation	Electricity/MJ		Oil fuels/MJ			Other fuels/MJ			Total energy /MJ
	Fuel production and delivery	Energy content of fuel	Fuel production and delivery	Energy content of fuel	Feedstock energy	Fuel production and delivery	Energy content of fuel	Feedstock energy	
(a) Delivery of 1 kg scrap to treatment plant	0.02	0.01	0.08	0.39	neg	0.01	0.04	–	0.55
(b) Treatment of 1 kg scrap	1.25	0.42	0.50	2.40	–	0.02	0.19	0.03	4.81
(c) Delivery of detinned steel	neg	neg	0.01	0.07	neg	neg	0.01	–	0.09
(d) Total energy to detin 1 kg of scrap	1.27	0.43	0.59	2.86	neg	0.03	0.24	0.03	5.45
(e) Energy associated with the production of 1 kg of detinned steel in the detinning works	0.43	0.14	0.20	0.96	–	neg	0.03	0.03	1.79
(f) Energy associated with the production of 1 kg of ingot tin in the detinning works	71.60	23.87	25.13	120.94	–	1.13	13.74	0.03	256.44

kWh of electricity and 0.005 gallons of medium fuel oil. Using the data of Table 9.1 for sodium nitrite production, then the total energy requirements for the production of detinned steel and tin ingot will be as given in entries (e) and (f) respectively in Table 11.17.

11.12.3 Tin-free steel recycling
Unlike tinplate where the resale value of recovered tin is sufficiently high to make scrap recycling economic, tin-free steel scrap does not possess this high value coating. Tin-free steel scrap is recyled but the presence of the chromium can be deleterious impurity if it is used in the production of high grade steel. Apart from shredding and cleaning to remove any coating or decoration, no further treatment is given. The considerably lower value of tin-free steel scrap means that recycling is not practised with the same vigour as is the case with tinplate scrap. In the later calculations zero recycling of this material has been assumed.

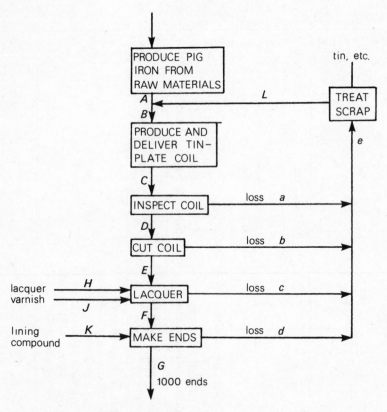

Fig. 11.14 – System used to produce tinplate ends from raw materials in the ground. See Table 11.18 for numerical data.

11.13 TOTAL ENERGY TO PRODUCE TINPLATE ENDS

The flow sheet for the complete production sequence for the manufacture of tinplate can ends from raw materials in the ground is shown in Fig. 11.14 and the materials flows are given in Table 11.18. The flows in the steel producing sub-systems are based on the data of Chapter 8. It is assumed that all scrap produced within the can making system is recycled and, after detinning, replaces pig iron. For convenience, this has been shown as replacing pig iron in the tinplate production sequence.

Table 11.18

Materials flows in the production of 1000 tinplate ends from raw materials in the ground using the system of Fig. 11.12

Flow (see Fig. 11.12)	Can size			
	6 fl. oz.	10, 12, 16 fl. oz.	4, 5 pint	7 pint
A (kg)	9.722	13.771	57.049	98.043
B (kg)	12.407	16.922	77.366	132.708
C (kg)	10.094	13.768	62.945	107.972
D (kg)	9.589	13.080	59.798	102.573
E (kg)	9.123	12.444	56.892	97.588
E (sheets)	7.045	7.691	33.818	33.838
F (kg)	9.077	12.382	56.608	97.100
F (sheets)	7.009	7.653	33.649	33.669
G (kg)	7.354	10.553	42.213	72.600
H (kg)	0.075	0.095	0.376	0.355
J (kg)	0.013	0.017	0.067	0.060
K (kg)	0.178	0.070	0.133	0.168
L (kg)	2.685	3.151	20.317	34.665
a (kg)	0.505	0.688	3.147	5.399
b (kg)	0.466	0.636	2.906	4.985
c (kg)	0.046	0.062	0.284	0.488
d (kg)	1.723	1.829	14.395	24.500
e (kg)	2.740	3.215	20.732	35.372

The inspection stage in Fig. 11.14, immediately before coil cutting, represents the point at which defective delivered coil is rejected by the can maker. This is estimated as 5% by the can maker and has been shown as passing through the scrap treatment works. In fact, some of this reject coil may well be resold by the primary metal producer after further treatment. However, any energy savings likely to accrue from this practice are probably balanced by the other assumptions, namely that *all* scrap generated in the can making process is recycled.

Lining compound masses include losses and refer to residual mass after evaporation of the solvent. Lacquer masses refer to residual mass and have been calculated as follows. The internal surface is coated with epoxy and vinyl with application rates of 5.4 gm^{-2} and 10.05 gm^{-2} respectively. The external surface is given in a single coat of epoxy at a rate of 2.7 gm^{-2}. The lacquer requirements have been derived from the sheet areas given earlier. A 5% lacquer loss has been included.

The detailed calculation for the energies associated with the production of ends for 6 fl. oz. cans is given in Table 11.19 and the results for all tinplate ends are summarised in Table 11.20.

Table 11.20

Gross energy in MJ required to produce 1000 tinplate ends of different sizes from raw materials in the ground

Operation	Can sizes			
	6 fl. oz.	10, 12 and 16 fl. oz.	4, 5 pint	7 pint
Production of pig iron	244.12	345.79	1432.49	2461.85
Production and delivery of tinplate	193.92	264.49	1247.60	2074.13
Coil cutting	2.15	2.35	10.34	10.35
Lacquering	6.89	7.51	33.06	33.07
End making	66.89	66.89	66.89	66.89
Scrap treatment	6.25	7.34	47.34	80.77
Production of lacquer	28.25	35.79	141.64	133.73
Production of varnish	3.50	4.57	17.96	16.10
Production of lining compound	32.05	12.64	24.00	30.33
Totals/MJ	584.02	747.37	3021.32	4907.22

11.14 TOTAL ENERGY TO PRODUCE TIN-FREE STEEL ENDS

The major difference between the production of tinplate ends and the production of tin-free steel ends is the absence of any scrap recycling loop in the tin-free steel system. Thus although the total production system is similar to that shown in Fig. 11.14 the flows L and e are zero. The flows in the remainder of the system are as shown in Table 11.21 and the energy requirements are given in Table 11.22. See Fig. 11.15.

Table 11.19
Total energy required to produce 1000 tinplate ends for 6 fl. oz. cans

Contribution	Electricity/MJ		Oil fuels/MJ			Other fuels/MJ			Total energy /MJ
	Fuel production and delivery energy	Energy content of fuel	Fuel production and delivery energy	Energy content of fuel	Feedstock energy	Fuel production and delivery energy	Energy content of fuel	Feedstock energy	
Production of pig iron	11.76	4.57	6.32	30.14	nil	27.71	163.62	nil	244.12
Production and delivery of tinplate	49.66	19.99	13.63	66.22	neg	1.82	42.60	nil	193.92
Coil cutting	0.52	0.17	0.25	1.21	nil	nil	nil	nil	2.15
Lacquering	1.72	0.57	nil	nil	nil	0.35	4.25	nil	6.89
End-making	34.52	11.51	2.68	14.86	nil	0.25	3.07	nil	66.89
Scrap treatment	1.21	0.40	0.78	3.54	nil	0.03	0.21	0.08	6.25
Supply of lacquer	1.38	0.46	1.53	7.54	11.37	0.47	5.50	nil	28.25
Supply of varnish	0.20	0.07	0.23	1.12	1.29	0.05	0.54	nil	3.50
Supply of lining compound	1.51	0.50	2.16	10.71	17.17	nil	nil	nil	32.05
Totals/MJ	102.48	38.24	27.58	135.34	29.83	30.68	219.79	0.08	584.02

© Crown Copyright 1981

Table 11.21

Materials flow in the production of 1000 tin-
free steel ends for 10, 12 and 16 fl. oz. cans
using the system shown in Fig. 11.12

Materials flow	Quantity
A and B	Not applicable
C (kg)	13.400
D (kg)	12.730
E (kg)	12.111
E (sheets)	7.744
F (kg)	12.051
F (sheets)	7.706
G (kg)	10.200
H (kg)	0.095
J (kg)	0.017
K (kg)	0.070
L	zero
a (kg)	0.670
b (kg)	0.619
c (kg)	0.060
d (kg)	1.851
e	zero

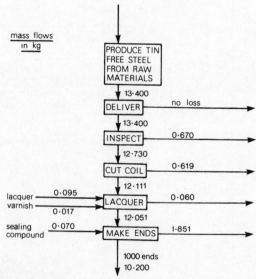

Fig. 11.15 – Production system for the manufacture of 1000 tin-free steel ends
for 10, 12 and 16 fl. oz. cans.

Table 11.22
Total energy required to produce 1000 tin-free steel ends for 10, 12 and 16 fl. oz. cans

Contribution	Electricity/MJ		Oil fuels/MJ			Other fuels/MJ			Total energy /MJ
	Fuel production and delivery energy	Energy content of fuel	Fuel production and delivery energy	Energy content of fuel	Feedstock energy	Fuel production and delivery energy	Energy content of fuel	Feedstock energy	
Production of tin-free steel	82.14	32.96	26.93	129.44	nil	48.37	324.41	nil	644.25
Delivery of tin-free steel	0.13	0.04	0.67	3.75	nil	0.03	0.40	nil	5.02
Coil cutting	0.57	0.19	0.28	1.33	nil	nil	nil	nil	2.37
Lacquering	1.89	0.63	nil	nil	nil	0.38	4.67	nil	7.57
End-making	34.52	11.51	2.68	14.86	nil	0.25	3.07	nil	66.89
Supply of lacquer	1.75	0.58	1.94	9.56	14.40	0.59	6.97	nil	35.79
Supply of varnish	0.26	0.09	0.30	1.46	1.69	0.06	0.71	nil	4.57
Supply of lining compound	0.60	0.20	0.85	4.22	6.77	nil	nil	nil	12.64
Totals/MJ	121.86	46.20	33.65	164.62	22.86	49.68	340.23	nil	779.10

© *Crown Copyright 1981*

11.15 TOTAL ENERGY TO PRODUCE ALUMINIUM EASY-OPEN ENDS

The flow sheet for the production of aluminium easy-open ends from raw materials in the ground is shown in Fig. 11.16 and the materials flows are given in Table 11.23. In practice the internal lacquer systems for beer cans differ from those for soft drinks with the latter receiving a considerably thicker coat. The flow diagram assumes an average coat thickness midway between the two. The energies associated with the operation of these systems are given in Tables 11.24 and 11.25. The only assumption that has been made in the calculation of the tables is that tab coil is all produced from primary sources whereas the alloy for the shell contains all the energy saving arising from the recycling process.

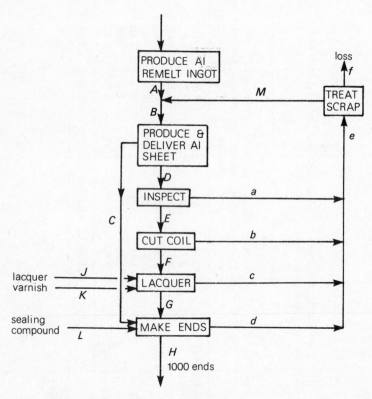

Fig. 11.16 – Production system for 1000 aluminium easy-open ends from raw materials in the ground. See Table 11.23 for numerical data.

Table 11.23
Materials flows for the production of 1000 easy-open aluminium ends using the production system shown in Fig. 11.16

Flow (Fig. 11.16)	Can size	
	6 fl. oz.	10, 12, 16 fl. oz.
A (kg)	3.796	5.4345
B (kg)	5.196	6.9743
C (kg)	0.746	0.7076
D (kg)	4.425	5.9985
E (kg)	4.204	5.6986
F (kg)	4.171	5.6536
F (sheets)	7.136	8.563
G (kg)	4.150	5.6253
G (sheets)	7.100	8.521
H (kg)	3.668	5.0525
J (kg)	0.0425	0.0552
K (kg)	0.0134	0.0174
L (kg)	0.0693	0.0913
M (kg)	1.400	1.5398
a (kg)	0.221	0.2999
b (kg)	0.033	0.0450
c (kg)	0.021	0.0283
d (kg)	1.228	1.2804
e (kg)	1.503	1.6536
f (kg)	0.103	0.1138

Table 11.24

Total energy required to produce 1000 aluminium easy-open ends for 6 fl. oz. cans

Contribution	Electricity/MJ		Oil fuels/MJ			Other fuels/MJ			Total energy /MJ
	Fuel production and delivery energy	Energy content of fuel	Fuel production and delivery energy	Energy content of fuel	Feedstock energy	Fuel production and delivery energy	Energy content of fuel	Feedstock energy	
Production of alloy for shells	389.46	195.85	30.74	153.75	77.95	5.65	67.74	nil	921.14
Production of aluminium for tabs	99.00	50.83	8.25	41.23	20.84	0.96	11.37	nil	232.48
Production of coil for shells	45.14	15.05	8.32	39.83	nil	3.41	41.37	nil	153.12
Production of coil for tabs	7.61	2.54	1.40	6.71	nil	0.57	6.98	nil	25.81
Coil cutting	0.50	0.17	0.25	1.23	nil	nil	nil	nil	2.15
Lacquering	1.74	0.58	nil	nil	nil	0.35	4.30	nil	6.97
End-making	34.78	11.59	nil	nil	nil	0.60	7.35	nil	54.32
Scrap recycling	0.57	0.20	0.20	1.08	nil	1.71	20.78	nil	24.54
Supply of lacquer	0.78	0.26	0.87	4.27	6.44	0.26	3.12	nil	16.00
Supply of varnish	0.21	0.07	0.24	1.15	1.33	0.05	0.56	nil	3.61
Supply of lining compound	0.59	0.20	0.84	4.18	6.70	nil	nil	nil	12.51
Totals/MJ	580.38	277.34	51.11	253.43	113.26	13.56	163.57	nil	1452.65

Table 11.25

Total energy required to produce 1000 aluminium easy-open ends for 10, 12 and 16 fl. oz. cans

Contribution	Electricity/MJ		Oil fuels/MJ			Other fuels/MJ			Total energy /MJ
	Fuel production and delivery energy	Energy content of fuel	Fuel production and delivery energy	Energy content of fuel	Feedstock energy	Fuel production and delivery energy	Energy content of fuel	Feedstock energy	
Production of alloy for shells	605.93	304.70	47.83	239.21	121.27	8.79	105.39	nil	1433.12
Production of aluminium for tabs	93.91	48.22	7.83	39.11	19.77	0.91	10.79	nil	220.54
Production of coil for shells	61.18	20.39	11.28	53.99	nil	4.62	56.09	nil	207.55
Production of coil for tabs	7.22	2.41	1.33	6.37	nil	0.54	6.62	nil	24.49
Coil cutting	0.61	0.20	0.31	1.47	nil	nil	nil	nil	2.59
Lacquering	2.09	0.70	nil	nil	nil	0.42	5.16	nil	8.37
End-making	34.78	11.59	nil	nil	nil	0.60	7.35	nil	54.32
Scrap treatment	0.63	0.22	0.22	1.19	nil	1.88	22.85	nil	26.99
Supply of lacquer	1.02	0.34	1.12	5.55	8.37	0.34	4.05	nil	20.79
Supply of varnish	0.27	0.09	0.31	1.49	1.72	0.06	0.72	nil	4.66
Supply of lining compound	0.78	0.26	1.11	5.51	8.83	nil	nil	nil	16.49
Totals/MJ	808.42	389.12	71.34	353.89	159.96	18.16	219.02	nil	2019.91

11.16 TOTAL ENERGY TO PRODUCE THREE-PIECE TINPLATE CANS

The complete system to produce three-piece open top tinplate cans from raw materials in the ground is shown in Fig. 11.17 and materials flows are given in Table 11.26. The detailed calculation of the total energy associated with the production of the 6 fl. oz. size is given in Table 11.27 and the energy required to produce the other sizes is summarised in Table 11.28.

These Tables also include the energy associated with the production of the loose end delivered separately to the filler so enabling the total energy to manufacture the completed can to be calculated.

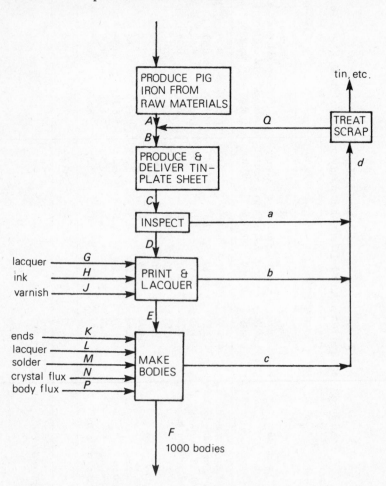

Fig. 11.17 — Total system required to produce 1000 bodies for three-piece tinplate cans. For materials flows see Table 11.26.

Table 11.26

Materials flows in the production of 1000 three-piece tinplate cans of different sizes using the system shown in Fig. 11.17

Flow (Fig. 11.17)	Can size						
	6 fl. oz.	10 fl. oz.	12 fl. oz.	16 fl. oz.	4 pint	5 pint	7 pint
A (kg)	25.628	32.615	37.947	49.108	203.786	255.678	299.385
B (kg)	27.486	34.916	40.623	52.605	219.625	275.656	321.567
C (kg)	22.397	28.408	33.051	42.800	178.688	224.275	261.628
D (kg)	21.277	26.988	31.398	40.660	169.754	213.061	248.547
E (kg)	21.171	26.853	31.241	40.457	168.905	211.996	247.304
E (sheets)	25.470	32.175	36.331	50.864	128.017	170.689	254.219
F (kg)	27.885	36.634	40.894	49.806	205.159	246.522	311.666
G (kg)	0.152	0.192	0.224	0.290	0.383	0.654	1.171
H (kg)	0.053	0.067	0.075	0.105	0.263	0.351	0.527
J (kg)	0.081	0.102	0.119	0.154	0.204	0.347	0.623
K (number)	1004	1002	1002	1002	1010	1010	1001
L (kg)	0.005	0.005	0.007	0.010	0.012	0.016	0.015
M (kg)	0.420	0.638	0.432	0.908	nil	nil	1.259
N (kg)	0.04	0.04	0.04	0.04	nil	nil	0.04
P (kg)	0.06	0.06	0.06	0.06	nil	nil	0.06
Q (kg)	1.858	2.301	2.676	3.497	15.839	19.978	22.182

a, b, c, d can be calculated by subtraction

Table 11.27

Total energy required to produce 1000 three-piece tinplate cans of 6 fl. oz. capacity

Contribution	Electricity/MJ		Oil fuels/MJ			Other fuels/MJ			Total energy /MJ
	Fuel production and delivery energy	Energy content of fuel	Fuel production and delivery energy	Energy content of fuel	Feedstock energy	Fuel production and delivery energy	Energy content of fuel	Feedstock energy	
Production of pig iron	31.01	12.06	16.66	79.45	nil	73.04	431.32	nil	643.54
Production and delivery of tinplate sheet	110.64	44.57	30.24	147.15	nil	4.03	94.52	0.67	431.82
Printing and lacquering	14.67	4.89	nil	nil	nil	3.10	37.72	nil	60.38
Body-making	88.92	21.02	nil	nil	nil	7.29	88.56	nil	205.79
Scrap treatment	0.84	0.28	0.54	2.45	nil	0.02	0.15	0.05	4.33
Supply of lacquer	2.81	0.93	3.10	15.29	23.03	0.95	11.15	nil	57.26
Supply of ink	0.11	0.04	0.23	1.12	1.94	nil	nil	nil	3.44
Supply of varnish	1.25	0.42	1.42	6.95	8.03	0.29	3.36	nil	21.72
Supply of ends	102.89	38.39	27.69	135.88	29.95	30.80	220.67	0.08	586.35
Supply of lacquer (sidestripe)	0.09	0.03	0.10	0.50	0.76	0.03	0.37	nil	1.88
Supply of tin solder	7.56	2.52	2.94	15.54	nil	1.59	19.32	nil	49.47
Supply of crystal flux	0.31	0.10	0.08	0.32	0.24	0.06	0.72	nil	1.83
Supply of body soldering flux	0.21	0.07	0.28	1.34	1.83	neg	neg	nil	3.73
Totals for bodymaking/MJ	361.31	125.32	83.28	405.99	65.78	121.20	907.86	0.80	2071.54
Loose end	580.38	277.34	51.11	253.43	113.26	13.56	163.57	nil	1452.65
Totals/MJ	941.69	402.66	134.39	659.42	179.04	134.76	1071.43	0.80	3524.19

© Crown Copyright 1981

Table 11.28

Summary of the total energy in MJ required to produce 1000 tinplate three-piece cans of various sizes

Contribution	Can size						
	6 fl. oz.	10. fl. oz.	12 fl. oz.	16 fl. oz.	4 pint	5 pint	7 pint
Production of pig iron	643.54	818.96	952.87	1233.10	5117.07	6420.07	7517.56
Production and delivery of tinplate	431.82	547.70	623.23	825.18	3445.10	4324.03	5044.19
Print and lacquer	60.38	76.29	86.14	120.59	303.52	404.70	602.75
Bodymaking	205.79	228.65	228.65	261.32	770.10	943.71	1036.54
Scrap treatment	4.33	5.34	6.23	8.13	36.92	46.55	51.68
Production of lacquer	57.26	72.33	84.39	109.25	144.29	246.37	441.14
Production of ink	3.44	4.35	4.86	6.81	17.05	22.75	34.17
Production of varnish	21.72	27.33	31.91	41.30	54.70	93.03	167.04
Production of end	586.35	749.23	749.23	749.23	3051.54	3051.54	4912.12
Production of lacquer (sidestripe)	1.88	1.88	2.62	3.76	4.51	6.03	5.65
Production of solder	49.47	34.19	50.87	48.67	nil	nil	148.27
Production of crystal flux	1.83	1.83	1.83	1.83	nil	nil	1.83
Production of body solder flux	3.73	3.73	3.73	3.73	nil	nil	3.73
Totals for body making	2071.54	2572.01	2840.56	3412.90	12944.80	15558.78	19966.67
Loose end production	1452.65	2019.91	2019.91	2019.91	3021.32	3021.32	4907.22
Totals/MJ	3524.19	4591.72	4860.47	5432.81	15966.12	18580.10	24873.89

11.17 TOTAL ENERGY TO PRODUCE THREE-PIECE TIN-FREE STEEL CANS

The complete system required to produce tin-free steel cans is as shown in Fig. 11.18 and the materials flows are given in Table 11.29. The total energies associated with all operations involved in can making are given in Tables 11.30 and 11.31 and, as before, include the energy associated with the provision of the loose end.

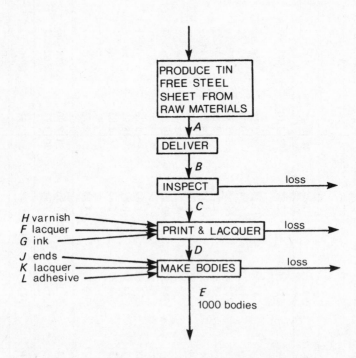

Fig. 11.18 – Total system required to produce 1000 bodies for tin-free steel cans from raw materials in the ground. See Table 11.29 for numerical data.

Table 11.29

Materials flows required to produce 1000 tin-free steel cans
using the system of Fig. 11.18

Flow (Fig. 11.15)	Can size	
	12 fl. oz.	16 fl. oz.
A (kg)	33.294	41.478
B (kg)	33.294	41.478
C (kg)	31.629	39.404
C (sheets)	36.514	51.119
D (kg)	31.471	39.207
D (sheets)	36.331	50.864
E (kg)	40.300	47.700
F (kg)	0.290	0.330
G (kg)	0.280	0.360
H (kg)	0.200	0.300
J (number)	1002	1002
K (kg)	0.008	0.010
L (kg)	0.132	0.176

Table 11.30
Total energy required to produce 1000 three-piece tin-free steel 16 fl. oz. cans

Contribution	Electricity/MJ		Oil fuels/MJ			Other fuels/MJ			Total energy /MJ
	Fuel production and delivery energy	Energy content of fuel	Fuel production and delivery energy	Energy content of fuel	Feedstock energy	Fuel production and delivery energy	Energy content of fuel	Feedstock energy	
Production of tin-free steel	204.09	81.90	66.92	321.62	nil	120.19	806.05	nil	1600.77
Delivery of tin-free steel	1.00	0.33	1.66	9.66	nil	0.09	1.00	1.00	14.74
Printing and lacquering	20.93	6.98	nil	nil	nil	4.42	53.81	nil	86.14
Bodymaking	123.44	41.15	5.91	29.17	43.95	9.56	116.30	nil	290.45
Supply of lacquer	5.35	1.78	1.22	5.93	10.24	1.81	21.28	nil	109.25
Supply of ink	0.58	0.19	3.52	17.17	19.83	nil	nil	nil	18.16
Supply of varnish	3.09	1.03	33.72	164.95	22.91	0.70	8.30	nil	53.64
Supply of ends	122.10	46.29	0.16	0.80	1.21	49.78	340.91	nil	780.66
Supply of lacquer (sidestripe)	0.15	0.05	0.92	4.47	5.94	0.05	0.59	nil	3.01
Supply of adhesive	0.33	0.11				nil	nil	nil	11.77
Totals for bodymaking/MJ	481.06	179.81	114.03	553.77	104.08	186.60	1348.24	1.00	2968.59
Loose ends	808.42	389.12	71.34	353.89	159.96	18.16	219.02	nil	2019.91
Totals/MJ	1289.48	568.93	185.37	907.66	264.04	204.76	1567.26	1.00	4988.50

Table 11.31

Total energy required to produce 1000 three-piece tin-free steel 16 fl. oz. cans

Contribution	Electricity/MJ		Oil fuels/MJ			Other fuels/MJ			Total energy /MJ
	Fuel production and delivery energy	Energy content of fuel	Fuel production and delivery energy	Energy content of fuel	Feedstock energy	Fuel production and delivery energy	Energy content of fuel	Feedstock energy	
Production of tin-free steel	254.26	102.04	83.37	400.68	nil	149.74	1004.18	nil	1994.27
Delivery of tin-free steel	1.24	0.41	2.07	12.03	nil	0.11	1.24	1.24	18.34
Printing and lacquering	29.30	9.77	nil	nil	nil	6.19	75.33	nil	120.59
Bodymaking	123.44	41.15	nil	nil	nil	9.56	116.30	nil	290.45
Supply of lacquer	6.09	2.03	6.73	33.19	50.01	2.06	24.22	nil	124.33
Supply of ink	0.74	0.25	1.56	7.62	13.17	nil	nil	nil	23.34
Supply of varnish	4.63	1.54	5.28	25.75	29.74	1.06	12.45	nil	80.45
Supply of ends	122.10	46.29	33.72	164.95	22.91	49.78	340.91	nil	780.66
Supply of lacquer (sidestripe)	0.18	0.06	0.21	1.01	1.52	0.06	0.73	nil	3.77
Supply of adhesive	0.44	0.15	1.22	5.96	7.92	nil	nil	nil	15.69
Totals for bodymaking/MJ	542.42	203.69	134.16	651.19	125.27	218.56	1575.36	1.24	3451.89
Loose ends	808.42	389.12	71.34	353.89	159.96	18.16	219.02	nil	2019.91
Totals/MJ	1350.84	592.81	205.50	1005.08	285.23	236.72	1794.38	1.24	5471.80

11.18 TOTAL ENERGY TO PRODUCE TWO-PIECE TINPLATE CANS

The total system required to produce bodies for two-piece tinplate cans is shown in Fig. 11.19. As before, a 5% coil rejection rate has been included and it is assumed that all scrap is returned for detinning. Lacquer use is based on actual production data but varnish and ink consumption has been estimated assuming the same use rate as for three-piece tinplate body cylinders after making an allowance for the different levels of spoilage. The energy associated with the operation of this system is shown in Table 11.32 and when the energy associated with the production of the loose end is included, this gives the total energy of the complete can.

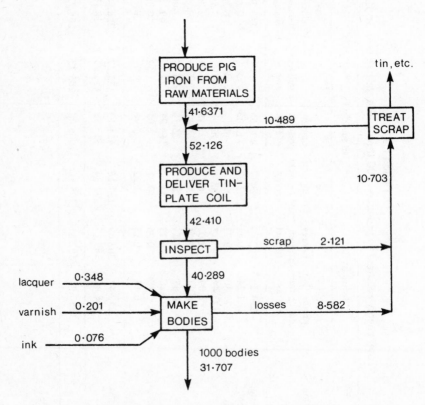

Fig. 11.19 – Total system required to produce 1000 bodies for two-piece tinplate cans of capacity 12 fl. oz. All flows are in kg.

Table 11.32

Total energy required to produce 1000 tinplate two-piece 12 fl. oz. cans

Contribution	Electricity		Oil fuels/MJ			Other fuels/MJ			Total energy /MJ
	Fuel production and delivery energy	Energy content of fuel	Fuel Production and delivery energy	Energy content of fuel	Feedstock energy	Fuel production and delivery energy	Energy content of fuel	Feedstock energy	
Production of pig iron	50.38	19.57	27.06	129.07	nil	118.67	700.75	nil	1045.50
Production and delivery of tinplate	208.66	83.97	57.25	278.21	nil	7.63	178.97	nil	814.69
Bodymaking	385.53	128.51	30.69	147.77	nil	30.39	369.52	nil	1092.41
Scrap treatment	4.72	1.57	3.04	13.85	nil	0.10	0.84	0.31	24.43
Supply of lacquer	6.42	2.14	7.09	35.00	52.74	2.17	25.54	nil	131.10
Supply of varnish	3.10	1.03	3.54	17.25	19.26	0.71	8.34	nil	53.23
Supply of ink	0.16	0.05	0.33	1.61	2.78	nil	nil	nil	4.93
Totals for bodymaking/MJ	658.97	236.84	129.00	622.76	74.78	159.67	1283.96	0.31	3166.29
Supply of loose ends	808.42	389.12	71.34	353.89	159.96	18.16	219.02	nil	2019.91
Totals/MJ	1467.39	625.96	200.34	976.65	234.74	177.83	1502.98	0.31	5186.20

© Crown Copyright 1981

11.19 TOTAL ENERGY TO PRODUCE TWO-PIECE ALUMINIUM CANS

The total system required to produce two-piece aluminium cans is shown in Fig. 11.20 and the materials flows in this system are given in Table 11.33. System energy requirements are given in Tables 11.34 and 11.35.

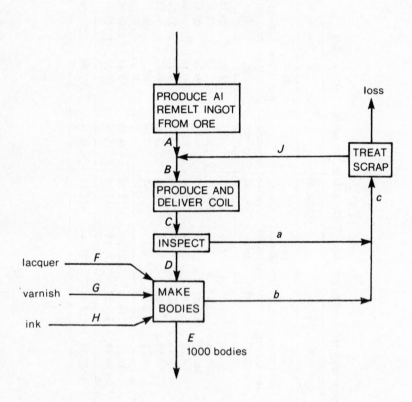

Fig. 11.20 – Total system required to produce 1000 bodies for two-piece aluminium cans. See Table 11.33 for numerical data.

Table 11.33

Materials flows to produce 1000 two-piece aluminium cans
using the system of Fig. 11.20. All values are in kilogrammes

Flow (Fig. 11.20)	Can size	
	12 fl. oz.	16 fl. oz.
A	17.153	18.969
B	23.049	25.494
C	22.162	24.513
D	21.054	23.287
E	15.830	17.506
F	0.147	0.184
G	0.167	0.210
H	0.063	0.079
J	5.896	6.525
a	1.108	1.226
b	5.224	5.781
c	6.332	7.007

Table 11.34
Total energy required to produce 1000 12 fl. oz. two-piece aluminium cans

Contribution	Electricity/MJ		Oil fuels/MJ			Other fuels/MJ			Total energy /MJ
	Fuel production and delivery energy	Energy content of fuel	Fuel production and delivery energy	Energy content of fuel	Feedstock energy	Fuel production and delivery energy	Energy content of fuel	Feedstock energy	
Production of primary aluminium remelt ingot	2191.12	1115.63	178.73	892.81	451.30	23.67	279.77	nil	5133.03
Production and delivery of aluminium coil	226.05	75.35	41.66	199.46	nil	17.06	207.21	nil	766.79
Bodymaking	399.54	133.18	35.88	172.76	nil	36.90	448.75	nil	1227.01
Scrap treatment	2.42	0.83	0.83	4.54	nil	7.19	87.50	nil	103.31
Supply of lacquer	2.71	0.90	3.00	14.79	22.28	0.92	10.79	nil	55.39
Supply of varnish	2.58	0.86	2.94	14.33	16.55	0.59	6.93	nil	44.78
Supply of ink	0.13	0.04	0.27	1.33	2.30	nil	nil	nil	4.07
Totals for bodymaking	2824.55	1326.79	263.31	1300.02	492.43	86.33	1040.95	nil	7334.38
Loose ends	808.42	389.12	71.34	353.89	159.96	18.16	219.02	nil	2019.91
Totals/MJ	3632.97	1715.91	334.65	1653.91	652.39	104.49	1259.97	nil	9354.29

Table 11.35

Total energy required to produce 1000 16 fl. oz. two-piece aluminium cans

Contribution	Electricity/MJ		Oil fuels/MJ			Other fuels/MJ			Total energy /MJ
	Fuel production and delivery energy	Energy content of fuel	Fuel production and delivery energy	Energy content of fuel	Feedstock energy	Fuel production and delivery energy	Energy content of fuel	Feedstock energy	
Production of primary aluminium remelt ingot	2423.10	1233.74	197.66	987.34	499.07	26.18	309.38	nil	5676.47
Production and delivery of aluminium coil	250.03	83.34	46.08	220.62	nil	18.88	229.20	nil	848.15
Bodymaking	399.54	133.18	35.88	172.76	nil	36.90	448.75	nil	1227.01
Scrap treatment	2.68	0.91	0.91	5.02	nil	7.96	96.83	nil	114.31
Supply of lacquer	3.40	1.13	3.75	18.51	27.88	1.15	13.50	nil	69.32
Supply of varnish	3.24	1.08	3.69	18.02	20.82	0.74	8.71	nil	56.30
Supply of ink	0.16	0.05	0.34	1.67	2.89	nil	nil	nil	5.11
Totals for bodymaking	3082.15	1453.43	288.31	1423.94	550.66	91.81	1106.37	nil	7996.67
Loose ends	808.42	389.12	71.34	353.89	159.96	18.16	219.02	nil	2019.91
Totals/MJ	3890.57	1842.55	359.65	1777.83	710.62	109.97	1325.39	nil	10016.58

CHAPTER 12

Production of PET Bottles

12.1 PRODUCTION OF POLYETHYLENE TEREPHTHALATE (PET)

The polyester, PET, is produced by the condensation reaction between ethylene glycol and terephthalic acid. The production route from crude oil is complex and involves the production of a number of chemical intermediates. From information available from a number of industrial sources, the total average production energy requirement from raw materials in the ground is as shown in Table 12.1.

Table 12.1

Total energy required to produce 1 kg of polyethylene terephthalate from crude oil

Fuel type	Fuel production and delivery energy/MJ	Energy content of fuel /MJ	Feedstock energy /MJ	Total energy /MJ
Electricity	16.17	5.39	nil	21.56
Oil fuels	14.58	71.18	46.56	132.32
Other fuels	nil	nil	nil	nil
Totals/MJ	30.75	76.57	46.56	153.88

12.2 THE PET BOTTLE SYSTEM

At the present time, the PET bottle used in Europe consists of a transparent body, blow moulded in the solid state from virgin PET resin. The body is welded or glued to an injection moulded base made from PET or from some less expensive polymer such as low density polyethylene. Waste PET generated during body production cannot be recycled back to the beginning of the operation because of contamination and degradation. If PET bases are used, body waste can be granulated and used as part of the feed to base production since this process can accept material of a lower specification than body production. However, insufficient waste is generated during body production for the whole of the input to base production to be derived from waste polymer. Hence the difference must be made up using virgin polymer. In contrast, the use of polyethylene bases requires that the whole input to base production be virgin polymer so that any waste PET produced represents a net loss from the system. These alternative processing possibilities can be summarised in the composite flowsheet of Fig. 12.1.

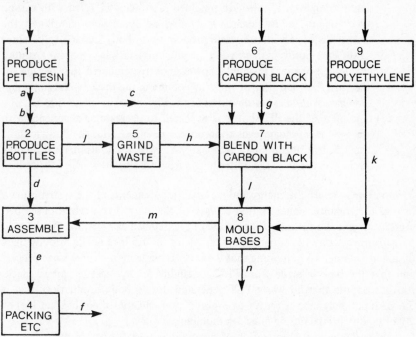

Fig. 12.1 – System used to produce PET bottles. Note that if polyethylene bases are used, the loss of PET from bottle making is represented by flow *j*.

12.3 ENERGY REQUIREMENTS FOR BOTTLE PRODUCTION – A TYPICAL EXAMPLE

Evaluation of the energy required to produce PET bottles using the system of Fig. 12.1 is complicated by four main factors. These are:

   (a) Bottles are currently produced in two different sizes; 1.5 litre and 2.0 litre. The principal effect of this is on the bottle mass. The 1.5 litre bottle typically weighs 60 g with a base (PET) of mass 5 g. The 2.0 litre bottle typically weighs 65 g with a base (PET) of mass 15 g.

(b) Bottles are currently produced by two different methods. The most widely used process is the two stage process in which a preform is injection moulded and the finished bottle is solid state blow moulded from this preform. The alternative method is a three stage process in which a transparent tube is extruded. The preform is thermoformed from this tube and the finished bottle is solid state blow moulded from this preform as in the two stage process. From the data available to date, there appears to be little difference in the energy used directly by these two methods but the materials efficiencies are markedly different. The two stage process exhibits overall PET losses of the order of 1% whereas the three stage process gives typical losses of the order of 5%.

(c) Bases are currently injection moulded in either PET filled with carbon black or unfilled low density polyethylene. As already mentioned, the use of PET bases allows waste production in body making to be taken up in base production whereas the polyethylene bases require a supply of virgin polymer. It should however be remembered that because of the different densities of the two polymers, the mass of the polyethylene base will be lower than that of the PET base (by about 20%).

(d) The mass of the PET bottle is expected to decrease as experience with the new technology leads to improved practice. Thus the body mass of the 1.5 litre bottle might reasonably be expected to decrease from 60 g to 55 g and that of the 2 litre bottle from 65 g to 60 g.

A proper analysis of the energy and materials requirements of the system shown in Fig. 12.1 requires consideration of each of the above likely variations. However, by way of illustration of the nature of the detailed calculations that have to be performed, consider initially the example of the 1.5 litre bottle of body mass 60 g and base mass 5 g. Assume that the body is made by the three stage process and that the base is made from PET containing 5% by mass of carbon black. Further assume that all waste PET generated during body production (that is 5% of input polymer) is fed to base production and that there is a net loss of 2% of input material during injection moulding.

Using the law of conservation of mass, the materials flows in the system of Fig. 12.1 can be calculated to give the values shown in Table 12.2 for the production of 1000 finished bottles.

To calculate the energy requirement of the system, it is necessary to assign an energy requirement to each of the sub-systems shown in Fig. 12.1; these are summarised in Table 12.3. Note that it has proved impossible to separate the energy associated with body production from that associated with assembly.

The total energy required to produce 1.5 litre PET bottles of the type specified above may therefore be calculated by multiplying the unit energy requirements in Table 12.3 by the appropriate mass flows in Table 12.2. This yields Table 12.4.

Table 12.2

Materials flows for the production of 1000 PET bottles of capacity 1.5 litre using the system shown in Fig. 12.1. Body mass = 60 g, Base mass = 5 g. Base material is PET containing 5% carbon black.

Mass flow in Fig. 12.1	Description of flow	Magnitude
a	Total PET input/kg	64.847
b	Virgin PET to bodymaking/kg	63.158
c	Virgin PET to base production/kg	1.689
d	Finished bodies/kg	60.000
e	Assembled bodies/kg	65.000
f	Packed bottles/kg	65.000
g	Carbon black input/kg	0.255
h	Granulated PET waste/kg	3.158
j	Waste PET to grinding/kg	3.158
k	Virgin polyethylene input/kg	nil
l	PET + carbon black blend/kg	5.102
m	Bases to assembly/kg	5.000
n	PET loss from base moulding/kg	0.102

Table 12.3

Energy requirements of unit operations in PET bottle production using the system of Fig. 12.1

Operation	Electricity/MJ		Oil fuels/MJ			Other fuels/MJ			Total energy /MJ
	Fuel production and delivery	Energy content of fuel	Fuel production and delivery	Energy content of fuel	Feedstock energy	Fuel production and delivery	Energy content of fuel	Feedstock energy	
Production of 1 kg PET resin from crude oil	16.17	5.39	14.58	71.18	46.56	nil	nil	nil	153.88
Production and assembly of 1000 PET bottles	3358.80	1119.60	nil	nil	nil	nil	nil	nil	4478.40
Production of 1 kg granulated PET from waste PET	0.99	0.33	nil	nil	nil	nil	nil	nil	1.32
Production of 1 kg carbon black from crude oil	21.48	7.16	16.00	51.21	32.79	nil	nil	nil	128.64
Blending operations to produce 1 kg of blended product	3.06	1.02	nil	nil	nil	nil	nil	nil	4.08
Injection moulding to produce 1 kg of product	23.44	7.81	nil	nil	nil	nil	nil	nil	31.25
Production of 1 kg of low density polyethylene resin	8.28	2.76	7.08	36.82	49.95	nil	nil	nil	104.89

Table 12.4

Energy required to produce 1000 PET bottles of capacity 1.5 litre. Body mass = 60 g. Base mass = 5 g. Production using the system of Fig. 12.1

| Operation | Electricity/MJ | | Oil fuels/MJ | | | Total energy /MJ |
	Fuel production and delivery	Energy content of fuel	Fuel production and delivery	Energy content of fuel	Feedstock energy	
PET resin production	1048.58	349.53	945.47	4615.81	3019.28	9978.67
Bottle production	3358.80	1119.60	–	–	–	4478.40
Grinding PET waste	3.13	1.04	–	–	–	4.17
Carbon black production	5.48	1.83	4.08	13.06	8.36	32.81
Blending	15.61	5.20	–	–	–	20.81
Injection mould bases	117.20	39.05	–	–	–	156.25
Totals/MJ	4548.80	1516.25	949.55	4628.87	3027.64	14671.11

12.4 ENERGY REQUIREMENTS OF BOTTLE PRODUCTION – SYSTEM VARIATIONS

Using detailed calculations of the type presented in the previous section, the effect of changing bottle size, bottle mass, type of base material and type of production process can be evaluated. To demonstrate this, 16 separate cases have been identified as shown in Table 12.5. The materials flows within these various systems are given in Table 12.6 and the gross energy requirements are summarised as the final column in Table 12.5. These calculations assume that when a PET base is used, all waste from body production is fed to base production and when polyethylene bases are employed, any PET waste is regarded as a net materials loss from the system.

Table 12.5

Properties of the cases identified to illustrate the effect of system variables on the energy and materials requirements for PET bottle production

Case reference	Bottle capacity in litre	Production process	Base material	Body mass in gramme	Gross energy requirement in MJ/bottle
1	1.5	2 stage	PET	60	14.66
2				55	13.90
3			Peth	60	14.37
4				55	13.59
5		3 stage	PET	60	14.67
6				55	13.90
7			Peth	60	14.75
8				55	13.93
9	2.0	2 stage	PET	65	17.41
10				60	16.62
11			Peth	65	16.15
12				60	15.50
13		3 stage	PET	65	17.51
14				60	16.72
15			Peth	65	16.35
16				60	15.77

It is clear from Table 12.5 that the combined effects of changing the above variables can lead to an 8% variation in the bottle making energy. The data also demonstrate the need to define accurately the system being studied since quite small changes in the system parameters can lead to significant changes in the system energy requirement.

Table 12.6

Materials flows (Fig. 12.1) corresponding to the various systems defined in Table 12.5. All values are per 1000 finished bottles produced. All values are in kilogrammes

Flow (Fig. 12.1)	Case (Table 12.5) 1	2	3	4	5	6	7	8	9	10	11	12	13	14	15	16
a	64.84	59.84	60.61	55.56	64.85	59.85	63.16	57.90	79.54	74.54	65.66	60.61	80.31	75.04	68.42	63.16
b	60.61	55.56	60.61	55.56	63.16	57.90	63.16	57.90	68.42	63.16	65.66	60.61	68.42	63.16	68.42	63.16
c	4.23	4.28	–	–	1.69	1.95	–	–	11.12	11.38			11.89	11.89	–	–
d	60	55	60	55	60	55	60	55	65	60	65	60	65	60	65	60
e	65	60	64	59	65	60	64	59	80	75	76	71	80	75	76	71
f	65	60	64	59	65	60	64	59	80	75	76	71	80	75	76	71
g	0.26	0.26	–	–	0.26	0.26	–	–	0.77	0.77	–	–	0.77	0.77	–	–
h	0.61	0.56	–	–	3.16	2.90	–	–	3.42	3.16	–	–	3.42	3.16	–	–
j	0.61	0.56	–	–	3.16	2.90	–	–	3.42	3.16	–	–	3.42	3.16	–	–
k	–	–	4.08	4.08	–	–	4.08	4.08	–	–	11.22	11.22	–	–	11.22	11.22
l	5.10	5.10	–	–	5.10	5.10	–	–	15.31	15.31	–	–	15.31	15.31	–	–
m	5	5	4	4	5	5	4	4	15	15	11	11	15	15	11	11
n	0.10	0.10	0.08	0.08	0.10	0.10	0.08	0.08	0.31	0.31	0.22	0.22	0.31	0.31	0.22	0.22

CHAPTER 13

Distribution of Empty Glass Bottles

13.1 INTRODUCTION

Finished glass bottles are usually stacked onto pallets in the glass factory and pallet loads are delivered to customers by lorry. Vehicles with payloads of 20 tonne are invariably used and the energy associated with this delivery operation arises from two main contributions; the energy required to run the lorry and the energy associated with the provision of the ancillary packing materials.

Calculations of the total distribution energy are however complicated by five major factors:

13.1.1 Bottle size

Bottles are usually stacked so that the overall dimensions of a full pallet are the same or similar for all types of bottles. This allows uniform stacking of pallets on lorries carrying mixed loads. However the number of bottles that can be stacked within this overall pallet pack size will depend upon bottle size; the larger the bottle the lower will be the number which can be stacked on a pallet and since lorries are of fixed dimensions, this influences the number of bottles that can be carried on a single lorry load.

13.1.2 Pack type

Bottles may be stacked in a number of ways to give a stable pallet-load. The simplest is to stack them in layers on cardboard sheets and to enclose the whole pallet stack in shrinkwrap polyethylene film. This method of stacking (known as bulk palletising) usually permits the maximum number of bottles to be stacked on a pallet. However, customers may request bottle deliveries in returnable crates, cartons or trays and these packages are also stacked on pallets to facilitate bulk handling. Such packaging methods usually result in a lower number of bottles per pallet compared with bulk palletising and hence the number of bottles per lorry-load is also reduced. At the same time, the use of different ancillary materials in the form of crates, cartons and trays also influences the overall energy requirement of this component of the distribution system.

13.1.3 Lorry loading

Whilst every effort is made to ensure that deliveries of bottles employ fully loaded lorries for maximum running efficiency, it is not always possible to achieve this optimum loading. Part-empty lorries and sub-optimum loading because of mixed loads lead to a decrease in the overall distribution efficiency.

13.1.4 Delivery distance

Delivery distances are determined by the respective geographical locations of glass factory and customer. Any description of the precise distribution system from a factory is likely to vary from year to year as customers change and as demand for different types of containers changes. Furthermore almost all beer and cider bottles are produced in amber glass which is made at only a limited number of factories. This centralised production of beer and cider bottles contrasts sharply with the widespread production of the white flint glass used in the production of bottles for carbonated soft drinks and might be expected to influence the distribution distances of the two groups of containers.

13.1.5 Nature of the return load

The energy associated with operating a vehicle is significantly affected by the amount of empty running. In 1977, most glass bottle deliveries involved full loads in one direction only. The return load was usually empty apart from transport of those returnable components of packaging such as pallets and layer pads.

13.2 ANALYSIS OF A SAMPLE OF BOTTLES DISTRIBUTED

Detailed data have been obtained for the distribution of some 214 million beer, cider and carbonated soft drink bottles (that is approximately 20% of the total number produced in 1977). Within this sample 22% were returnable beer/cider bottles, 10% were non-returnable beer/cider bottles, a further 10% were returnable bottles for carbonated soft drinks and the remaining 58% were non-returnable bottles for carbonated soft drinks. The overall delivery characteristics of this sample are shown in Table 13.1 and it is on the basis of this sample that distribution energies have been calculated. There is no correlation between bottle size and type and one-way delivery distance; that is the effect of a limited number of factories producing amber bottles is not detectable.

Table 13.1 shows that the average delivery distances for the four groups of containers lie in the range 109 to 137 miles with an overall average, weighted by the number of containers of different sizes, of 117 miles (RMS variation ± 50 miles). This value of 117 miles has been used throughout the remainder of this chapter as the average one-way delivery distance for all beverage containers.

In order to calculate delivery energies we also need to consider the loadings of the delivery lorries on both the outward and return journeys.

Table 13.1

Overall delivery characteristics of the distribution sample examined

Container type	Total number of containers moved	Total number of loads	Total vehicle-miles (one way)	Average containers per load	Average delivery distance (miles)	Vehicle-miles per 1000 bottles
Returnable beer/cider	46,234,519	902	98,172	51,258	109	2.12
Non-returnable beer/cider	22,433,645	491	67,285	45,690	137	3.00
Returnable soft drinks	22,278,341	806	105,145	27,641	130	4.72
Non-returnable soft drinks	123,225,050	2325	256,972	53,000	111	2.09
Whole sample	214,171,555	4524	527,574	47,341	117	2.46

13.3 LORRY LOADINGS IN PRACTICE

The most efficient form of packaging for glass bottles is bulk palletisation and the number of bottles per pallet may be readily calculated from the pack specification. This is shown in Table 13.2. Lorries typically carry between 36 and 44 pallets depending upon the precise dimension of the palletised pack.

Table 13.2

Number of bottles per pallet and per lorry calculated from packing specifications and expected lorry loadings

Capacity		Specified number of bottles per pallet
fl. oz.	litre	
Returnable beer/cider bottles		
6.34	0.18	2712
6.5	0.19	2772
9.68	0.28	1805
11.6	0.33	1734
11.6	0.33	1620
19.4	0.55	932
35.2	1.00	652
39.1	1.11	652
78	2.22	195
Non-returnable beer/cider bottles		
8.8	0.25	2226
9.68	0.28	2082
15.5	0.44	1792
17.6	0.50	1320
19.4	0.55	900
26.4	0.75	704
35.2	1.00	712
35.2	1.00	676
Returnable carbonated soft drink bottles		
6.34	0.18	1855
6.5	0.19	2706
6.5	0.19	2718
8.8	0.25	2260
25.7	0.73	812

Table 13.2 – *continued*

Capacity		Specified number of bottles per pallet
fl. oz.	litre	
Returnable carbonated soft drink bottles – contd.		
25.7	0.73	784
26.4	0.75	812
35.2	1.00	652
35.2	1.00	704
39.1	1.11	474
39.1	1.11	600
Non-returnable carbonated soft drinks bottles		
3.7	0.11	Not bulk palletised
6.5	0.19	3164
6.5	0.19	2712
8.8	0.25	2226
8.8	0.25	2968
9.68	0.28	1855
17.6	0.50	1280
26.4	0.75	704
35.2	1.00	676

Using data from the sample, actual lorry loadings have been calculated for each container size and type. This enables a comparison with the theoretical maximum load which can be carried on a lorry. Non-returnable beer/cider containers exhibit the best packing (90%) followed by returnable beer/cider bottles (80%), non-returnable bottles for carbonated soft drinks (80%) and finally returnable bottles for carbonated soft drinks (70%). The main source of these lower packaging efficiencies for returnable beer bottles and for all carbonated soft drinks bottles appears to be the use of packaging systems other than bulk palletised packs. In the case of returnable beer/cider bottles, a proportion are packed in the customers own crates so that they can be fed to the filling lines in the same way as the returns from their retail customers. A small number are also delivered in cartons. The same is true of returnable bottles for carbonated soft drinks, but an additional factor here is the predominance of the larger sizes of bottles in this sector, compared with returnable beer bottles so that the packing efficiency is considerably lower. Non-returnable bottles for carbonated soft drinks are despatched in a variety of different sizes and packs (predominantly trays and cartons) with some of the packs being subsequently used after filling to form the outer packaging for retail sale.

Return journeys also require brief consideration since, as we have noted before, lorries will be loaded well below capacity. For lorries delivering bulk packs, the return journey typically involves the return of pallets and layer pads. When packaging materials are supplied by the customer, the return journey will carry pallets and supplies of new packaging materials. In all cases, the total mass of packaging materials for the return load is unlikely to exceed 1 tonne (5% of capacity).

These load factors for the outward and return journeys are shown in Table 13.3. If we apply them to data on the effect of loading on vehicle performance given in [1] we produce the factors shown in columns 4 and 5 of this table. These in turn may be combined to yield the multiplier given in the final column. Since the difference between these values is only 3% we have assumed a constant multiplier for all container types of 1.66.

Table 13.3

Performance characteristics of delivery vehicles for glass bottles

Container type	Outward loading factor	Return loading factor	Outward energy as a proportion of that at max. loading	Return energy as a proportion of that at max. loading	Total load factor
Returnable beer/ cider	80%	5%	0.95	0.71	1.66
Non-returnable beer/cider	90%	5%	0.98	0.71	1.69
Returnable carbonated soft drink	70%	5%	0.93	0.71	1.64
Non-returnable carbonated soft drink	80%	5%	0.95	0.71	1.66

The energy to deliver 1 lorry load of glass containers (including the return journey) is now given by 1.66 $\times$ 117 $\times$ Table 3.1. This yields Table 13.4.

Table 13.4

Energy required to run the delivery vehicle in the transport of one lorry load of glass bottles from the glass factory to the bottler

Fuel type	Fuel production and delivery energy/MJ	Energy content of fuel /MJ	Feedstock energy /MJ	Total energy /MJ
Electricity	266.08	83.51	nil	349.59
Oil fuels	1237.18	6300.50	15.54	7553.22
Other fuels	108.76	666.17	nil	774.93
Totals/MJ	1612.02	7050.18	15.54	8677.74

© *Crown Copyright 1981*

13.4 DELIVERY ENERGY PER CONTAINER

If the number of bottles per lorry load is known, then the vehicle energy associated with their delivery can be calculated from Table 13.4. The most convenient method of calculating an average for bulk palletising is to plot the reciprocal of the number of bottles on a pallet, N, against the volume capacity of the bottle, V. This is shown in Fig. 13.1 for all glass bottles and gives the empirical ralationship.

$$\frac{10^6}{N} = 1.32\,V + 130$$

where volume capacity is measured in millilitres. Using this relationship together with the loading factors of table 13.3 and assuming an average lorry loading of 40 pallets, the average number of bottles per lorry will be as shown in Table 13.5.

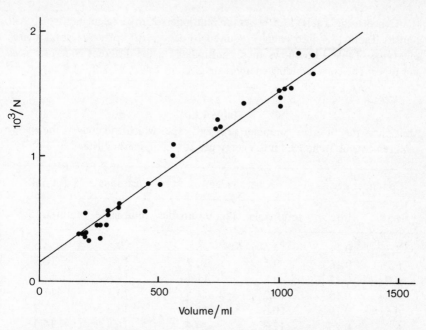

Fig. 13.1 – Relationship between the number of glass bottles per pallet, N, and the volume, V, of the bottle. Plotted points represent data calculated from actual pack specifications.

Table 13.5

Average number of bottles per lorry as a function of the bottle capacity V (ml.)

Bottle type	Number of bottles per lorry
Returnable beer/cider	$\dfrac{36 \times 10^6}{1.32V + 130}$
Non-returnable beer/cider	$\dfrac{32 \times 10^6}{1.32V + 130}$
Returnable carbonated soft drink	$\dfrac{28 \times 10^6}{1.32V + 130}$
Non-returnable carbonated soft drink	$\dfrac{32 \times 10^6}{1.32V + 130}$

Hence using Table 13.5, a set of multipliers can be calculated which will convert Table 13.4 into energy required to deliver 10^6 bottles of various sizes and types. Table 13.6 shows these multipliers for the different container sizes and types. The corresponding energies are given in Tables 13.7 to 13.9.

Table 13.6

Multipliers for different container sizes and types which will convert the energy requirements of Table 13.4 into energy per 10^6 containers delivered

Container size		beer/cider		carbonated soft drink	
fl. oz.	litre	returnable	non-returnable	returnable	non-returnable
3.7	0.11	7.6	8.6	9.8	8.6
5.5	0.16	9.5	10.7	12.1	10.7
6.0	0.17	9.8	11.1	12.6	11.1
6.34	0.18	10.2	11.5	13.1	11.5
6.5	0.19	10.6	11.9	13.6	11.9
8.8	0.25	12.8	14.4	16.4	14.4
9.68	0.28	13.9	15.7	17.8	15.7
11.6	0.33	15.7	17.7	20.2	17.7
15.5	0.44	19.7	22.3	25.3	22.3
17.6	0.50	21.9	24.7	28.2	24.7
19.4	0.55	23.8	26.8	30.5	26.8
25.0	0.71	29.6	33.4	38.1	33.4
25.7	0.73	30.4	34.2	39.0	34.2
26.4	0.75	31.1	35.0	40.0	35.0
34.5	0.98	39.5	44.5	50.8	44.5
35.2	1.00	40.3	45.4	51.7	45.4
38.0	1.08	43.2	48.7	55.5	48.7
39.1	1.11	44.3	49.9	56.9	49.9
78.0	2.22	85.0	95.7	109.3	95.7

Table 13.7

Average vehicle energy to deliver 1000 returnable beer/cider bottles

Container size		Electricity/MJ		Oil fuels/MJ			Other fuels/MJ			Total energy /MJ
fl. oz.	litre	Fuel production and delivery energy	Energy content of fuel	Fuel production and delivery energy	Energy content of fuel	Feedstock energy	Fuel production and delivery energy	Energy content of fuel	Feedstock energy	
3.7	0.11	2.02	0.63	9.40	47.88	0.12	0.83	5.06	—	65.94
5.5	0.16	2.53	0.79	11.75	59.85	0.15	1.03	6.33	—	82.43
6.0	0.17	2.61	0.82	12.12	61.74	0.15	1.07	6.53	—	85.04
6.34	0.18	2.71	0.85	12.62	64.26	0.16	1.11	6.79	—	88.50
6.5	0.19	2.82	0.89	13.11	66.78	0.16	1.15	7.06	—	91.97
8.8	0.25	3.40	1.07	15.83	80.64	0.20	1.40	8.52	—	111.06
9.68	0.28	3.70	1.16	17.19	87.57	0.22	1.52	9.26	—	120.62
11.6	0.33	4.18	1.31	19.42	98.91	0.24	1.71	10.46	—	136.23
15.5	0.44	5.24	1.65	24.37	124.11	0.31	2.15	13.12	—	170.95
17.6	0.50	5.83	1.83	27.09	137.97	0.34	2.38	14.59	—	190.03
19.4	0.55	6.33	1.99	29.44	149.94	0.38	2.59	15.85	—	206.52
25.0	0.71	7.87	2.49	36.62	186.48	0.46	3.23	19.71	—	256.86
25.7	0.73	8.09	2.54	37.60	191.52	0.47	3.31	20.25	—	263.78
26.4	0.75	8.27	2.60	38.47	195.93	0.48	3.38	20.71	—	269.84
34.5	0.98	10.51	3.30	48.86	248.85	0.61	4.29	26.31	—	342.73
35.2	1.00	10.72	3.37	49.85	253.89	0.62	4.39	26.84	—	349.68
38.0	1.08	11.49	3.61	53.44	272.16	0.67	4.71	28.77	—	374.85
39.1	1.11	11.78	3.70	54.80	279.09	0.71	4.83	29.50	—	384.41
78.0	2.22	22.61	7.10	105.15	535.54	1.32	9.27	56.61	—	737.60

Table 13.8

Vehicle energy required to deliver 1000 non-returnable bottles for beer, cider or carbonated soft drinks

Container size		Electricity/MJ		Oil fuels/MJ			Other fuels/MJ			Total energy /MJ
fl. oz.	litre	Fuel production and delivery energy	Energy content of fuel	Fuel production and delivery energy	Energy content of fuel	Feedstock energy	Fuel production and delivery energy	Energy content of fuel	Feedstock energy	
3.7	0.11	2.29	0.72	10.64	54.18	0.13	0.94	5.73	–	74.63
5.5	0.16	2.85	0.89	13.24	67.41	0.17	1.17	7.13	–	92.86
6.0	0.17	2.95	0.93	13.73	69.93	0.17	1.21	7.39	–	96.31
6.34	0.18	3.06	0.96	14.23	72.45	0.18	1.25	7.66	–	99.79
6.5	0.19	3.16	0.99	14.72	74.97	0.18	1.29	7.93	–	103.24
8.8	0.25	3.83	1.20	17.81	90.72	0.22	1.56	9.59	–	124.93
9.68	0.28	4.18	1.31	19.42	98.91	0.24	1.71	10.46	–	136.23
11.6	0.33	4.71	1.48	21.89	111.51	0.27	1.93	11.79	–	153.58
15.5	0.44	5.93	1.86	27.59	140.49	0.35	2.43	14.85	–	193.50
17.6	0.50	6.57	2.06	30.55	155.61	0.38	2.68	16.45	–	214.30
19.4	0.55	7.13	2.24	33.15	168.84	0.42	2.92	17.85	–	232.55
25.0	0.71	8.88	2.79	41.32	210.42	0.52	3.64	22.24	–	289.81
25.7	0.73	9.10	2.86	42.31	215.46	0.53	3.72	22.78	–	296.76
26.4	0.75	9.31	2.92	43.30	220.50	0.54	3.81	23.31	–	303.69
34.5	0.98	11.84	3.72	55.05	280.35	0.69	4.84	29.64	–	386.13
35.2	1.00	12.08	3.79	56.16	286.02	0.70	4.95	30.24	–	393.94
38.0	1.08	12.95	4.07	60.24	306.81	0.76	5.30	32.43	–	422.56
39.1	1.11	13.27	4.17	61.73	314.37	0.77	5.43	33.23	–	432.97
78.0	2.22	25.46	7.99	118.38	602.91	1.49	10.43	63.74	–	830.40

Table 13.9
Vehicle energy required to deliver 1000 returnable bottles for carbonated soft drinks

Container size		Electricity/MJ		Oil fuels/MJ			Other fuels/MJ			Total energy /MJ
fl. oz.	litre	Fuel production and delivery energy	Energy content of fuel	Fuel production and delivery energy	Energy content of fuel	Feedstock energy	Fuel production and delivery energy	Energy content of fuel	Feedstock energy	
3.7	0.11	2.61	0.82	12.12	61.74	0.15	1.07	6.53	–	85.04
5.5	0.16	3.22	1.01	14.97	76.23	0.19	1.32	8.06	–	105.00
6.0	0.17	3.35	1.05	15.59	79.38	0.20	1.37	8.39	–	109.33
6.34	0.18	3.48	1.09	16.20	82.53	0.20	1.43	8.72	–	113.65
6.5	0.19	3.62	1.14	16.82	85.68	0.21	1.48	9.06	–	118.01
8.8	0.25	4.36	1.37	20.29	103.32	0.25	1.79	10.92	–	142.30
9.68	0.28	4.73	1.49	22.02	112.02	0.28	1.94	11.85	–	154.45
11.6	0.33	5.37	1.69	24.99	127.26	0.31	2.20	13.45	–	175.27
15.5	0.44	6.73	2.11	31.30	159.39	0.39	2.76	16.85	–	219.53
17.6	0.50	7.50	2.35	34.88	177.66	0.44	3.07	18.78	–	244.68
19.4	0.55	8.11	2.55	37.73	192.15	0.47	3.32	20.31	–	264.64
25.0	0.71	10.13	3.18	47.13	240.03	0.59	4.15	25.37	–	330.58
25.7	0.73	10.37	3.26	48.24	245.70	0.61	4.25	25.97	–	338.40
26.4	0.75	10.64	3.34	49.48	252.00	0.62	4.35	26.64	–	347.07
34.5	0.98	13.51	4.24	62.84	320.04	0.79	5.54	33.83	–	440.79
35.2	1.00	13.75	4.32	63.95	325.71	0.80	5.64	34.43	–	448.60
38.0	1.08	14.76	4.63	68.65	349.65	0.86	6.04	36.96	–	481.55
39.1	1.11	15.14	4.75	70.39	358.47	0.88	6.14	37.90	–	493.67
78.0	2.22	29.07	9.13	135.20	688.59	1.69	11.91	72.81	–	948.40

13.5 PACKAGING MATERIALS FOR BOTTLE DELIVERY

The other contributor to the total energy required to deliver glass bottles is the provision of the packaging materials. For bulk palletised packs, each pallet load requires one pallet (returnable), one polyethylene shroud (non-returnable) and a number of layer pads (returnable). The polyethylene is typically of mass 1.1 kg per pallet. The amount of board employed as layer pads depends upon the size and shape of the bottles. From an examination of the detailed pack specifications for a number of different beer, cider and carbonated soft drinks bottles, the total mass of board when plotted as a function of bottle size gives Fig. 13.2. There is a gradual increase in the board requirement per container as the bottle capacity increases. The plotted points in Fig. 13.2 relate to all types of beverage containers and no difference can be detected between bottles with different end uses.

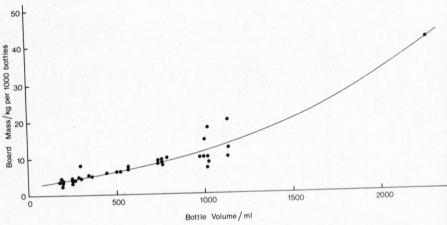

Fig. 13.2 – Board required to pack 1000 glass bottles in bulk packs as a function of bottle capacity.

Fig. 13.2 has been used to construct Table 13.10 which shows the typical packaging materials requirements for different sizes of bottles when bulk pallet-ised. Table 13.10 makes no allowance for the fact that some of the packaging materials are returnable. In general, polyethylene used in packaging is discarded but pallets and layer pads are returnable. None of the factories examined keep detailed records of the lifetimes of pallets or layer pads. An average lifetime of 8 trips has been assumed for pallets. Layer pads are thought to last for approxi-mately two trips although many customers regard them as disposable items; a lifetime of 2 trips has been assumed in calculations. The importance of having returnable items of packaging is that the energy associated with their manu-facture is spread over more than one trip and consequently their contribution to the delivery energy is correspondingly reduced.

Table 13.10

Typical materials requirements for the bulk palletisation of 1000 glass bottles of different sizes

Container size		Packaging materials requirements		
fl. oz.	litre	Pallets	Polyethylene	Board/kg
3.7	0.11	0.30	0.33	3.0
5.5	0.16	0.38	0.42	3.5
6.0	0.17	0.39	0.43	3.5
6.34	0.18	0.40	0.44	3.5
6.5	0.19	0.42	0.46	3.7
8.8	0.25	0.50	0.55	4.2
9.68	0.28	0.54	0.59	4.5
11.6	0.33	0.61	0.67	5.0
15.5	0.44	0.76	0.84	6.0
17.6	0.50	0.84	0.92	6.6
19.4	0.55	0.92	1.01	7.0
25.0	0.71	1.13	1.24	8.7
25.7	0.73	1.16	1.28	9.0
26.4	0.75	1.19	1.31	9.1
34.5	0.98	1.52	1.67	12.0
35.2	1.00	1.54	1.69	12.2
38.0	1.08	1.66	1.83	13.2
39.1	1.11	1.69	1.86	13.6
78.0	2.22	5.00	5.50	40.0

© *Crown Copyright 1981*

The energy associated with the supply of these materials may therefore be calculated using Table 5.2 and the energy requirements are shown in Table 13.11.

Table 13.11

Energy required to provide packaging materials for 1000 bottles of different sizes in bulk palletised packs. An allowance has been made for returnable items

Container size		Electricity/MJ		Oil fuels/MJ			Other fuels/MJ			Total energy /MJ
fl. oz.	litre	Fuel production and delivery energy	Energy content of fuel	Fuel production and delivery energy	Energy content of fuel	Feedstock energy	Fuel production and delivery energy	Energy content of fuel	Feedstock energy	
3.7	0.11	70.26	23.43	13.28	65.59	16.81	0.14	25.73	43.48	258.72
5.5	0.16	84.84	28.28	16.05	79.32	21.40	0.17	30.09	52.33	312.48
6.0	0.17	85.65	28.55	16.21	80.11	21.91	0.17	30.11	52.77	315.48
6.34	0.18	86.46	28.82	16.36	80.90	22.42	0.18	30.13	53.21	318.48
6.5	0.19	91.18	30.40	17.24	85.25	23.44	0.18	31.86	56.31	335.86
8.8	0.25	105.77	35.26	20.02	98.98	28.02	0.22	36.22	65.16	389.65
9.68	0.28	113.51	37.84	21.47	106.19	30.06	0.23	38.82	70.03	418.15
11.60	0.33	126.98	42.33	24.05	118.96	34.14	0.26	43.15	78.01	467.88
15.5	0.40	155.34	51.77	29.43	145.65	42.80	0.31	51.85	95.27	572.42
17.6	0.50	170.83	56.95	32.34	160.06	46.87	0.35	57.05	105.02	629.47
19.4	0.55	184.01	61.34	34.86	172.52	51.46	0.38	60.58	112.98	678.13
25.0	0.71	227.34	75.78	43.07	213.12	63.18	0.46	75.25	139.57	837.77
25.7	0.73	234.77	78.26	44.49	220.15	65.22	0.48	77.83	144.01	865.21
34.5	0.98	310.18	103.99	58.69	290.68	85.09	0.62	103.70	190.54	1142.89
35.2	1.00	314.93	104.98	59.63	295.01	86.11	0.64	105.43	193.65	1160.38
38.0	1.08	340.54	113.51	64.49	319.14	93.24	0.68	114.06	209.16	1254.82
39.1	1.11	348.56	116.18	66.02	326.64	94.77	0.70	117.44	214.05	1284.36
78.0	2.22	1028.04	342.68	194.69	963.34	280.23	2.06	345.51	631.49	3788.04

13.6 ALTERNATIVE METHODS FOR PACKAGING

In addition to bulk palletisation, alternative packs in the form of cartons and trays are in use in the delivery of empty glass bottles. Most of these alternative packs are used for carbonated soft drinks bottles although a small number are used for beer and cider bottles. The important features of these alternative packaging methods are:

(a) They usually reduce the number of bottles that can be stacked on a pallet.

(b) Often the polyethylene shroud may be eliminated. However, if protection of the pallet contents is still necessary as for non-returnable bottles which may be simply rinsed or air blown at the filling plant, the plastic covering can often be made of finer gauge film than is used in bulk packs where it imparts mechanical rigidity to the pack.

(c) They usually require significantly higher quantities of board then bulk packs; typically by factors ranging from 2 to 10.

An analysis of the sample described earlier shows that the average materials requirements for bottles packed in these alternative ways are such that:

(i) The pallet requirement should be increased by a factor of 1.1 compared with bulk palletising.

(ii) The polyethylene film requirement should be the same as for bulk palletising.

(iii) The board requirement should be increased by a factor of 2.5 compared with bulk palletising.

Hence for bottles packed in these alternative packs, the estimated average materials requirements will be as shown in Table 13.12. As before an allowance has been made for the returnable components on the basis of a pallet lifetime of 8 trips and a board lifetime of 2 trips. The energy associated with the provision of these materials has been calculated using Table 5.2 to give Table 13.13. One further variation should be noted. Frequently empty bottles are despatched in packs which will subsequently be used in the retail delivery of the container after filling. It is important when calculating system energies to ensure that the energy associated with such packaging is not counted twice.

13.7 TOTAL DELIVERY ENERGY FOR EMPTY GLASS BOTTLES

The total energy associated with packaging and delivering empty glass bottles is the sum of the vehicle energy (Tables 13.7 to 13.9) and the energy required to provide the packaging materials (Tables 13.11 or 13.13).

Table 13.12

Average materials requirements for bottles packed in alternative packs when allowance is made for the returnable components. These packs are used solely for delivering the empty bottle from the glass factory to the bottler

Container size		Packaging materials required		
fl. oz.	litre	Pallets	Polyethylene/kg	Board/kg
3.7	0.11	0.042	0.33	3.75
5.5	0.16	0.053	0.42	4.38
6.0	0.17	0.054	0.43	4.38
6.34	0.18	0.055	0.44	4.38
6.5	0.19	0.058	0.46	4.63
8.8	0.25	0.069	0.55	5.25
9.68	0.28	0.075	0.59	5.63
11.6	0.33	0.084	0.67	6.25
15.5	0.44	0.105	0.84	7.50
17.6	0.50	0.116	0.92	8.25
19.4	0.55	0.127	1.01	8.75
25.0	0.71	0.155	1.24	10.88
25.7	0.73	0.160	1.28	11.25
26.4	0.75	0.164	1.31	11.38
34.5	0.98	0.209	1.67	15.00
35.2	1.00	0.212	1.69	15.25
38.0	1.08	0.229	1.83	16.50
39.1	1.11	0.232	1.86	17.00
78.0	2.22	0.688	5.50	50.00

Table 13.13

Energy associated with the provision of packaging materials for delivery of glass bottles in alternative packs when the materials are used solely for delivery of the empty bottle.

Container size		Electricity/MJ		Oil fuels/MJ			Other fuels/MJ			Total energy /MJ
fl. oz.	litre	Fuel production and delivery energy	Energy content of fuel	Fuel production and delivery energy	Energy content of fuel	Feedstock energy	Fuel production and delivery of fuel	Energy content of fuel	Feedstock energy	
3.7	0.11	134.66	44.89	25.24	123.98	16.81	0.26	63.23	85.48	494.55
5.5	0.16	160.22	53.41	30.05	147.64	21.40	0.31	73.93	101.57	588.53
6.0	0.17	161.03	53.67	30.20	148.43	21.91	0.31	73.96	102.01	591.52
6.34	0.18	161.84	53.94	30.36	149.22	22.42	0.31	73.98	102.45	594.52
6.5	0.19	170.77	56.92	32.02	157.41	23.44	0.33	78.20	108.23	627.32
8.8	0.25	196.05	65.35	36.77	180.81	28.02	0.38	88.73	124.14	720.25
9.68	0.28	210.58	70.19	39.48	194.10	30.06	0.41	95.18	133.56	773.56
11.6	0.33	234.73	78.24	44.03	216.53	34.14	0.45	105.67	148.59	862.38
15.5	0.44	284.78	94.93	53.43	262.81	42.80	0.56	126.90	180.16	1046.37
17.6	0.50	313.22	104.41	58.73	288.93	46.87	0.61	139.60	198.39	1150.76
19.4	0.55	335.14	111.71	62.86	309.26	51.46	0.66	148.13	212.15	1231.37
25.0	0.71	415.01	138.34	77.87	383.03	63.18	0.80	184.13	262.52	1524.88
25.7	0.73	428.93	142.98	80.47	395.88	65.22	0.83	190.39	271.33	1576.03
26.4	0.75	435.32	145.10	81.67	401.78	66.74	0.85	192.64	275.41	1599.51
34.5	0.98	568.75	189.58	106.70	524.81	85.09	1.11	253.77	359.87	2089.68
35.2	1.00	577.69	192.57	108.36	532.99	86.11	1.12	257.99	365.66	2122.49
38.0	1.08	625.00	208.33	117.26	576.71	93.24	1.22	279.13	395.47	2296.36
39.1	1.11	641.43	213.81	120.35	591.89	94.77	1.24	287.50	405.73	2356.72
78.0	2.22	1889.82	529.94	354.54	1743.71	280.24	3.67	845.71	1195.78	6943.41

CHAPTER 14

Distribution of Empty Cans

14.1 INTRODUCTION

Finished cans are invariably bulk palletised. Usually a wooden top frame is placed over the stack and the whole pack is held rigidly together with steel banding. Pallet stacks are built so that the overall external dimensions are similar for all sizes of can to facilitate stacking onto delivery lorries. Hence the number of cans per pallet load varies with the size of the can. Typical packing details are given in Table 14.1. The components of the pallet stack are similar for all companies although there are some variations in the masses of the components. The greatest variations are in the masses of the pallets (21 to 30 kg) and in the masses of the top frame (4.5 to 5.0 kg). Nevertheless, the values shown in Table 14.1 are thought to be reasonable averages.

Table 14.1
Typical average can packaging detail

| Can type | Number per pallet | Mass in pallet stack/kg | | | | | |
		Pallet	Layer pads	Top frame	Steel strapping	Cans	Total
6 fl. oz.	6072	25.00	12.00	4.75	0.54	175.36	217.65
10 fl. oz.	3872	25.00	12.00	4.75	0.50	170.40	212.65
12 fl. oz.	3520	25.00	11.00	4.75	0.52	166.85	208.12
16 fl. oz.	2464	25.00	8.00	4.75	0.50	147.84	186.09
4 pint	510	25.00	6.00	4.75	0.53	105.85	142.13
5 pint	425	25.00	5.00	4.75	0.54	105.95	141.24
7 pint	300	25.00	5.00	4.75	0.53	94.84	130.12

The energy associated with the delivery of empty cans from the can maker to the filler comprises three components; that associated with warehousing, that associated with can transport and that associated with the packaging materials.

Table 14.2
Energy per 1000 cans associated with warehousing

Can size	Electricity/MJ		Oil fuels/MJ			Other fuels/MJ			Total energy /MJ
	Fuel production and delivery energy	Energy content of fuel	Fuel production and delivery energy	Feedstock energy	Energy content of fuel	Fuel production and delivery energy	Energy content of fuel	Feedstock energy	
6 fl. oz.	18.93	6.31	0.37	nil	1.96	0.89	10.86	nil	39.32
10 fl. oz.	29.69	9.90	0.59	nil	3.07	1.40	16.98	nil	61.63
12 fl. oz.	32.66	10.89	0.65	nil	3.39	1.53	18.66	nil	67.78
16 fl. oz.	46.67	15.56	0.92	nil	4.83	2.19	26.68	nil	96.85
4 pint	225.45	75.15	4.45	nil	23.37	10.61	129.06	nil	468.09
5 pint	270.53	90.18	5.34	nil	28.05	12.73	154.79	nil	561.62
7 pint	383.26	127.75	7.57	nil	39.74	18.03	219.32	nil	795.67

© Crown Copyright 1981

14.2 WAREHOUSING

Warehousing is important in can making because the product is likely to deteriorate if stored in cold damp conditions because of rusting in the case of tinplate and tin-free steel and pitting in the case of aluminium. Also, unlike glass bottles, cans are already decorated (labelled) and so very large stocks of finished cans must often be stored in reserve to satisfy the demand for a particular design.

Data have been obtained from a number of can warehousing operations and although practices differ, typical average fuel requirements per pallet stored are 10.646 kWh of electricity, 0.624 therms of natural gas and 0.0718 gallons of kerosine. The fuel requirements per 1000 cans may be calculated from the data in Table 14.1 for the number of cans per pallet, and the corresponding energy requirements are as shown in Table 14.2.

14.3 VEHICLE DELIVERY ENERGY

Cans are invariably delivered by road using 20 tonne payload articulated vehicles. We have examined the distribution pattern for some 10^9 cans (approximately one third of the total produced in 1977) and using the same procedure as outlined in Chapter 13 for glass bottles, the average distribution distance has been calculated as 100 miles.

Unlike the distribution of glass bottles, the mass of a full lorry load of metal cans is considerably lower than the maximum mass which the lorry may carry. A typical loading is 36 pallets for which the actual loading as a fraction of the maximum permitted load varies from 0.23 for 7 pint cans to 0.39 for 6 fl. oz. cans. Using the data of [1], the multipliers which take account of this loading will be 0.83 for the outward journey and 0.71 for the return journey. Hence the total average energy per lorry load of cans is shown in Table 14.3. Using the data of Table 14.1, the vehicle energy per 1000 cans is shown in Table 14.4.

Table 14.3
Average energy required to deliver one lorry-load of cans

Fuel type	Fuel production and delivery energy/MJ	Energy content of fuel /MJ	Feedstock energy /MJ	Total energy /MJ
Electricity	210.98	66.22	nil	277.20
Oil fuels	980.98	4995.76	12.32	5989.06
Other fuels	86.24	528.22	nil	614.46
Totals/MJ	1278.20	5590.20	12.32	6880.72

Table 14.4
Average vehicle energy required to deliver 1000 cans

Can size	Electricity/MJ		Oil fuels/MJ			Other fuels/MJ			Total energy /MJ
	Fuel production and delivery energy	Energy content of fuel	Fuel Production and delivery energy	Energy content of fuel	Feedstock energy	Fuel production and delivery energy	Energy content of fuel	Feedstock energy	
6 fl. oz.	0.97	0.30	4.49	22.85	0.06	0.39	2.42	nil	31.48
10 fl. oz.	1.51	0.48	7.04	35.84	0.09	0.62	3.79	nil	49.37
12 fl. oz.	1.66	0.52	7.74	39.42	0.10	0.68	4.17	nil	54.29
16 fl. oz.	2.38	0.75	11.06	56.32	0.14	0.97	5.95	nil	77.57
4 pint	11.94	3.61	53.43	272.10	0.67	4.70	28.77	nil	374.77
5 pint	13.79	4.33	64.12	326.52	0.81	5.64	34.52	nil	449.73
7 pint	19.54	6.13	90.83	462.57	1.14	7.99	48.91	nil	637.11

© Crown Copyright 1981

14.4 ENERGY ASSOCIATED WITH PACKAGING MATERIALS

Of the packaging materials used, pallets, layer pads and top frames are returnable items. The steel strapping is disposable. The average reported lifetimes of the returnable components are 12 trips for pallets, 11 trips for layer pads and 10 trips for top frames. From the data in Table 14.1 the mass attributable to each pallet load per trip will be as shown in Table 14.5.

Table 14.5

Masses of packaging materials attributable to each pallet-pack per trip when allowance is made for returnable items of packaging

Can size	Component mass/kg			
	Pallet	Layer pads	Top frames	Steel strapping
6 fl. oz.	2.08	1.09	0.475	0.54
10 fl. oz.	2.08	1.09	0.475	0.50
12 fl. oz.	2.08	1.00	0.475	0.52
16 fl. oz.	2.08	0.73	0.475	0.50
4 pint	2.08	0.55	0.475	0.53
5 pint	2.08	0.45	0.475	0.54
7 pint	2.08	0.45	0.475	0.53

© Crown Copyright 1981

The energy associated with the provision of pallets may be derived from Table 5.1 and that for the provision of layer pads from Table 5.2. The energy required to manufacture steel strapping is taken from Table 8.7 Top frames are essentially four softwood battens fixed to form a 'picture-frame' and the energy associated with their production has been calculated from Table 5.1 assuming a 5% materials loss during production. Hence the total energy associated with packaging materials is as shown in Table 14.7.

Table 14.6

Energy required to produce a top-frame of mass 4.75 kg used in can packing

Fuel type	Fuel production and delivery energy/MJ	Energy content of fuel /MJ	Feedstock energy /MJ	Total energy /MJ
Electricity	58.81	19.61	nil	78.42
Oil fuels	7.53	36.79	neg	44.32
Other fuels	0.01	0.13	90.10	90.24
Totals/MJ	66.35	56.53	90.10	212.98

© Crown Copyright 1981

Table 14.7

Energy required to package 1000 cans for delivery

Can size	Electricity/MJ		Oil fuels/MJ			Other fuels/MJ			Total energy /MJ
	Fuel production and delivery energy	Energy content of fuel	Fuel production and delivery energy	Energy content of fuel	Feedstock energy	Fuel production and delivery energy	Energy content of fuel	Feedstock energy	
6 fl. oz.	10.87	3.62	1.62	7.92	neg	0.27	4.57	10.60	39.47
10 fl. oz.	16.99	5.67	2.54	12.39	neg	0.40	7.02	16.62	61.63
12 fl. oz.	18.01	6.00	2.66	12.99	neg	0.45	7.38	17.83	65.32
16 fl. oz.	22.61	7.54	3.22	15.73	neg	0.62	8.60	23.51	81.83
4 pint	99.67	33.22	13.76	67.14	neg	3.16	36.55	107.27	360.77
5 pint	113.13	37.72	15.27	74.59	neg	3.81	40.26	124.52	409.30
7 pint	160.07	53.37	21.63	105.57	neg	5.33	56.57	176.40	578.94

14.5 DELIVERY OF CAN ENDS

Loose ends are delivered to the filler for seaming onto the can body after filling. Two methods of packaging the ends for delivery are in use. For the smaller sizes (up to 16 fl. oz.) the ends are packed in paper bags and stacked on pallets with reinforced paper interleaving. The ends for 4, 5 and 7 pint cans are packed in cartons which are subsequently stacked on pallets. Typical packaging detail is given in Table 14.8.

Table 14.8
Packaging detail for can ends

Item	Ends for can size:			
	6 fl. oz.	10 fl. oz. 12 fl. oz. 16 fl. oz.	4 pint 5 pint	7 pint
Mass of pallet/kg	30	30	30	30
Pallet lifetime/trips	20	20	20	20
Number of ends per pallet	132,000	97,240	19,200	16,200
Number of bags per pallet	600	442	–	–
Mass of bags per pallet/kg	7.351	5.415	–	–
Lifetime of bags/trips	1	1	–	–
Mass of interleaving paper per pallet/kg	2.5	2.5	–	–
Number of cartons per pallet	–	–	30	30
Mass of cartons per pallet/kg	–	–	52.5	52.5
Number of trips per carton	–	–	3	3
Number of pallets per lorry	18	18	18	18

The energy associated with warehousing loose ends may be calculated from the data of section 14.2.

Eighteen pallets stacked on a 10 tonne lorry corresponds to a full loading. Using an average delivery distance of 100 miles as before, the energy required to deliver ends may be calculated from Table 3.1.

The energy required to produce paper bags has been taken from Table 5.2 and the energy of pallet manufacture is derived from Table 5.1. The interleaving paper is reinforced with polypropylene to the extent of 20% by mass and so the data of Table 5.2 can be used to calculate the energy requirement. The energy associated with cartons has also been taken from Table 5.2. The total energy associated with warehousing, packing and delivering can ends is therefore as shown in Table 14.9.

Table 14.9
Total energy required to warehouse, pack and deliver 1000 can ends

Can size	Electricity/MJ		Oil fuels/MJ			Other fuels/MJ			Total energy /MJ
	Fuel production and delivery energy	Energy content of fuel	Fuel production and delivery energy	Energy content of fuel	Feedstock energy	Fuel production and delivery energy	Energy content of fuel	Feedstock energy	
6 fl. oz.	2.48	0.83	0.65	3.25	0.20	0.06	1.83	1.46	10.76
10, 12, 16 fl. oz.	3.00	1.00	0.79	4.03	0.27	0.12	2.15	1.63	12.99
4, 5 pint	33.04	10.99	7.07	35.11	0.02	0.48	19.66	17.65	124.02
7 pint	39.18	13.06	8.41	41.63	0.03	0.59	23.30	20.92	147.12

14.6 TOTAL ENERGY TO PACK AND DELIVER CANS

The total energy required to warehouse, pack and deliver 1000 cans together with the loose ends may be calculated as the sum of Tables 14.2, 14.4 and 14.7. These energies are shown in Table 14.10.

Table 14.10

Total energy associated with the warehousing, packing and delivery of 1000 cans with loose ends

Can size	Electricity/MJ		Oil fuels/MJ			Other fuels/MJ			Total energy /MJ
	Fuel production and delivery energy	Energy content of fuel	Fuel production and delivery energy	Energy content of fuel	Feedstock energy	Fuel production and delivery energy	Energy content of fuel	Feedstock energy	
6 fl. oz.	33.25	11.06	7.13	35.98	0.26	1.61	19.68	12.06	121.03
10 fl. oz.	51.19	17.05	10.96	55.33	0.36	2.54	29.94	18.25	185.62
12 fl. oz.	55.33	18.41	11.84	59.83	0.37	2.78	32.36	19.46	200.38
16 fl. oz.	74.66	24.85	15.99	80.91	0.41	3.90	43.38	25.14	269.24
4 pint	369.65	122.97	78.71	397.72	0.69	18.95	214.04	124.92	1327.65
5 pint	430.49	143.22	91.74	464.27	0.83	22.66	249.23	142.17	1544.61
7 pint	602.05	200.31	128.44	649.51	1.17	31.94	348.10	197.32	2158.84

CHAPTER 15

Distribution of Empty PET Bottles

15.1 PACKAGING PET BOTTLES

Empty PET bottles are invariably bulk palletised for distribution to the filling plant. Pack specifications vary slightly from one supplier to another but the energy associated with the provision of packaging materials (that is, pallets, layer pads and shrinkwrap film are as shown in Table 15.1). These are based on Tables 5.1 and 5.2 and assume a pallet lifetime of 10 trips and a lifetime for layer pads of 2 trips.

15.2 DELIVERING PET BOTTLES

Delivery energies have been calculated assuming that 20 tonne payload lorries carrying 40 pallets, are used. An average one-way journey distance of 100 miles with empty return loads has been assumed. The vehicle energy requirement is also shown in Table 15.1.

15.3 TOTAL PACKING AND DELIVERY ENERGY

The total energy associated with packing and delivery of empty PET bottles is simply the sum of the above two components and this is also shown in Table 15.1.

Table 15.1
Energy associated with packing and delivering 1000 empty PET bottles

Contribution	Electricity/MJ		Oil Fuels/MJ			Other fuels/MJ			Total energy /MJ
	Fuel production and delivery energy	Energy content of fuel	Fuel production and delivery energy	Energy content of fuel	Feedstock energy	Fuel production and delivery energy	Energy content of fuel	Feedstock energy	
1.5 litre bottle									
Packaging	171	57	33	161	45	neg	59	105	631
Delivery	6	2	27	137	neg	2	15	nil	189
Total	177	59	60	298	45	2	74	105	820
2.0 litre bottle									
Packaging	187	62	35	175	49	neg	65	114	687
Delivery	8	3	39	199	neg	3	21	nil	273
Total	195	65	74	374	49	3	86	114	960

CHAPTER 16

Closures for Glass Bottles

16.1 INTRODUCTION

A variety of closures are used for sealing bottles with the precise design depending in part upon bottle design and in part upon the preferences of the bottler. As with metal can production, it is expected that a significant proportion of the total energy required to produce closures from raw materials in the ground arises from the production of the metal used and this in turn depends upon the mass of the closure. Energies have therefore been calculated (from manufacturers data) for the production of two specific closures, one made from tinplate (crown closure) and one made from aluminium (standard aluminium screw closure). The energies associated with other types of closure have been calculated by increasing or decreasing these energies in proportion to the relative finished masses.

16.2 PRODUCTION OF CROWN CLOSURES

The materials flows for the production of crown closures of finished mass 0.00214 kg (tinplate content) are given in Fig. 16.1. A rejection rate of 5% of delivered tinplate has been included and, as in the case of metal can production, this has been shown as all passing through the detinning process, even though a proportion is retreated by the primary metal producer for resale.

Flows in the scrap treatment plant are based on Chapter 11. Output tinplate from the primary metal producer has been converted into an iron requirement using the demand of 1.2292 kg of pig iron to produce 1 kg of finished tinplate (see Chapter 8).

The energy associated with closure production has been calculated from a detailed factory analysis and includes contributions for heating, lighting and administrative functions. The total energy is therefore as shown in Table 16.1.

Finished crown closures are packed in cartons; typically 6000 crowns are packed in a returnable carton of mass 1 kg and the expected lifetime of the carton is 3 trips. The delivery distance obviously varies with the location of the filler but we have assumed an average delivery distance of 100 miles using 20 tonne payload vehicles. This packing and delivery energy is also shown in Table 16.1.

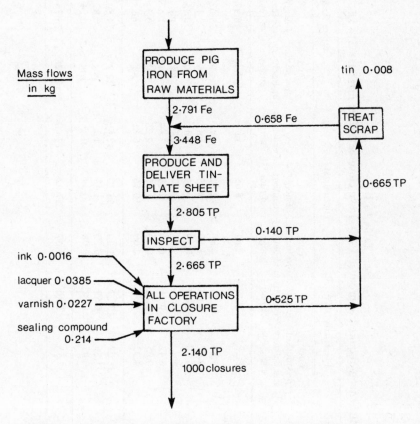

Fig. 16.1 – Total system required to produce 1000 crown closures from raw materials in the ground. (TP = tinplate).

Table 16.1

Total energy required to produce 1000 crown closures each of mass 0.00214 kg

	Electricity/MJ		Oil fuels/MJ			Other fuels/MJ			Total energy /MJ
	Fuel production and delivery energy	Energy content of fuel	Fuel production and delivery energy	Energy content of fuel	Feedstock energy	Fuel production and delivery energy	Energy content of fuel	Feedstock energy	
Production of pig iron	3.38	1.31	1.81	8.65	nil	7.95	46.97	nil	70.07
Production and delivery of tinplate sheet	15.51	6.23	4.66	22.61	nil	4.07	25.02	0.08	78.18
Closure factory operation	12.09	4.03	1.69	8.25	nil	0.03	0.15	nil	26.24
Scrap treatment	0.30	0.10	0.19	0.87	nil	0.01	0.05	0.02	1.54
Ink production	neg	neg	0.01	0.03	0.06	nil	nil	nil	0.10
Lacquer production	0.42	0.14	0.46	2.28	3.44	0.14	1.67	nil	8.55
Varnish production	0.35	0.12	0.40	1.95	2.25	0.08	0.94	nil	6.09
Sealing compound production	4.02	1.34	2.08	10.56	6.85	nil	nil	nil	24.85
Totals	36.07	13.27	11.30	55.20	12.60	12.28	74.80	0.10	215.62
Packaging	1.56	0.52	0.29	1.43	nil	neg	0.92	0.99	5.71
Delivery	0.02	0.01	0.08	0.39	neg	neg	0.04	nil	0.54
Totals	37.65	13.80	11.67	57.02	12.60	12.28	75.76	1.09	221.87

16.3 PRODUCTION OF ALUMINIUM SCREW CLOSURES

The aluminium screw closure examined had an aluminium content of 0.00131 kg (finished mass) and the materials flows in the production system are as shown in Fig. 16.2. Inputs to the closure factory are based on actual operations as also are the scrap and spoilage levels. Scrap treatment follows the data of Chapter 11 and all scrap, including reject plate, is assumed to pass through the recycling loop. The conversion of aluminium remelt ingot to finished sheet is based on a requirement of 1.04 kg of ingot to produce 1 kg of sheet. The energy associated with each of the operations in Fig. 16.2 is given in Table 16.2.

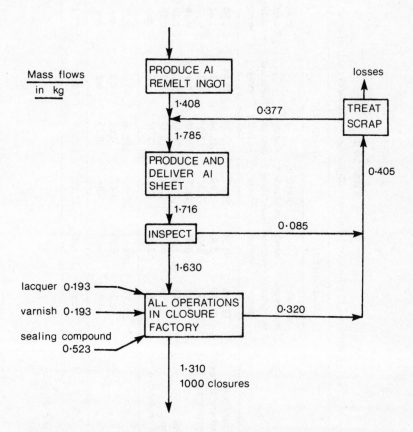

Fig. 16.2 — Total system required to produce 1000 aluminium screw closures from raw materials in the ground.

Table 16.2

Energy required to produce 1000 aluminium screw closures each of mass 0.00131 kg

	Electricity/MJ		Oil fuels/MJ			Other fuels/MJ			Total energy /MJ
	Fuel production and delivery energy	Energy content of fuel	Fuel production and delivery energy	Energy content of fuel	Feedstock energy	Fuel production and delivery energy	Energy content of fuel	Feedstock energy	
Production of primary ingot	179.86	91.58	14.67	73.29	37.04	1.94	22.96	nil	421.34
Production and delivery of sheet	17.50	5.83	3.23	15.44	nil	1.32	16.04	nil	59.36
Closure factory operation	19.20	6.40	3.27	15.97	nil	0.03	0.32	nil	45.19
Scrap treatment	0.15	0.05	0.05	0.29	nil	0.46	5.59	nil	6.59
Lacquer production	3.56	1.19	3.93	19.41	29.24	1.20	14.16	nil	72.69
Varnish production	2.98	0.99	3.39	16.57	19.13	0.68	8.01	nil	51.75
Sealing compound production	6.86	2.29	3.55	17.97	11.65	nil	nil	nil	42.32
Totals	230.11	108.33	32.09	158.94	97.06	5.63	67.08	nil	699.24
Packaging	2.24	0.75	0.42	2.05	nil	neg	1.33	1.43	8.22
Delivery	0.02	0.01	0.08	0.39	neg	neg	0.04	nil	0.54
Totals	232.37	109.09	32.59	161.38	97.06	5.63	68.45	1.43	708.00

The finished closures are packed at the rate of 5000 per returnable carton of mass 1.2 kg and the carton is expected to last 3 trips. Delivery is assumed in 20 tonne lorries with a one-way delivery distance of 100 miles. Table 16.2 also shows the energy associated with these packing and delivery operations.

16.4 SCREW CLOSURES FOR WIDE MOUTH BOTTLES

The 8.8 fl. oz. (0.25 litre) non-returnable bottle for carbonated soft drinks is available with a wide mouth. Large aluminium screw closures are used to seal such bottles and they typically weigh 0.00214 kg compared with 0.00131 kg for the standard variety. It has been assumed that the energy associated with the production of the closure can be scaled up in proportion to the relative masses. Such an assumption is reasonable since the major contributors to the closure energy are the energies required to supply the aluminium and the lining compound, both of which are proportional to mass. The estimated energy required to produce such closures is therefore given in Table 16.3.

Table 16.3

Total energy required to produce 1000 large aluminium closures for wide necked bottles of capacity 8.8. fl. oz. (0.25 litre)

Fuel type	Fuel production and delivery energy/MJ	Energy content of fuel /MJ	Feedstock energy /MJ	Total energy /MJ
Electricity	379.60	178.21	nil	557.81
Oil fuels	53.24	263.63	158.56	475.43
Other fuels	9.20	111.82	2.34	123.36
Totals/MJ	442.04	553.66	160.90	1156.60

© Crown Copyright 1981

16.5 RIP-CAP ALUMINIUM CLOSURES

A recent innovation for wide mouth bottles is the use of rip-cap aluminium closures which are similar in concept to the ring pull easy-open ends for cans. Most rip cap closures are currently made in the U.S.A. and the energy associated with the production of this type of closure has been calculated by scaling up the energy associated with the standard aluminium screw closure in proportion to the relative masses. The calculated energy requirement is given in Table 16.4.

Table 16.4

Total energy associated with the provision of 1000 rip-cap closures for wide necked bottles.

Fuel type	Fuel production and delivery energy/MJ	Energy content of fuel /MJ	Feedstock energy /MJ	Total energy /MJ
Electricity	342.35	160.72	nil	503.07
Oil fuels	48.01	237.76	143.00	428.77
Other fuels	8.29	100.85	2.11	111.25
Totals/MJ	398.65	499.33	145.11	1043.09

CHAPTER 17

Inputs to Beverage Filling Lines

17.1 LABELS

A wide variety of different sizes and shapes of labels are in use in beverage packaging and it is impossible to define an 'average label'. We have therefore examined the output from a factory producing beverage labels only. The total consumption by the factory of materials and fuels for a fixed period has been divided by the total number of saleable labels, to provide average data which are thought to be representative of the industry as a whole.

Typical materials requirements per 1000 labels are 0.4559 kg of coated paper, 0.031 kg ink, 0.0036 kg toluene and 0.0034 kg methanol. The energy associated with paper production is taken from Table 5.2 and the energy associated with the other inputs are taken from Table 9.1. The total energy associated with label production is as shown in Table 17.1.

17.2 MULTIPACK CARTONS

Non-returnable bottles are frequently packed in multipack cartons after filling. As with labels, the variety of designs in use is large and so we have chosen as typical, a printed pack of finished mass 0.025 kg (board content) able to contain four bottles. The production of 250 cartons to pack 1000 bottles requires 6.875 kg board, 0.124 kg ink, 0.014 kg methanol and 0.014 kg toluene.

The energy associated with their production is shown in Table 17.2. It is assumed that the input board is delivered an average one way distance of 100 miles. The delivery energy of the other commodities is negligible. The delivery energy of finished packs to the filler has also been calculated assuming a delivery distance of 100 miles.

17.3 GLUE

The glue for labelling is assumed to be either animal glue or starch based. The production processes are predominantly concerned with the physical separation

Table 17.1

Energy required to produce 1000 beer labels

	Electricity/MJ		Oil fuels/MJ			Other fuels/MJ			Total energy /MJ
	Fuel production and delivery energy	Energy content of fuel	Fuel production and delivery energy	Energy content of fuel	Feedstock energy	Fuel production and delivery energy	Energy content of fuel	Feedstock energy	
Factory operations	2.28	0.76	0.30	1.47	nil	nil	nil	nil	4.81
Paper production	10.59	3.53	2.67	13.03	nil	0.02	7.58	8.16	45.58
Ink production	0.06	0.02	0.13	0.66	1.13	nil	nil	nil	2.00
Toluene production	0.01	neg	0.02	0.11	0.16	nil	nil	nil	0.30
Methanol production	neg	neg	0.01	0.04	0.10	nil	nil	nil	0.15
Delivery of paper	−0.01	neg	0.03	0.16	neg	neg	0.02	nil	0.22
Totals/MJ	12.95	4.31	3.16	15.47	1.39	0.02	7.60	8.16	53.06

Table 17.2

Energy required to produce and deliver multipacks for 1000 bottles

	Electricity/MJ		Oil fuels/MJ			Other fuels/MJ			Total energy /MJ
	Fuel production and delivery energy	Energy content of fuel	Fuel production and delivery energy	Energy content of fuel	Feedstock energy	Fuel production and delivery energy	Energy content of fuel	Feedstock energy	
Factory operations	9.12	3.04	1.20	5.88	nil	nil	nil	nil	19.24
Board production	192.84	64.28	36.09	176.21	nil	0.34	114.33	122.99	707.08
Ink production	0.26	0.09	0.54	2.63	4.53	nil	nil	nil	8.05
Toluene production	0.05	0.02	0.09	0.43	0.63	nil	nil	nil	1.22
Methanol production	0.01	neg	0.03	0.16	0.39	nil	nil	nil	0.59
Delivery of board	0.08	0.03	0.37	1.90	neg	0.02	0.20	nil	2.60
Totals/MJ	202.36	67.46	38.32	187.21	5.55	0.36	114.53	122.99	738.78
Delivery	0.07	0.03	0.33	1.71	neg	0.02	0.18	nil	2.34
Totals/MJ	202.43	67.49	38.65	188.92	5.55	0.38	114.71	122.99	741.12

of the required components from naturally occurring materials and the detailed processing is complex. The energy requirement shown in Table 17.3 has been calculated from the data of Sundstrom [7].

Table 17.3
Energy required to produce 1 kg of glue

Fuel type	Fuel production and delivery energy/MJ	Energy content of fuel /MJ	Feedstock energy /MJ	Total energy /MJ
Electricity	26.25	8.75	nil	35.00
Oil fuels	7.17	35.00	nil	42.17
Other fuels	nil	nil	nil	nil
Totals/MJ	33.42	43.75	nil	77.17

17.4 PLASTIC CRATES

The energy associated with the manufacture of polypropylene crates has been derived from the energy to produce the polymer resin and from that to injection mould the crate. The energy required to injection mould polypropylene products has been calculated elsewhere [22] and the total energy required to produce 1 kg of injection moulded crate is as shown in Table 17.4. The energy has been expressed in this form because, as will be seen later, crates of different masses are in use.

Table 17.4
Energy required to produce 1 kg of injection moulded beverage crate from raw materials in the ground

Fuel type	Fuel production and delivery energy/MJ	Energy content of fuel /MJ	Feedstock energy /MJ	Total energy /MJ
Electricity	28.35	9.45	nil	37.80
Oil fuels	7.72	40.39	52.65	100.76
Other fuels	nil	nil	nil	nil
Totals/MJ	36.07	49.84	52.65	138.56

CHAPTER 18

Beer Filling

18.1 INTRODUCTION

In 1977, consumption of beer in the U.K. amounted to some 6.6×10^9 litres of which approximately 78% was sold as draught beer and the remainder was packaged in returnable or non-returnable bottles and metal cans. Trends in packaged beer consumption by container type since 1972 are shown in Table 18.1.

Over this period, the amount of beer packed in both returnable and non-returnable bottles declined whilst that packed in cans increased. Returnable bottle sales showed the greatest decline from 1.3×10^9 litres in 1973 to 0.9×10^9 litres in 1977. Nevertheless, returnable bottles still dominate the packaged beer market. Note that the growth of the can market did not replace the volume lost by returnable and non-returnable bottles and consequently the total volume of packaged beer declined.

The use of non-returnable bottles in the brewery trade appears to have increased in the early 1970's, peaked in 1974 and declined thereafter. The reasons for the lack of success of non-returnable bottles are complex and relate to the structure of the beer trade with its division into on-licence and off-licence sales. In the on-licence trade there is virtually a closed loop involving the filler and the retailer. Bottles are emptied on the premises and not handed to the customer. This ensures a high rate of bottle returns and, since brewers adopted standard bottles many years ago, there are no losses when a bottle filled by one brewer is returned to another. Since returnable bottles satisfy the needs of the on-trade, there is no requirement for non-returnable bottles in this sector. In the off-licence trade, the major development in the recent past has been the growth of licensed supermarkets. They have only limited storage space and need a container which can be densely stacked and is not subject to breakage. Cans fulfil this requirement much better than non-returnable bottles.

For canned beer, recent growth has come in the smaller sizes and in particular the 15.5 fl. oz. size, which in 1976 increased in unit terms by 19%. Sales in Scotland are almost exclusively in this can size although the 9.68 fl. oz. size

Table 18.1

Packaged beer consumption in the United Kingdom (Source: H.M. Customs and Excise and the Brewers Society)

Year	Returnable bottles		Non-returnable bottles		Cans		Total packaged		Total beer mbb
	%	mbb‡	%	mbb‡	%	mbb‡	%	mbb‡	
1972	21.9	8.03	0.5†	0.18	4.2†	1.54	26.6	9.75	36.65
1973	21.2	8.11	0.5†	0.19	5.3†	2.03	27.0	10.33	38.25
1974	19.7	7.70	0.6	0.23	5.9	2.31	26.2	10.24	39.09
1975	16.8	6.76	0.5	0.20	6.9	2.78	24.2	9.74	40.22
1976	14.6	5.93	0.4	0.16	8.0	3.25	23.0	9.34	40.65
1977	13.1	5.27	0.4	0.16	8.3	3.34	21.8	8.77	40.26

† Denotes estimates as separate analysis between cans and non-returnable bottles is not available.

‡ mmb = millions of bulk barrels; 1 bulk barrel = 36 imperial gallons = 163.656 litres.

© Crown Copyright 1981

Table 18.2

Sales of packaged beer by container size. Data relates to the U.K. in 1976 (Source: Brewers Society)

Container size	Returnable glass bottles			Non-returnable glass bottles			Metal cans		
	Volume litre $\times 10^6$	Bulk barrels $\times 10^3$	Number of containers $\times 10^6$	Volume litre $\times 10^6$	Bulk barrels $\times 10^3$	Number of containers $\times 10^6$	Volume litre $\times 10^6$	Bulk barrels $\times 10^3$	Number of containers $\times 10^6$
(1) up to 7 fl. oz.	34	210	181.3	4	23	19.8	–	–	–
(2) 7–9 fl. oz.	–	–	–	3	17	11.0	–	–	–
(3) 9–10 fl. oz.	829	5067	2918.8	6	36	3.4	173	1060	610.5
(4) 10–13.3 fl. oz.	–	–	–	3	17	8.4	24	150	71.9
(5) 13.3–16 fl. oz.	–	–	–	–	–	–	287	1753	631.2
(6) 16–20 fl. oz.	97	593	170.8	5	30	8.7	–	–	–
(7) 20–40 fl. oz;	10	60	8.6	6	36	5.2	–	–	–
(8) 78 fl. oz.	–	–	–	–	–	–	27	167	12.1
(9) 98 fl. oz.	–	–	–	–	–	–	17	99	5.7
(1) 136 fl. oz.	–	–	–	–	–	–	3	21	0.9
Totals	970	5930	3279.5	27	159	56.5	531	3250	1332.3

retains a significant share of the English market. In 1976, the number of cans sold in other small sizes (11.6 fl. oz. and 0.33 litre) was negligible although significant numbers of 11.6 fl. oz. cans were filled primarily for export.

Cans of capacity 78, 98 and 136 fl. oz. serve a restricted market. Moreover, because of the seasonal demand and limited shelf life of the product, they present problems. Most brewers report a static or slightly declining market for these larger can sizes although in total, the market shows some growth due to their increased use by smaller brewers.

Table 18.2 shows the sales of packaged beer broken down by container type and size. The exact numbers of sales by container size are not known in some instances, hence the need to quote ranges. A number of assumptions have been made in constructing this table. Entry (1) in Table 18.2 can be assumed to be all 6.34 fl. oz. (Nip) size bottles. Entry (2) relates to 0.25 litre non-returnable bottles. Entry (3) is the 9.68 fl. oz. bottle or can. Entry(4) for non-returnable bottles relates to 0.33 litre bottles whereas the value for cans may be

Table 18.3
Container sizes considered in this section

| Container name | Capacity as stated on label or can | | Container type | Conventional number of containers per barrel |
	fl. oz.	litre		
nip bottle	6.34	0.18	RB/NRB	864
small bottle	9.68	0.275	RB/NRB	576
large bottle	19.4	0.55	RB/NRB	288
flagon bottle	39.1	1.11	RB	144
25 cl bottle	8.8	0.25	NRB	633
33 cl bottle	11.6	0.33	NRB	480
440 ml bottle	15.5	0.44	NRB	360
50 cl. bottle	17.6	0.50	NRB	316
Northern extra large bottle	25.7	0.73	NRB	217
1 litre bottle	35.2	1.00	NRB	158
small can	9.68	0.275	C	576
33 cl can	11.6	0.33	C	480
large can	15.5	0.44	C	360
party can	78	2.22	C	72
party can	98	2.78	C	58
party can	136	3.86	C	42

assumed to be the 11.6 fl. oz. size filled for export. Entry (5) is the 15.5 fl. oz. can. Entry (6) relates to the pint returnable and non-returnable bottles and includes the 0.5 litre size. Entry(7) relates to the 39.1 fl. oz. returnable flagon as well as the 1 litre non-returnable bottle.

The present study examines the filling of containers in the sizes shown in Table 18.3. Containers are frequently referred to in terms of their approximate capacity, e.g. ½ pint or 10 fl. oz. whereas the actual declared contents usually differs from this value. In the example quoted, the content on the label would be 9.68 fl. oz. or 0.275 litre. Table 18.3 shows the relationship between the approximate container size, stated contents on the label or can, and the conventional number of containers that can be filled from unit volume of beer. In the remainder of this Chapter, containers are referred to by their stated label contents and the abbreviations RB, NRB and C are used for returnable bottle, non-returnable bottle and can respectively.

18.2 BEER FILLING LINES

The operations included in a filling line depend primarily on the type of container being filled and Figs. 18.1, 18.2 and 18.3 show simplified flowsheets for the three principal types of filling lines. Most lines tend to be highly mechanised with minimal manual handling but practices vary considerably from plant to plant. Most plants employ several filling lines and usually handle a number of container sizes. Since individual filling lines are seldom monitored separately for fuel and raw materials consumption, it is usually necessary to apportion total consumption between the different filling lines and between container sizes when a number of containers are filled on the same line.

A further complication is that many filling plants are attached to breweries and it is rare for the fuel consumption of different areas of the site, e.g. brewery, filling plant, warehouses, etc. to be monitored separately. Hence before apportionment between container sizes and types within a filling plant can be effected, some breakdown of the total site consumption is needed. In this work, this was usually achieved in consultation with the site engineer.

The predominant feed to a returnable bottle filling line such as that shown in Fig. 18.1 is the incoming supply of empty bottles. These may be dirty bottles returned for refilling or new bottles added to the line as replacements or to allow for increased production. All bottles are given a hot detergent wash followed by a rinse. The function of the washing operation is not only to clean the bottle but also to remove labels. Water and detergent solution are usually steam heated. The second major consumer of thermal energy on the filling line is the pasteuriser and in all plants examined, bottles were tunnel pasteurised at a temperature of 336 K.

For non-returnable bottle lines such as that shown in Fig. 18.2, the incoming bottles are usually bulk palletised. After manual removal of any shrink-wrap film and top cover, depalletisation is usually automatic or semi-automatic.

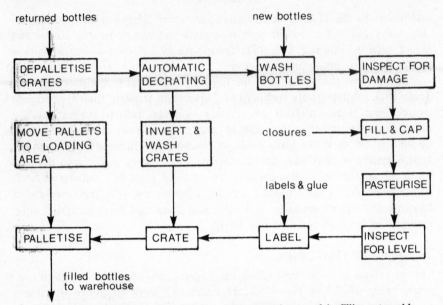

Fig. 18.1 – Schematic flow diagram of the operations used in filling returnable bottles. The diagram does not show the bottle sorting operation which removes foreign or wrong sized bottles. Similarly, any decanting or reprocessing operations have been omitted. The energy associated with both of these functions has however been taken into account in the text.

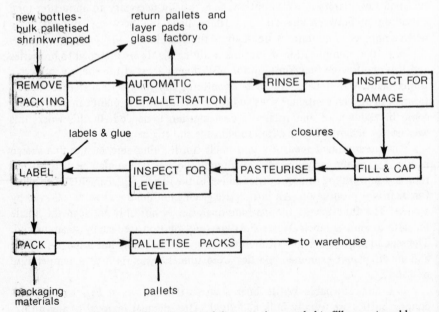

Fig. 18.2 – Schematic flow diagram of the operations needed to fill non-returnable bottles.

Before filling, bottles require only a rinse, although in many of the filling plants examined, the bottles were frequently given a hot wash. As with returnable bottles, current practices favour tunnel pasteurisation after filling.

Fig. 18.3 shows that after depalletisation, cans are given a cold rinse so that thermal fuels are used only for pasteurisation. This is strictly true only for can sizes up to 15.5 fl. oz., since the large party cans are sterilised with hot water prior to filling. Tunnel pasteurisation is used for the smaller can sizes (up to 15.5 fl. oz.) but flash pasteurisation is usual on lines filling 78, 98 and 136 fl. oz. cans because the much larger sizes of container would require excessively long residence times in a tunnel pasteuriser.

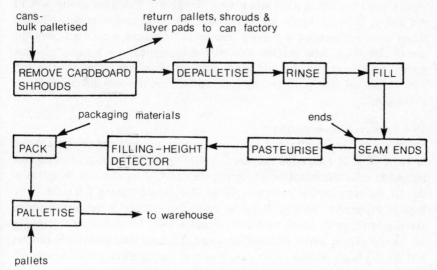

Fig. 18.3 – Schematic flow diagram of the operations used to fill cans.

18.3 ENERGY OF FILLING

The energy required by beer filling lines is the sum of the electrical energy to drive the individual machines and conveyors plus the thermal energy supplied to the washer, where appropriate, and the pasteuriser. A minor complication is that some beers are not pasteurised; such products represent only a very small proportion of packaged beers and have been ignored.

In addition to the energy required to drive the filling line itself, further energy is used in the provision of lighting and space heating in the bottling store and for refrigeration of beer prior to packaging. It is arguable that the energy associated with refrigeration, excepting that following flash pasteurisation, should be excluded. However, in order that the beer retain its condition prior to packaging and pasteurisation, refrigeration is essential. That proporttion of refrigeration energy which can be attributed to packaged beer has therefore been included.

During this work, only one plant, a can filling plant, was discovered in which detailed monitoring of fuels was carried out. This is reported in section 18.4 to provide some basis for comparison and because it also illustrates some interesting features. Subsequent sections are concerned with evaluating energy requirements which are thought to be representative of more general practices.

18.4 ANALYSIS OF A CANNING PLANT

The plant consisted of a single line, filling 9.68 and 11.6 fl. oz. cans. 90% of the filling line throughput was 9.86 fl. oz. cans. In addition to the can filling operation, a small keg filling plant was present on the site. However, almost 80% of the output from the whole site was canned beer. Fuel consumption in the keg filling area was metered separately and could be readily subtracted from the overall site total. Fuel consumption was also metered at a number of other points throughout the site and electricity and steam use in the canning hall were known accurately. Electricity and fuel oil were the only direct fuel inputs to the site.

18.4.1 Oil fuels consumption

Steam use within the plant was monitored at four locations and these are shown in Table 18.4. In the period examined, 294,179 gallons of medium fuel oil were burned to raise this steam, which corresponds to 47.1 kg of steam per gallon of oil. Of the consumption shown in Table 18.4, meter reading 1 is solely attributable to canning: reading 2 can be subtracted out and some proportion of the overheads as indicated by meters 3 and 4 must be attributed to canning. In the absence of better information, overheads have been charged to canning and kegging in proportion to the volume of beer passing through each operation.

Table 18.4
Steam use at the canning site

Meter number	Indicated consumption of steam/kg	Operational area and use of steam
1	4,675,020	Canning plant — pasteurisation and space heating
2	4,993,728	Keg plant — sterilisation and space heating
3	2,865,330	Offices and laboratories — space heating only
4	1,318,961	Offices and warehouse — space heating only

Hence the total 'charge' to canning is meter reading 1 plus 78.55% of meter readings 3 and 4 which corresponds to a total steam consumption of 7.96×10^6 kg (169,074 gallons of medium fuel oil). For a total production of 92,015,904 cans during this period, the fuel requirement per 1000 cans of capacity 9.68 fl. oz. is 1.84 gallons of fuel oil.

This quantity is the sum of two contributions; 1.08 gallons per 1000 cans arising from the consumption for pasteurisation and space heating in the canning hall and 0.76 gallons per 1000 cans for space heating in the offices, warehouses, etc. Although this plant is well metered compared with all the other plants examined, it is still not possible to separate steam use in pasteurisation from that used in space heating. A consumption of 1.08 gallons of oil per 1000 cans represents an energy requirement of approximately 201.2 MJ/1000 cans. As shown later, the design running load of a can pasteuriser is of the order of 70 MJ/1000 cans. If this value is doubled to take account of the efficiency of steam generation and the energy consumption during warm-up periods, then the thermal energy used *within the canning hall* can be split approximately as 70% to pasteurisation and 30% to space heating. When ancilliary site operations are taken into account, pasteurisation represents only 40% of the total thermal fuels consumed.

18.4.2 Electricity consumption

Electricity consumption was examined for two separate 12 month periods and the data are presented in Table 18.5. Note that overheads in the form of lighting, fans, pumps and air compression, comprise some 31% of the total. Beer refrigeration requires a further 40% and the filling operation itself takes up the remaining 30%.

Production in 1973/74 was 82,988,640 cans and in 1976/77 was 92,015,904 cans so that the electricity consumption per 1000 cans decreased from 16.72 kWh in 1973/74 to 13.71 kWh in 1976/77.

In 1973/74, all cans were packed in multipacks (4 cans per pack) and 6 multipacks were then shrinkwrapped together. In 1976/77, all production was in Hicone, packed in fours, with 24 cans placed on a board tray. The effect of this change on the electricity demand can be judged by comparing the installed loads of the two lines shown in Tables 18.6 and 18.7. It is clear that packing in Hicone and cardboard trays is much less energy intensive than packaging with shrinkwrap because of the high electrical consumption of the shrinkwrapping tunnel. This is one of the major contributors to the reduced electricity consumption noted above.

A further interesting feature of this analysis concerns the running load of a filler/seamer which dropped from 15.2 kW in 1973/74 to 11.7 kW in 1976/77 because of a decrease in the load factor from 80% to 62%. The reason for this was that in 1973/74, empty cans were delivered to the plant with the easy-open aluminium end fixed, so that the steel end was seamed on after filling. This situation was reversed in 1976/77 so that the aluminium end was seamed on in the filling plant.

Table 18.5

Metered electricity consumption in a canning site

Meter number	Recorded consumption kWh	Operational area	Proportion attributed to canning	'Canning' Electricity usage/kWh	% of total for canning
1973/74					
1	467,378	Lighting, fans, etc.	30%	140,213	10.1
2	453,334	Canning hall	100%	453,334	32.6
3	258,400	Air compressor	85%	219,640	15.8
4	213,150	Cold rooms	90%	191,835	13.8
5	107,300	Keg plant	0	—	—
6	140,100	Lighting, fans, etc.	50%	70,050	5.1
7	390,700	Refrigeration	80%	312,560	22.6
			Totals	1,387,632	100.0
1976/77					
1	390,598	Lighting, fans, etc.	30%	117,177	9.3
2	363,132	Canning hall	100%	363,132	28.8
3	229,200	Air compressor	85%	194,820	15.4
4	185,400	Cold rooms	90%	166,860	13.2
5	118,400	Keg plant	0	—	—
6	161,900	Lighting, fans, etc.	50%	80,950	6.4
7	423,400	Refrigeration	80%	338,720	26.9
			Totals	1,261,659	100.0

Table 18.6

Electrical data for a canning line with final packaging in multipacks with shrink-wrapping for final trays

Operation	Installed load/kW	Load factor %	Running load/kW	Line demand %
Depalletiser and empty can conveyors	7.5	60	4.5	3.42
Filler and seamer	19.0	80	15.2	11.56
Conveyors from filler to pasteuriser	6.0	60	3.6	2.73
Pasteuriser	30.0	90	27.0	20.54
Conveyors from exit post to packing	9.0	60	5.4	4.10
Multipack sleeving and indexing	14.6	75	10.9	8.29
Shrinkwrapping	108.0	60	64.8	49.36
Totals	194.1	—	131.4	100.00

© *Crown Copyright 1981*

Table 18.7

Electrical data for a canning line with final packaging in Hicone and cardboard trays

Operation	Installed loads/kW	Load factor %	Running load/kW	Line demand %
Depalletiser and empty can conveyors	7.5	60	4.5	7.52
Filler and seamer	19.0	62	11.7	19.56
Conveyors from filler to pasteuriser	6.0	60	3.6	6.02
Pasteuriser	30.0	90	27.0	45.17
Conveyors from exit post to packing	9.0	75	6.7	11.20
Hicone and tray loader	9.0	70	6.3	10.53
Totals	80.5	—	59.8	100.00

© *Crown Copyright 1981*

18.5 APPORTIONMENT OF ENERGY WHEN PLANT IS NOT METERED

All of the plants, with the exception of that described in section 18.4, were not metered to any significant extent and the data supplied for the filling operation inevitably included fuel consumptions in ancillary areas such as offices and laboratories, although fuels used in brewery operations were excluded. The problem therefore was one of analysing the total energy consumption to derive energy requirements for containers of specific type and size.

One method of energy apportionment is to divide the consumption of all fuels by the total number of containers filled within the plant. Although simple, this procedure provides only an approximate energy requirement for all containers since it takes no account of:

(a) container type; that is returnable bottle, non-returnable bottle or can, a factor which determines for example whether or not containers are hot washed, or rinsed.

(b) container size,

(c) the relative filling speeds of different lines within the plant,

(d) the degree of mechanical handling on the line which will in turn affect the installed electrical capacity of the line.

A second possibility is to relate fuel consumption to the throughput of a particular container size and type by taking relative line speeds into account. This effectively partitions energy on the basis of the time for which the plant is running. This is an acceptable method for apportioning electrical and thermal fuels used in **ancillary** operations but it is not the best technique for apportioning **line** consumption of thermal fuels, which depend upon whether or not a container is hot washed and/or pasteurised. Some other means of apportionment is therefore required to take account of this type of variation and the method applied is discussed in the following sections.

18.6 THERMAL ENERGY CONSUMPTION OF FILLING LINES

Thermal fuels are used in filling plants for washing, pasteurising and space heating. This section is primarily concerned with the thermal energy use in pasteurisers and washers.

18.6.1 Pasteurisation energy

The minimum energy which must be supplied during pasteurisation is that required to raise the temperature of the container and contents from the input temperature to the pasteurisation temperature (337 K). The minimum energy requirements of different container types and sizes may be compared by calculating the energy required to raise the temperature of the container and contents by 1 K. These values can be calculated from the masses and specific heats of the container and contents, and are given in Table 18.8 and clearly demonstrate that it is the volume of the contents which dominates the minimum energy input required. For bottles the energy use in pasteurisation is approximately proportional to container volume because the specific heat capacity of the contents is 8.38 times greater than that of glass. A second factor to note is that there is no significant difference between all steel and all aluminium cans of the same volume. The reason for this is that although the mass of an aluminium can is typically only half that of the equivalent steel can, the specific heat capacity of aluminium is approximately twice that of steel.

Table 18.8

Theoretical minimum energies required to raise the temperature of various containers and contents by 1 K

Container type	Volume of beer fl. oz.	Energy J
returnable bottle	9.68	1357
non-returnable bottle	9.68	1307
returnable bottle	19.40	2590
non-returnable bottle	19.40	2548
all steel can	9.68	1201
all aluminium can	9.68	1203
all steel can	15.5	1916

© *Crown Copyright 1981*

The energies given in Table 18.8 represent the minimum values required to raise the temperature of the container plus contents. They do not allow for factors such as heat losses, machine efficiencies and regeneration, or the fact that containers of different types and sizes must be processed through the same or similar pasteurisers at different speeds.

Operational factors can be accommodated by examining the designed operating loads of can and bottle pasteurisers. A pasteuriser must be designed to operate at different speeds for different container sizes and must also take in sufficient energy to overcome heat losses, etc. from the container heating medium. By calculating the thermal energy input per 1000 containers pasteurised, differences in expected performance can be illustrated and a comparison with the theoretical predictions of Table 18.8 can be made. Table 18.9 therefore lists the designed thermal energy requirements of some typical pasteurisers for a number of different containers.

Table 18.9

Thermal loadings for pasteurisers under normal running conditions. Values based on machine manufacturers' data

Container size fl. oz.	Type[†]	Electrical energy per 1000 containers kWh	Thermal energy per 1000 containers MJ	Volume ratio	Energy ratio
9.68	C	0.85	70.79	1:1.6	1:1.15
15.5	C	0.94	81.76		
9.68	C	0.60	67.68	1:1.6	1:1.12
15.5	C	0.70	75.71		
9.68	RB	1.014	69.17	1:2	1:1.99
19.4	RB	2.028	137.55		
9.68	RB	not known	72.72	1:2	1:1.98
19.4	RB	not known	143.99		
9.68	RB	not known	77.90	1:2	1:1.95
19.4	RB	not known	151.80		

†C = can; RB = returnable bottle.

© Crown Copyright 1981

The values in Table 18.9 are grouped in pairs with each pair relating to the operation of one particular pasteuriser for different container sizes. Electricity consumption is also given for completeness where known, but electricity use on filling lines is considered in detail later. The data for returnable bottles are plotted in Fig. 18.4.

The values in Table 18.9 are of interest for a number of reasons. If we examine the thermal inputs for bottles, then the energies are approximately in the ratios of the liquid volumes and in agreement with the calculations shown in Table 18.8. Cans however are different; instead of the ratio being 1:1·6 for the 9.68 and 15.5 fl. oz. cans, it is 1:1.15. The reason for this different behaviour between cans and bottles is that as the liquid volume in a bottle increases, both the height and diameter of the container increases. For small cans however, only the height increases with volume. Consequently the heating zone of the pasteuriser can accommodate fewer bottles as the container size increases but the same number of cans. The reduction in line speed for cans as a function of container size is therefore less than that for bottles and so the differences in

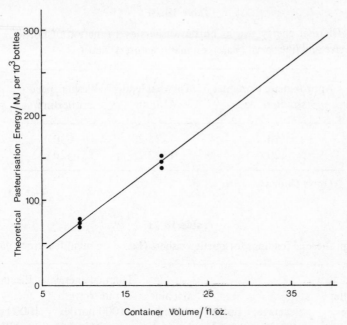

Fig. 18.4 – Variation in the theoretical pasteurisation energy with container volume based on manufacturers' data.

pasteurisation energy between containers of different size is much smaller for cans.

Note however that the pasteurisation energy for a can of a given size is very similar to that for a bottle of the same size. This is again the consequence of the high specific heat capacity of the liquid compared with those of the various container materials. These general conclusions are used later.

18.6.2 Bottle washing

Returnable bottles need a hot detergent wash to remove labels and to clean and sterilise them. Frequently new bottles are added to the line as replacements and these also go through the washer as a matter of convenience. Non-returnable bottles require only a rinse but because these containers are frequently filled on lines designed for returnable bottles, they are often processed through the washer to avoid installing separate rinsing equipment. Cans with the exception of party cans usually receive only a cold rinse or air cleaning.

Like pasteurisers, bottle washers run at slower speeds when treating larger containers but now the thermal energy input per unit is also lower as shown in Table 18.10. The consequence of this is shown in Table 18.11 where the designed thermal energy input per 1000 bottles washed is calculated for a number of bottle washers installed in plants visited. Data from both beer and soft drinks filling plants have been included.

Table 18.10

Thermal energy use of bottle washers as a function of bottle
size and line speed (Based on manufacturers' data)

Approximate container size in fl. oz.	Thermal load MJ/hour	Washer speed bottles/min
10	2274.2	630
20	1672.2	420

© *Crown Copyright 1981*

Table 18.11

Design thermal loadings for bottle washers (Based on manufacturers' data)

Bottle size fl. oz.	Beverage filled*	Bottle type†	Machine speed b.p.m.‡	Thermal energy input to wash 1000 bottles MJ	Electrical energy per 1000 bottles kWh
(1) 3.7	CSD	RB	675	49.96	1.77
(2) 6.34	CSD	RB	690	59.08	1.26
(3) 9.68	B	RB	330	63.90	3.77
(4) 19.40	B	RB	170	71.20	7.32
(5) 9.68	B	RB	630	60.16	1.28
(6) 19.40	B	RB	420	66.35	1.93
(7) 9.68	B	NRB	330	23.96	1.05
(8) 19.40	B	NRB	170	27.91	2.04
(9) 35.2	CSD	RB	200	116.85	2.75

*CSD = carbonated soft drink; B = beer.
†NRB = non-returnable bottle; RB = returnable bottle.
‡b.p.m. = bottles per minute.

© *Crown Copyright 1981*

Considering entries (3) and (4) or (5) and (6) in Table 18.11, then within
any pair of numbers, the thermal energy consumption per 1000 bottles washed
is similar (to within 10%) for the two container sizes. Entries (7) and (8) in
Table 18.11 represent the energy consumption of the same machine as entries
(3) and (4) but with the washer downrated to handle non-returnable bottles.

This thermal energy reduction is effected by decreasing the temperature of the soaking liquor which in turn reduces the steam requirement of the washer. Appreciable savings in electrical power can also be achieved by isolating certain pump and drive units. The use of detergent is also less when using the washer for non-returnable bottles but water consumption is unaffected.

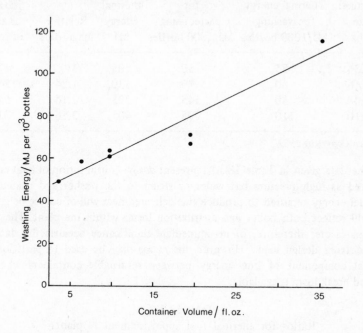

Fig. 18.5 — Variation in the theoretical bottle washing energy with container volume, based on manufacturers' data.

Table 18.11 and Fig. 18.5 demonstrate that although the washing energy increases with container size, the relative change is much less than in pasteurisation; that is as the container volume increases by almost a factor of 10 from 3.7 to 35.2 fl. oz., the washing energy increases only by a factor of 2. Remember that pasteurisation energy is approximately directly proportional to volume (Fig. 18.4).

18.6.3 Thermal energy use on returnable bottle lines
Figs. 18.4 and 18.5 may be used to estimate the minimum theoretical energy to be supplied as thermal fuels to wash and pasteurise returnable bottles. These are shown in Table 18.12 and have been obtained by interpolation from the two graphs.

Table 18.12
Estimated thermal energy requirement for washing and pasteurising returnable bottles

Container size fl. oz.	Thermal energy for washing MJ/1000 bottles	Thermal energy for pasteurising MJ/1000 bottles	Total thermal energy MJ	Relative magnitude	Pasteurisation as a % of total
6.34	55	50	105	1	47
9.68	60	70	130	1.24	54
19.40	80	145	225	2.14	64
39.10	110	295	405	3.86	73

© *Crown Copyright 1981*

The data given in Table 18.13 represent *design* consumption of heat energy delivered as high pressure hot water or steam to the washer and pasteuriser. The fuel energy required to produce this delivered heat will of course be greater and will reflect both boiler and distribution losses within the plant. The absolute values are, therefore, of no immediate significance because the data are derived from design loads. However, the *ratios* may be used to partition the thermal component of line energy between returnable containers when no detailed breakdown is available.

Table 18.13
Ratios for thermal fuel apportionment in plants filling only non-returnable bottles

Container volume fl. oz.	Ratio for apportionment
6.34	1
8.8 (0.25 l)	1.39
9.68	1.53
11.6 (0.33 l)	1.83
15.5 (0.44 l)	2.44
17.6 (0.50 l)	2.78
19.4	3.06
25.7 (0.75 l)	4.05
35.2 (1.00 l)	5.55
39.10	6.17

© *Crown Copyright 1981*

18.6.4 Thermal energy use on non-returnable bottle lines

The above analysis applies to returnable bottles and the ratios derived in Table 18.12 are not expected to be applicable to non-returnable bottles because of differing practices. Table 18.11 showed that hot rinsing in a downrated washer, lowers the energy by more than 60%. In plants where a rinser, designed specifically for non-returnable bottles, is used, the rinsing energy will be even lower. Consequently, it is assumed that pasteurising energy is dominant on non-returnable bottle lines and, to a first approximation that the consumption of thermal fuels will be in the ratio of the container volumes. Thermal fuels may therefore be apportioned between different bottle sizes in plants filling only non-returnable bottles, as shown in Table 18.13.

18.6.5 Thermal energy use on can lines

Cans also require their own set of partitioning parameters. Table 18.9 demonstrated that the pasteurisation energy for 15.5 fl. oz. cans compared with that for 9.68 fl. oz. cans is in ratio 1.15:1 rather than 1.6:1 as might have been expected from the relative volumes of liquid. Evidence from a plant filling 9.68 and 11.6 fl. oz. cans on the same line suggests that there is no significant difference in the pasteurisation energies for these two container sizes.

Beer packed in 78, 98 or 136 fl. oz. cans is usually flash pasteurised and in these circumstances pasteurisation energy is strictly proportional to the volume of beer treated. Estimated values, together with known data for small can sizes are given in Table 18.14 and allow calculation of the apportionment ratios shown. They may be used to allocate thermal fuel energy between different can sizes in filling plants where no detailed fuel breakdown is available.

Table 18.14

Notional pasteurisation energies for cans and ratios for thermal fuel apportionment

Can size fl. oz.	Pasteurisation energy MJ/1000 cans	Ratio for apportionment
9.68	70.79	1
11.6	70.79	1
15.5	81.76	1.155
78	605.83	8.56
98	757.29	10.70
136	1056.31	14.92

© Crown Copyright 1981

18.6.6 Summary of thermal energy use

The method of apportionment described above for each type of container is based on a study of the use of thermal fuels in canning and bottling lines and should strictly be applied only to that proportion of the total energy attributable to bottle washing and/or pasteurisation. Section 18.4 showed that for a can line, pasteurisation accounts for 70% of the thermal energy used in the canning hall but only for some 40% of the total requirement of thermal fuels. A similar situation is likely to occur in the filling of non-returnable bottles. However, in plants filling returnable bottles, the thermal energy required for washing, raises the proportion of the total energy that is directly attributable to the container.

Thermal energy, other than that consumed on the line, should arguably be apportioned on the basis of the relative times that the plant is filling each type of container. This would take account of both the relative line speeds and the total number of containers of each type and size filled.

The ideal approach therefore, is to apportion **line usage** using the ratios calculated for the different container types and sizes and **overheads** as described above. When the breakdown between these two functions is not known the alternative is to apportion *total* energy use on the basis of line energy use or relative filling times, recognising that neither approach is ideal. Where necessary, the method based on line usage has been chosen because it is related to the size and type of container being filled. Table 18.15 shows the values obtained for thermal fuels consumption in a plant filling a range of container sizes and types apportioned on the basis of the alternatives described. The 'true' values will lie somewhere between the two sets and will probably be closer to the values in column (1).

Table 18.15

Heavy fuel oil consumption in a filling plant as a function of the method of apportionment

Container size fl.oz.	Container type	Gallons of oil per 1000 bottles with apportionment based on line energy use (1)	Gallons of oil per 1000 bottles with apportionment based on the time for which the plant was filling each type of container (2)
6.34 fl. oz.	R	2.99	4.90
9.68 fl. oz.	R	3.71	3.38
19.40 fl. oz.	R	6.40	5.87
39.10 fl. oz.	R	11.54	11.76
0.5 litre	NR	4.10	5.36
1 litre	NR	8.43	10.70

© *Crown Copyright 1981*

18.7 ELECTRICAL ENERGY CONSUMPTION ON FILLING LINES

For a returnable bottle filling line similar to that shown in Fig. 18.1, typical electrical ratings for the equipment on the line will be as shown in Table 18.16. This line can fill 630 bottles of capacity 9.68 fl. oz. or 430 bottles of capacity 19.4 fl. oz. per minute. In both cases, the average load factor is 90%, so that the electricity consumption per 1000 bottles filled will be 8.57 kWh for 9.68 fl. oz. bottles and 12.86 kWh for 19.4 fl. oz. bottles. In both cases consumption is proportional to throughput and the electricity used to fill a given number of containers is related to the plant running time. This is also true of course for non-returnable bottles and can filling lines where the overall electricity consumption is usually lower. For example, Table 18.17 shows the ratings for a can line suitable for filling either 9.68 or 15.5 fl. oz. cans and packing the product in Hicone and cardboard trays.

Remember that the total installed load for non-returnable bottle and can lines is sensitive to the type of outer packaging used. Shrinkwrapping tunnels for example are significantly higher energy consumers than other forms of packaging machinery (section 18.4).

If the line described in Table 18.17 fills 9.68 fl. oz. cans at the rate of

Table 18.16

Electrical ratings for the components of a returnable bottle filling line

Equipment	Rating/kW
Depalletiser	26
Unpacker	12
All bottle conveyors	49
Case conveyors	67
Case washer	11
Washer	54
Bottle inspector	2
Filler/capper	8
Closure feed system	6
Pasteuriser	52
Level detector	2
Labeller	10
Packer	16
Palletiser	45
Total installed load	360

© *Crown Copyright 1981*

Table 18.17
Electrical ratings of the components in a
typical can filling line

Equipment	Rating/kW
Depalletiser	18
Filler/seamer	42
Level detector	5
Can conveyors	30
Pasteuriser	63
Can coder	2
Hicone machine	4
Tray packer	8
Tray conveyor	37
Palletiser	45
Total installed load	254

© *Crown Copyright 1981*

1050 cans per minute and 15.5 fl. oz. cans at the rate of 950 cans per minute, then the electricity consumption per 1000 cans filled will be 3.43 kWh and 3.79 kWh respectively. Again, the electricity consumption is expected to be proportional to throughput and therefore to the time for which the plant is filling each type of can.

The above method of apportionment is obviously most satisfactory for plant where containers of only one type are filled. It is also applicable to plant where mixed returnable and non-returnable bottles are filled on the same line. Fortunately, these two situations cover most of the practices encountered.

In all the plants examined, *total* electricity use has been apportioned on the basis of relative throughput taking account of relative speeds and the total number of containers of each type filled, although once again the above discussion relates specifically to *line* consumption. This assumption is felt to be reasonable because the analysis in section 18.4 suggests that approximately 30% of the total electrical energy is used on the can line itself and is therefore directly attributable to the container and proportional to throughput. (This proportion is likely to be higher for bottle lines because of the higher rating of the lines and the requirement for more mechanical handling equipment.) A proportion of the 31% overheads will also be directly attributable in this way since they include power to provide services to the line. The other major contributor to the total electrical consumption, namely refrigeration energy, will also be proportional to the volume throughput of the plant. This leaves lighting as the only charge to the system which is not in any way proportional to throughput.

18.8 FILLING ENERGIES AS DETERMINED BY PROCESS ANALYSIS

A total of 10 beer filling plants have been examined in some detail and the data obtained using the procedures outlined in the preceding Sections is presented in Table 18.18. Reasonable average fuel requirements can be obtained by plotting these known requirements as a function of container size and obtaining a best fit line which satisfies the plotted points. In this way, the fuel requirements for intermediate container sizes may be obtained by interpolation. When converted to energy requirements using Table 2.2, the results as as shown in Table 18.19.

Table 18.18
Direct fuel consumption per 1000 containers for beer filling

Container size fl. oz.	Electricity (kWh)	Oil fuels (gallons)	Natural gas (MJ)
Returnable bottles			
9.68	31.34	–	623.00
9.68	29.51	3.24	–
9.68	29.98	3.68	2.39
9.68	30.31	0.15	1277.40
9.68	56.86	4.53	6.54
9.68	23.23	3.71	–
6.34	33.65	2.99	–
19.40	50.07	–	983.50
19.40	54.60	5.12	–
19.40	59.96	5.81	3.78
19.40	89.99	0.24	2018.29
19.40	40.38	6.40	–
39.10	80.81	11.54	–
Non-returnable bottles			
9.68	25.05	3.30	–
17.6 (½ litre)	36.82	4.10	–
35.2 (1 litre)	73.51	8.43	–
Cans			
9.68	13.71 to 16.72	1.84	–
9.68	25.14	1.82	–
9.68	18.16	–	350.00
15.5	22.89	–	367.50
78	158.12	25.86	–
78	266.55	14.22	–
136	463.80	24.79	–

Table 18.19
Total energy required to fill 1000 beer containers of different types and sizes

Container size in fl. oz. and type	Electricity/MJ		Oil fuels/MJ			Other fuels/MJ			Total energy /MJ
	Fuel production and delivery energy	Energy content of fuel	Fuel production and delivery energy	Energy content of fuel	Feedstock energy	Fuel production and delivery energy	Energy content of fuel	Feedstock energy	
6.34 RB	363.42	121.14	81.38	389.93	nil	13.74	167.11	nil	1136.72
9.68 RB	361.80	120.60	103.82	497.49	nil	17.53	213.21	nil	1314.45
19.40 RB	637.20	212.40	151.24	724.71	nil	25.54	310.59	nil	2061.68
39.10 RB	872.76	290.92	283.31	1357.58	nil	47.84	581.82	nil	3434.23
6.34 NRB	194.40	64.80	58.43	280.00	nil	9.87	120.00	nil	727.50
8.80 NRB	245.16	81.72	73.04	350.00	nil	12.33	150.00	nil	912.25
9.68 NRB	270.54	90.18	89.81	430.36	nil	15.16	184.44	nil	1080.49
11.60 NRB	281.88	93.96	87.65	420.00	nil	14.80	180.00	nil	1078.29
15.50 NRB	374.76	124.92	111.39	533.75	nil	18.81	228.75	nil	1392.38
17.60 NRB	397.65	132.55	111.58	534.66	nil	18.84	229.14	nil	1424.42
19.40 NRB	470.88	156.96	137.31	658.00	nil	23.19	282.00	nil	1728.34
25.70 NRB	591.84	197.28	171.64	822.50	nil	28.98	352.50	nil	2164.74
35.20 NRB	793.92	264.64	229.43	1099.35	nil	38.73	471.15	nil	2897.22
9.68 C	181.77	60.59	50.25	240.80	nil	8.49	103.20	nil	645.10
11.60 C	203.04	67.68	52.04	249.38	nil	8.79	106.87	nil	687.80
15.50 C	247.20	82.40	53.68	257.25	nil	9.07	110.25	nil	759.85
78.0 C	2293.17	764.39	545.39	2613.45	nil	92.10	1120.05	nil	7428.55
98.0 C	3207.60	1069.20	588.70	2821.00	nil	99.41	1209.00	nil	8994.91
136.0 C	5009.04	1669.68	674.65	3232.88	nil	113.93	1385.52	nil	12085.70

RB = returnable bottle; NRB = non-returnable bottle; C = can.

CHAPTER 19

Soft Drinks Filling

19.1 SALES OF CARBONATED SOFT DRINKS

Production of carbonated soft drinks in recent years is shown in Table 19.1 and demonstrates an annual growth rate of some 3 to 4%. It has proved more difficult than in the case of beer to obtain sales data broken down by container size and type because the Trade Federation does not collect this information in sufficient detail. Three separate estimates are given in Tables 19.2 to 19.4.

Table 19.1

Production of carbonated soft drinks in the U.K. (Source: Food Manufacturers Federation Annual Report (1977))

Year	Production	
	million gallons	million litres
1968	254.0	1154.7
1969	264.0	1200.1
1970	267.8	1217.4
1971	274.0	1245.6
1972	291.4	1324.7
1973	359.0	1632.0
1974	374.6	1702.9
1975	398.9	1813.4
1976	438.2*	1992.1*
1977	427.3	1942.5

*1976 was not a typical year because of the very hot summer.

© Crown Copyright 1981

Table 19.2

Estimated split of the carbonated soft drinks trade based on Trade Sources

Year	Returnable bottles 10^6 litre	Non-returnable bottles 10^6 litre	Cans 10^6 litre	Bulk 10^6 litre	Total 10^6 litre
1973	996	260	324	50	1630
1974	1008	274	361	60	1703
1975	1091	290	362	70	1813
1976	1157	313	447	75	1992
1977	1101	301	460	80	1942

© Crown Copyright 1981

Table 19.3

Estimated consumption of carbonated soft drinks by type and size of pack in the U.K. in 1976. Values are in millions of units. (Source: private communications with bracketed values taken from Table 11.1)

Container size	Cans	Non-returnable bottles	Returnable bottles	Total
under 7 fl. oz.	(46)	95	1723	1818
0.25 litre/10 fl. oz.	–	468	36	504
11.6 fl. oz.	1420	24	–	1444
13 to 20 fl. oz.	(15)	119	–	119
26 fl. oz.	–	27	526	553
litre/quart	–	121	409	530
Totals	1420	854	2694	4968

© Crown Copyright 1981

Plastic (PET) bottles for carbonated soft drinks have recently entered the market and are considered later. At present, returnable bottles dominate the soft drinks market in both on-trade (principally 3.7, 6.0 and 6.5 fl. oz. sizes) and in the take-home sector (26 and 38 fl. oz. and 1 litre sizes). Cans are almost exclusively restricted to the 11.6 fl. oz. size with small numbers of 5.5 fl. oz. cans being used for mixer drinks.

Trends in the industry are for the volume in returnable bottles to remain fairly constant with the growth stemming from convenience packaging sectors of cans and, to a lesser extent, non-returnable bottles. There is also a move towards larger bottle sizes. This is true both for mixers where upgrading to the 0.5 litre size is occurring and for beverages destined for home consumption where, for example, the 26 fl. oz. bottle is being substantially replaced by the 1 litre size. It is as yet difficult to estimate the long term impact of the 1.5 and 2.0 litre PET bottles. The present examination considers filling of the container sizes given in Table 19.5.

Table 19.4

U.K. consumption of carbonated soft drinks in 1977 by type and size of pack. (Source: British Soft Drinks Council). Values are in millions of litres

Container size	Cans	Non-returnable bottles	Returnable bottles	Total	Plastic bottles	Grand total
under 7 fl. oz.	—	10	251	261		
0.25 litres	—	110				
11.6 fl. oz.	459	8	50	682	45	1913
0.5 litre	—	55				
0.75 litre	—	20	386	406		
1 litre	—	110	409	519		
Totals	459	313	1096	1868	45	1913

© *Crown Copyright 1981*

Table 19.5

Containers considered in Chapter 19

Approximate capacity	Capacity as stated on the label fl. oz.	Type of container	Volume in litres assumed for conversion
4.0 fl. oz.	3.7	RB	0.105
6 to 7 fl. oz.	6.0/6.5	RB	0.177
10 fl. oz.	9.68	RB	0.275
26 fl. oz.	25.0	RB	0.710
1 litre	34/35	RB	0.990
40 fl. oz.	38.0	RB	1.080
6.0 fl. oz.	6.0	NRB	0.170
0.25 litre	8.8	NRB	0.250
0.50 litre	17.6	NRB	0.500
0.75 litre	26.4	NRB	0.750
1.00 litre	34/35.2	NRB	0.986
6.0 fl. oz.	5.5	C	0.156
12.0 fl. oz.	11.6	C	0.327
16.0 fl. oz.	15.5	C	0.440
1.5 litre PET non-returnable plastic bottle			1.500

RB = returnable bottle; NRB = non-returnable bottle; C = can.

© *Crown Copyright 1981*

19.2 FILLING LINES FOR CARBONATED SOFT DRINKS

Filling lines for carbonated soft drinks are essentially similar to those for beer filling (Figs. 18.1 to 18.3) although only relatively small volumes of product are pasteurised. Those which are pasteurised (e.g. shandy) usually contain beer. This Chapter is concerned only with those beverages where no pasteurisation is applied. In general soft drinks filling lines operate at slower speeds than beer filling lines.

19.3 FILLING ENERGY

As with beer filling, energy consumption is an aggregate of line usage and over-heads such as heating, lighting and power to ancilliary operations. In general, overheads depend upon the type of site which may range from small one-line bottling operations which buy in concentrate and employ the minimum of support facilities, to large multi-line plants which make their own syrup and concentrates and have substantial office, laboratory, garaging and warehouse accommodation. A much wider range of filling energies is therefore expected for carbonated soft drinks than for beer.

Because of the nature of the industry, it is much easier to identify plant filling only one type of container or a restricted range of types than was the case in the brewing industry. The problem of apportionment of energy is therefore much less severe in the case of carbonated soft drinks. For example, in plants filling only non-returnable containers, apportionment was based on the relative times for which the plant was filling different sizes. This is readily justified because, without washing or pasteurising the line consumption of thermal fuels is negligible compared with the 'overheads'. In plants filling only returnable bottles or where both returnable and non-returnable bottles or cans are filled, an estimate of the fuel energy to wash the particular container size was made and attributed to the relevant container. These data were interpolated from Fig. 18.5 and the efficiency of steam generation was taken into account. The remaining thermal fuels use was partitioned as for non-returnable bottles and cans. Electricity consumption was in all cases apportioned as in section 18.7.

19.4 FILLING ENERGIES AS DETERMINED BY PROCESS ANALYSIS

24 filling plants for carbonated soft drinks were examined in some detail and the data relating to their filling energies are presented in Tables 19.6 and 19.7. Many of the operators were able to estimate the electrical and thermal consumption with some confidence. These data are also given in Tables 19.6 and 19.7 where known and are identified by the relevant note. They include the heating and lighting of the bottling store but not energy consumptions due to offices, garaging, warehousing, etc. and therefore represent the lower limit of the range of energy requirements. Data are most comprehensive for returnable bottles and least detailed for cans.

Table 19.6

Fuel consumed in filling 1000 returnable bottles with carbonated soft drinks

Container size fl. oz.	Electricity (kWh)	Oil fuels (gallons)	Natural gas (MJ)	Notes
3.7	15.20	1.73	4.20	2
3.7	22.56	1.88	3.37	2, 3
3.7	11.05	0.38	261.29	1
6.0†	13.40	1.53	3.16	2
6.0†	24.44	3.57	1.05	2,3
6.0†	10.98	0.38	259.72	1
6.0†	15.86	1.51	4.21	2, 3, 4
6.0†	19.84	2.64	17.92	2, 3
6.0†	18.78	2.15	4.22	2
6.0†	21.11	3.47	2.11	2
6.0†	14.74	2.44	25.11	2, 3, 4
9.6	12.94	0.44	306.03	1
9.6	25.66	3.52	1.05	2, 3
25.0	18.50	3.21	–	2, 3
25.0	20.04	2.72	18.97	2
25.0	25.42	1.66	–	2, 3
25.0	24.23	2.95	–	2, 3
25.0	23.87	2.48	5.27	2
25.0	14.60	1.67	4.22	2
25.0	40.65	3.24	–	2, 3
25.0	47.01	5.69	97.75	2, 3
25.0	40.65	4.65	9.49	2
25.0	45.69	4.21	5.27	2
25.0	11.70	–	338.80	1
35.2‡	22.13	3.73	–	2, 3
35.2‡	20.98	2.82	20.03	2
35.2‡	25.32	1.66	–	2, 3
35.2‡	24.14	2.94	–	2, 3
35.2‡	28.96	3.01	6.32	2
35.2‡	21.05	2.40	5.27	2
35.2‡	54.10	4.31	–	2, 3
35.2‡	58.14	5.36	6.32	2
35.2‡	25.12	0.86	594.32	1
35.2‡	23.77	–	351.47	1
35.2‡	29.78	0.86	621.66	2
38.0	12.50	2.43	–	1, 3, 4
38.0	13.89	3.40	–	2

Notes: (1) Bottling store only (including heating and lighting).
        (2) Bottling store, syrup making, offices, warehouses, etc.
        (3) One filling line only.
        (4) One container only filled in the plant.
† 6.0 fl. oz. includes the 6.5 fl. oz.
‡ 35.2 fl. oz. includes the 34.5 fl. oz.

© *Crown Copyright 1981*

Table 19.7

Fuel consumed in filling 1000 non-returnable containers with carbonated soft drinks

Container size fl. oz.	Electricity (kWh)	Oil fuels (gallons)	Natural gas (MJ)	Notes
Non-returnable bottles				
6.0	11.03	0.73	–	1
6.0	14.66	0.79	–	2
8.8	26.94	3.12	1.05	2, 3
8.8	17.78	1.05	2.64	2
17.6	23.30	1.88	–	2
17.6	46.00	3.12	–	2, 3
17.6	26.09	3.02	1.05	2, 3
17.6	33.26	4.71	65.30	2, 3
17.6	29.66	2.01	10.44	2, 3, 4
17.6	16.86	1.80	11.28	2, 3
35.2	22.11	3.17	–	2, 3
35.2	21.00	1.97	–	2
35.2	25.27	1.12	–	2, 3
35.2	60.84	4.30	–	2, 3
35.2	34.47	3.98	1.05	2, 3
35.2	31.00	3.32	20.77	2, 3
Cans				
5.5	8.82	0.59	–	1
5.5	14.26	0.65	–	2
11.5	6.06	0.60	–	1
11.5	11.50	0.66	–	2
15.5	6.92	0.68	–	1
15.5	12.36	0.74	–	2

Notes: (1) Bottling store only (including heating and lighting).
          (2) Bottling store, syrup making, offices, etc.
          (3) One filling line only.
          (4) One container only filled.

By plotting the fuel requirement from Tables 19.6 and 19.7 as a function of container size, it can be shown that there is a much greater scatter of values than occurs in beer filling. However, in all cases, the fuel requirements for carbonated soft drinks filling are lower than for the equivalent container filled with beer — as expected from the absence of refrigeration and pasteurisation.

Representative energies are later required for filling containers of different sizes and so values have been chosen from Tables 19.6 and 19.7 with intermediate values deduced by interpolation. These representative fuel requirements are given in Table 19.8 and the corresponding energy requirements, using Table 2.2, are shown in Table 19.9.

Table 19.8

Typical fuel energies required to fill and pack 1000 containers with carbonated soft drinks. Values indicated by and asterisk are interpolated values

Container size fl. oz.	Electricity (kWh)	Thermal energy (MJ)
Returnable bottles		
3.7	18.88	340.47
6.0	19.30	462.00
6.5	19.30	462.00
9.6	19.30	480.90
25.0	30.06	618.70
34.5	32.20	621.66
35.2	32.20	621.66
38.0*	33.50	633.42
Non-returnable bottles		
6.0	14.66	147.18
8.8	18.56	500.03
17.6	21.84	388.44
26.4*	24.50	512.00
34.2	26.77	554.55
35.2	26.77	554.55
Cans		
5.5	14.26	121.10
11.6	11.50	122.96
15.5	12.36	137.86

© *Crown Copyright 1981*

Table 19.9
Total energy required to fill 1000 containers with carbonated soft drinks

Container type and size in fl. oz.	Electricity/MJ		Oil fuels/MJ			Other fuels/MJ			Total energy /MJ
	Fuel production and delivery energy	Energy content of fuel	Fuel production and delivery energy	Energy content of fuel	Feedstock energy	Fuel production and delivery energy	Energy content of fuel	Feedstock energy	
3.7 RB	203.91	67.97	62.17	297.91	–	3.50	42.56	–	678.02
6.0 RB } 6.5 RB	208.44	69.48	84.36	404.25	–	4.75	57.75	–	829.03
6.5 RB	208.44	69.48	87.81	420.79	–	4.94	60.11	–	851.57
9.6 RB	324.66	108.22	112.97	541.36	–	6.36	77.34	–	1170.91
25.0 RB	347.76	115.92	113.51	543.95	–	6.39	77.71	–	1205.24
34.5 **RB** } 35.2 RB	361.80	120.60	115.66	554.24	–	6.51	79.18	–	1237.99
38.0 RB	158.34	52.78	26.87	128.78	–	1.51	18.40	–	386.68
6.0 NRB	200.46	66.82	70.93	339.88	–	3.99	48.56	–	730.64
8.8 NRB (¼ litre)	235.86	78.62	91.31	437.53	–	5.14	62.50	–	910.96
17.6 NRB (½ litre)	264.60	88.20	93.49	448.00	–	5.26	64.00	–	963.55
26.4 NRB (¾ litre)	289.11	96.37	101.26	485.23	–	5.70	69.32	–	1046.99
34.2 NRB } 35.2 NRB (1 litre)	154.02	51.34	22.11	105.96	–	1.24	15.14	–	349.81
5.5 Can	124.20	41.40	22.45	107.59	–	1.26	15.37	–	312.27
11.6 Can	133.50	44.50	25.17	120.63	–	1.42	17.23	–	342.45
15.5 Can									

CHAPTER 20

Cider Filling

20.1 SALES OF CIDER

Cider making in the U.K. is in the hands of three major producers together with a number of small independent companies. The total annual production amounts to some 48 million gallons which is very small compared with the production of beer (1400 million gallons) or carbonated soft drinks (430 million gallons). Data collection within the industry is not comprehensive and Table 20.1 gives the best breakdown of sales by container type that is available.

Table 20.1

Approximate consumption of cider by type and size of pack. 1976 data based on information from the National Association of Cider Makers and private communications from individual companies

Container size and type	% of total production	Volume in 10^6 gall.
Returnable bottles up to 1 litre	10.96	5.222
Non-returnable bottles up to 1 litre	23.95	11.411
Cans	1.81	0.862
Returnable flagons	25.29	12.05
Draught and Keg	37.99	18.10
Totals	100.00	47.645

© *Crown Copyright 1981*

Table 20.1 clearly demonstrates that the flagon bottle has the largest share of the packaged cider market. Trends in cider packaging are similar to those for beer and carbonated soft drinks. However, the technical difficulties of packaging cider in cans means that as returnable bottle use declines, the market share is likely to be taken by non-returnable bottles.

Compared with beer and soft drinks, cider is packed in a restricted range of containers. For the purposes of the present study, we have examined the filling of containers in the sizes given in Table 20.2. The small amount of cider packaged in cans has been ignored.

Table 20.2

Container sizes used to package cider and considered in this Section

Container capacity as stated on the label fl. oz.	Container type	Volume in litres assumed for conversion
3.33	RB and NRB	0.10
6.0	RB	0.17
9.5†	RB	0.27
38.0‡	RB	1.08
35.2	NRB	1.00

†This is in fact the ½ pint bottle which has a stated label capacity of 9.68 fl. oz. in the brewery trade.
‡This is the flagon bottle which has a stated label capacity of 39.1 fl. oz. in the brewery trade.

© *Crown Copyright 1981*

20.2 CIDER FILLING LINES

Cider, like beer, is biologically active and requires sterilisation to achieve adequate shelf life. Approximately two thirds of packaged cider is pasteurised and the remainder is sterilised by filtration, a technique not used by the brewing industry.

Although it is possible to achieve sterility by filtration or pasteurisation in both beer and cider, the level of preservative (SO_2) in beer must not exceed 70 parts per million whereas cider is permitted up to 200 parts per million [61]. This might be the reason for the use of pasteurisation in breweries as against filtration. In-bottle pasteurisation is the safest method because it takes care of any microbiological spoilage in the equipment or bottles with which sterilised bulk product may subsequently come into contact.

Within a given plant, the energy consumption will be a function of the type of container filled and the method used for sterilisation. This latter parameter will have an effect because plants which use filtration need not supply thermal fuels for pasteurisation or the electrical energy for refrigeration if flash pasteurisation were used. There will however be a small contribution to the system energy associated with the provision of filter medium.

As with beer and carbonated soft drinks filling, the system energy will be an aggregate of line usage and overheads relating to heating, lighting and powering ancillary machinery associated with the filling plant.

20.3 CIDER FILLING ENERGIES AS DETERMINED BY PROCESS ANALYSIS

Data obtained from cider companies are given in Table 20.3. These data are by no means comprehensive but they do show that plants using filtration instead of pasteurisation consume less fuels. Because the data of Table 20.3 do not relate to a sufficient number of filling lines to be truly representative, the following assumptions have been made:

(a) For 3.33 and 6.0 fl. oz. bottles, the values obtained for filling soft drinks in returnable bottles of capacities 3.7 and 6.0 fl. oz. have been used. This is thought to be acceptable because the small bottle sizes filled with cider use filtration and the direct consumption of fuels in filling is close to that for soft drinks in these sizes.

(b) The three large bottle sizes, which are all pasteurised, are assumed to have the same filling energies as beer.

Table 20.3
Fuels consumed in filling 1000 bottles with cider

Container size fl. oz.	Electricity (kWh)	Oil fuels (gallons)	Natural gas (MJ)	Notes
Returnable bottles				
3.33	13.88	1.49	–	(1)
3.33	12.82	1.38	–	(1)
6.0	44.06	–	186.63	(1)
9.68	23.37	–	231.07	(1)
39.1	32.49	–	667.44	(1)
39.1	73.87	10.53	–	(2)
Non-returnable bottles				
35.2	32.83	–	526.15	(1)
35.2	73.48	7.69	–	(2)

Notes: (1) Filtration used to sterilise.
         (2) Flash pasteurisation.

© *Crown Copyright 1981*

The energy requirement for cider filling is therefore as shown in Table 20.4. It is expected that any errors introduced by making the above assumptions will be small since the volume of cider packed is small in comparison with the other beverages. Furthermore, as will be seen later, the filling energy is only a small component of the total system energy.

Table 20.4
Energy required to fill 1000 cider containers of different types and sizes

Container size and type Size in fl. oz.	Electricity/MJ		Oil fuels/MJ			Other fuels/MJ			Total energy /MJ
	Fuel production and delivery energy	Energy content of fuel	Fuel production and delivery energy	Energy content of fuel	Feedstock energy	Fuel production and delivery energy	Energy content of fuel	Feedstock energy	
3.33 RB (100 ml)	203.91	67.97	62.17	297.91	—	3.50	42.56	—	678.02
6.0 RB	208.44	69.48	84.36	404.25	—	4.75	57.75	—	829.03
9.68 RB (9.5)	361.80	120.60	103.82	497.49	—	17.53	213.21	—	1314.45
39.10 (38.0)	872.76	290.92	283.31	1357.58	—	47.84	581.82	—	3434.23
35.2 NRB	793.92	264.64	229.43	1099.35	—	38.73	471.15	—	2897.22

CHAPTER 21

Filling PET Bottles

21.1 FILLING PET BOTTLES

Because PET bottles for carbonated beverages effectively entered the market only in 1979, there are few filling lines which can be analysed satisfactorily. None of the data supplied by manufacturers was sufficiently detailed to enable calculation of PET bottle filling energies by process analysis. However, there is nothing intrinsic to the container which makes its filling behaviour substantially different from that of glass bottles and so the data of Chapter 19 have been extrapolated to give PET filling energies. These energies are shown in Tables 21.1 and 21.2.

Table 21.1

Energy required to fill 1000 1.5 litre PET bottles with carbonated soft drinks

Fuel type	Fuel production and delivery energy/MJ	Energy content of fuel /MJ	Feedstock energy /MJ	Total energy /MJ
Electricity	432.00	144.00	nil	576.00
Oil fuels	132.36	634.37	nil	766.73
Other fuels	7.47	90.63	nil	98.10
Totals/MJ	571.83	869.00	nil	1440.83

Table 21.2
Energy required to fill 1000 2.0 litre PET bottles with carbonated soft drinks

Fuel type	Fuel production and delivery energy/MJ	Energy content of fuel /MJ	Feedstock energy /MJ	Total energy /MJ
Electricity	462.24	154.08	nil	616.32
Oil fuels	141.63	678.77	nil	820.40
Other fuels	7.99	96.97	nil	104.96
Totals/MJ	611.86	929.82	nil	1541.68

CHAPTER 22

Materials Associated with Filling Operations

22.1 INTRODUCTION

Chapters 18 to 21 calculated the energies associated with the fuels directly consumed in the filling plant, to fill and pack the range of container sizes and types with different beverages. Energy contributions arise also from the provision of a number of materials inputs to the filling plant which are necessary to produce the finished packaged product ready for despatch to the customer. Of these, three main groups of supplies may be identified:

(a) Inputs to the line, for example lubricating oils, grease, carbon dioxide, water, detergent etc.
(b) Inputs to the container, for example, closures, labels, glue.
(c) Packaging materials, for example all outer packaging whether returnable (that is crates) or non-returnable.

Each of these inputs is considered in the following sections.

22.2 INPUTS TO THE FILLING LINE

22.2.1 Lubricating oil

The use of lubricating oil was found to range from 0.007 to 0.018 gallons per 1000 containers filled. This contribution to the overall system energy is small and so an average consumption of 0.014 gallons per 1000 containers has been assumed for all lines. The associated energy may be calculated from Table 2.2.

22.2.2 Grease

Grease consumption in filling plants was found to range from 0.002 kg to 0.007 kg per 1000 containers filled, with plants filling returnable bottles exhibiting the greater consumption. An overall average of 0.0043 kg per 1000 containers filled has been assumed for all lines and the associated energy requirement can be calculated from Table 2.2.

22.2.3 Carbon dioxide

Carbon dioxide is used in filling plants for a number of functions. Most obviously it is dissolved in the liquid to provide the carbonation. Gas used in this way should not be charged to the system energy since it forms part of the product beverage, which is itself excluded from the system. The gas is also used to move liquid round the filling plant and for undercover gassing on can filling lines. In the cider and soft drinks trade, compressed air is widely used for this latter operation since the product is less susceptible to oxidation.

Carbon dioxide use in a number of plants has been examined and, not surprisingly, fillers do not meter the amounts of gas supplied to each of the functions mentioned above. To estimate the amount of carbon dioxide to be charged to each container size and type, we have apportioned total filling plant usage in terms of the number of gallons of beverage going to each type and size of container. An overall average per container size and type has been calculated by averaging the values obtained from a number of plants. If the theoretical amount of carbon dioxide dissolved by the liquid volume is then subtracted from these averages, the result is the plant usage of carbon dioxide for filling the container. These data are given in Table 22.1.

The theoretical values in Table 22.1 have been calculated assuming an average carbonation level of 2.5% by volume. The level does in fact differ from beverage to beverage but these differences are not significant.

The energy associated with the manufacture of carbon dioxide for each container size and type can be calculated from Table 9.1. Transport energies involved in the delivery of carbon dioxide are negligible.

22.2.4 Water

Water use, excluding that appearing in the final product, varies significantly from plant to plant covering the range 443 gallons per 1000 containers filled on a can filling line to 1415 gallons per 1000 containers filled on a returnable bottle line. An average of 1000 gallons per 1000 containers has been chosen in the present calculations and the energy associated with the supply of this water may be determined from Table 9.1.

While variations in water use with container size and type are not significant in energy terms, they are significant if water is regarded as a valuable resource. Non-returnable glass bottles require, on average, only one third to one half of the water volume required for washing and rinsing returnable glass bottles.

22.2.5 Detergent

Detergent is used for bottle washing and plant cleaning. Returnable bottle lines are the major consumers although some detergent is used on non-returnable bottle and can filling lines. For the purposes of the present study we have assumed that the detergent use on non-returnable bottle and can lines is negligible and that returnable bottle lines consume detergent (sodium hydroxide) at the rate of 1.85 kg per 1000 containers filled. The energy associated with the provision of this caustic soda may be calculated from Table 7.8.

Table 22.1
Carbon dioxide usage as a function of container size and type

Container size fl. oz.	Total average usage in kg per 1000 containers	Residual dissolved carbon dioxide in kg per 1000 containers	Carbon dioxide attributable to filling in kg per 1000 containers
Bottles			
3.7	0.76	0.49	0.27
6.0	1.24	0.79	0.45
6.34	1.31	0.83	0.48
6.5	1.34	0.85	0.49
8.8	1.82	1.15	0.67
9.68	2.00	1.27	0.73
11.6	2.40	1.52	0.88
17.6	3.64	2.31	1.33
19.40	4.01	2.55	1.46
26.40	5.45	3.46	1.99
34.2	7.13	4.53	2.60
35.2	7.27	4.62	2.65
38.0	7.85	4.99	2.86
39.10	8.08	5.13	2.95
Cans			
5.5	1.14	0.72	0.42
9.68	2.00	1.27	0.73
11.6	2.40	1.52	0.88
15.5	3.20	2.03	1.17
78	16.11	10.23	5.88
98	20.25	12.86	7.39
136	28.10	17.84	10.26
PET Bottle			
1.5 litre	10.91	6.92	3.99
2.0 litre	14.55	9.23	5.32

© *Crown Copyright 1981*

22.2.6 Summary of feeds to filling lines

Using the above data the total energy associated with the materials feeds to the filling lines may now be calculated. It should be remembered that oil, grease, carbon dioxide and water are fed to all lines whereas sodium hydroxide detergent is fed only to returnable bottle lines. The total energy associated with these inputs as a function of container size and type are shown in Table 22.2

Table 22.2

Energy requirement for the provision of oil, grease, carbon dioxide, water and detergent to fill 1000 containers

Container size (fl. oz.) and type	Electricity/MJ			Oil fuels/MJ			Total energy /MJ
	Fuel production and delivery energy	Energy content of fuel	Fuel production and delivery energy	Energy content of fuel	Feedstock energy	
3.7 RB	64.65	21.55	0.57	2.74	–	89.51
6.0 RB	65.04	21.68	0.57	2.74	–	90.03
6.34 RB	65.13	21.71	0.57	2.74	–	90.15
6.5 RB	65.16	21.72	0.57	2.74	–	90.19
9.68 RB	65.73	21.91	0.57	2.74	–	90.95
19.40 RB	67.50	22.50	0.57	2.74	–	93.31
25.0 RB	68.52	22.84	0.57	2.74	–	94.67
34.5 RB	70.29	23.43	0.57	2.74	–	97.03
35.2 RB	70.41	23.47	0.57	2.74	–	97.19
38.0 RB	70.92	23.64	0.57	2.74	–	97.87
39.1 RB	71.13	23.71	0.57	2.74	–	98.15
6.0 NRB	31.08	10.36	0.57	2.74	–	44.75
6.34 NRB	31.17	10.39	0.57	2.74	–	44.87
8.8 NRB	31.62	10.54	0.57	2.74	–	45.47
9.68 NRB	31.77	10.59	0.57	2.74	–	45.67
11.6 NRB	32.13	10.71	0.57	2.74	–	46.15
15.5 NRB	32.85	10.95	0.57	2.74	–	47.11
17.6 NRB	33.24	11.08	0.57	2.74	–	47.63
19.40 NRB	33.54	11.18	0.57	2.74	–	48.03
25.7 NRB	34.56	11.52	0.57	2.74	–	49.39
26.4 NRB	34.83	11.61	0.57	2.74	–	49.75
34.2 NRB	36.33	12.11	0.57	2.74	–	51.75
35.2 NRB	36.45	12.15	0.57	2.74	–	51.91
5.5 C	31.02	10.34	0.57	2.74	–	44.67
9.68 C	31.77	10.59	0.57	2.74	–	45.67
11.60 C	32.13	10.71	0.57	2.74	–	46.15
15.5 C	32.85	10.95	0.57	2.74	–	47.11
78 C	44.28	14.76	0.57	2.74	–	62.35
98 C	47.97	15.99	0.57	2.74	–	67.27
136 C	54.93	18.31	0.57	2.74	–	76.55
1.5 litre PET	39.66	13.22	0.57	2.74	–	56.19
2.0 litre PET	42.88	14.29	0.57	2.74	–	60.48

22.3 PACKAGING MATERIALS FOR RETURNABLE BOTTLES

Returnable bottles filled with all three beverages are packed almost exclusively in returnable crates. Small numbers of such bottles filled with carbonated soft drinks are packed in fibreboard cartons but the number is insignificant. We have therefore assumed that all returnable bottles are transported in returnable crates. Both wood and plastic crates are in use and all three industries are in the process of changing over from wood to plastic. The proportions of wood and plastic crates currently in service is unknown so it is assumed that all crates are plastic since this is the position to which all the industries are moving.

The lifetime of plastic crates is uncertain since they have not been in service long enough to establish their performance with any accuracy. There is however, general agreement within the industries that the lifetime is likely to be between 10 and 20 years. Assuming that a crate makes five trips per year for a total of 15 years, the total lifetime trippage is 65 for all bottle sizes.

Table 22.3 lists the returnable bottles covered by this study and gives typical values for the masses of the crates in service. The mass of polypropylene attributable to the delivery of 1000 bottles of each size is also shown, based on the above trippage.

Table 22.3
Crates for returnable bottles

Container size fl. oz.	Number of bottles per crate	Mass of crate /kg	Number of crates required to deliver 1000 bottles	Mass of polypropylene associated with the delivery of 1000 bottles /kg
3.7[†]	48	1.26	0.32	0.40
6.0	24	1.30	0.64	0.83
6.34	24	1.36	0.64	0.87
6.5	24	1.59	0.64	1.02
9.68	24	1.82	0.64	1.17
19.40	12	1.82	1.28	2.33
25.0	12	1.82	1.28	2.33
34.5	12	1.82	1.28	2.33
35.2	12	1.82	1.28	2.33
38.0	6	1.48	2.56	3.79
39.1	6	1.48	2.56	3.79

†The 3.33 fl. oz. cider bottle is packed 24 to a crate.

© *Crown Copyright 1981*

The energy required to produce crates is given in Table 17.4. Delivery of crates average 0.1 vehicle miles per crate on 20 tonne payload vehicles so that transport energy can be calculated from Table 3.1. Hence the total energy associated with the packaging of returnable bottles in crates can be calculated to give the values shown in Table 22.4.

Table 22.4
Total energy associated with the provision of crates for 1000 returnable bottles

Container size (fl. oz.) and type	Electricity/MJ		Oil fuels/MJ			Other fuels/MJ			Total energy /MJ
	Fuel production and delivery energy	Energy content of fuel	Fuel production and delivery energy	Energy content of fuel	Feedstock energy	Fuel production and delivery energy	Energy content of fuel	Feedstock energy	
3.7 RB	11.38	3.79	3.29	17.20	21.06	0.01	0.11	—	56.84
6.0 RB	23.61	7.87	6.82	35.60	43.71	0.02	0.22	—	117.85
6.34 RB	24.74	8.25	7.13	37.22	45.82	0.02	0.22	—	123.40
6.5 RB	29.00	9.67	8.28	43.28	53.71	0.02	0.22	—	144.18
9.68 RB	33.25	11.09	9.44	49.34	61.61	0.02	0.22	—	164.97
19.4 RB	66.22	22.08	18.81	98.27	122.69	0.04	0.44	—	328.55
25.0 RB	66.22	22.08	18.81	98.27	122.69	0.04	0.44	—	328.55
34.5 RB	66.22	22.08	18.81	98.27	122.69	0.04	0.44	—	328.55
35.2 RB	66.22	22.08	18.81	98.27	122.69	0.04	0.44	—	328.55
38.0 RB	107.77	35.94	30.90	161.40	199.58	0.08	0.88	—	536.55
39.1 RB	107.77	35.94	30.90	161.40	199.58	0.08	0.88	—	536.55

22.4 MATERIALS FOR PACKAGING NON-RETURNABLE BOTTLES

A variety of packaging methods are available for non-returnable bottles but the principal variations are:

(a) cardboard cartons (both open top and full flap).
(b) cardboard trays with shrinkwrap film, and
(c) multipacks with shrinkwrap film.

Method (a) is used by some fillers for all non-returnable bottles. Method (b) is used for most sizes and method (c) is widely used for 9.68 fl. oz. wide mouth bottles and for some 19.4 fl. oz. bottles.

22.4.1 Non-returnable bottles in cartons

Cartons are non-returnable and typical masses of board required as cartons to deliver 1000 containers are shown in Table 22.5. The energy associated with the manufacture of these quantities of board can be obtained from Table 5.2. It is assumed that the energy required to manufacture the carton is negligible compared with the energy required to make the board itself. Table 22.6 gives the energy requirements for board provision as a function of container size. The energy associated with the delivery of the board is sufficiently small to be neglected.

Table 22.5

Mass of board required as cartons to pack 1000 non-returnable bottles

Container size fl. oz.	Mass of board kg
6.0	8.20
6.34	8.20
8.8	13.00
9.68	13.00
11.6	17.30
15.5	23.60
17.6	23.60
19.4	23.60
25.7	26.00
26.4	26.00
34.2	36.50
35.2	36.50
1.5 litre PET	37.83

© *Crown Copyright 1981*

Table 22.6

Energy requirement for the provision of board to pack 1000 non-returnable containers in cartons

Container size (fl. oz.) and type	Electricity/MJ		Oil fuels/MJ			Other fuels/MJ			Total energy /MJ
	Fuel production and delivery energy	Energy content of fuel	Fuel production and delivery energy	Energy content of fuel	Feedstock energy	Fuel production and delivery energy	Energy content of fuel	Feedstock energy	
6.0 NRB	230.01	76.67	43.05	210.17	—	0.41	136.37	146.70	843.38
6.34 NRB	230.01	76.67	43.05	210.17	—	0.41	136.37	146.70	843.38
8.8 NRB	364.65	121.55	68.25	333.19	—	0.65	216.19	232.57	1337.05
9.68 NRB	364.65	121.55	68.25	333.19	—	0.65	216.19	232.57	1337.05
11.6 NRB	485.27	161.76	90.83	443.40	—	0.87	287.70	309.50	1779.33
15.5 NRB	661.98	220.66	123.90	604.87	—	1.18	392.47	422.20	2427.26
17.6 NRB	661.98	220.66	123.90	604.87	—	1.18	392.47	422.20	2427.26
19.4 NRB	661.98	220.66	123.90	604.87	—	1.18	392.47	422.20	2427.26
25.7 NRB	729.30	243.10	136.50	666.38	—	1.30	432.38	465.14	2674.10
26.4 NRB	729.30	243.10	136.50	666.38	—	1.30	432.38	465.14	2674.10
34.2 NRB	1023.83	341.28	191.63	935.50	—	1.83	607.00	652.99	3754.06
35.2 NRB	1023.83	341.28	191.63	935.50	—	1.83	607.00	652.99	3754.06
1.5 litre NRB (PET)	1061.13	353.71	198.61	969.58	—	1.89	629.11	676.78	3890.81

© *Crown Copyright 1981*

22.4.2 Non-returnable bottles in trays with shrinkwrap

Typical masses of shrinkwrap film and cardboard trays to pack 1000 containers are shown in Table 22.7. Since all container sizes are not packed in this way, some of the data in Table 22.7 have been interpolated from known data so that an approximate comparison can be made with packaging in cartons.

Table 22.7

Masses of cardboard (as trays) and shrinkwrap film required to pack 1000 non-returnable bottles of different sizes

Container size fl. oz.	Mass of board kg	Mass of shrinkwrap kg
6.0	2.91	0.83
6.34	2.91	0.83
8.8	2.91	0.83
9.68	2.91	0.83
11.6	3.49	0.99
15.5	5.83	1.66
17.6	5.83	1.66
19.4	5.83	1.66
25.7	8.73	2.49
26.4	8.73	2.49
34.2	11.66	3.32
35.2	11.66	3.32
1.5 litre PET	12.50	4.00

© *Crown Copyright 1981*

The energy associated with the provision of shrinkwrap and cardboard trays can be calculated from Table 5.2 using the masses given in Table 22.7. Transport energies for the delivery of both commodities are negligible so that the total energy associated with shrinkwrapped trays is as in Table 22.8.

22.4.3 Non-returnable bottles in multipacks

Only two bottle sizes have been considered:

(a) The 9.68 wide mouth bottle for beer for which a multipack containing 4 bottles requires 6.25 kg of board and 1.17 kg of shrinkwrap film, both per 1000 bottles packed.

(b) The 19.4 fl. oz. beer bottle which requires 16.67 kg board and 2.83 kg of shrinkwrap film per 1000 bottles packed.

The energy required to provide these materials are taken from Tables 5.2 and 17.4 and the total energy requirements for this form of packing are given in Table 22.9.

Table 22.8

Total energy required to pack 1000 non-returnable bottles in cardboard trays with shrinkwrap film

Container size fl. oz.	Electricity/MJ		Oil fuels/MJ			Other fuels/MJ			Total energy /MJ
	Fuel production and delivery energy	Energy content of fuel	Fuel production and delivery energy	Energy content of fuel	Feedstock energy	Fuel production and delivery energy	Energy content of fuel	Feedstock energy	
6.0	121.74	40.58	25.30	125.43	42.29	0.15	48.39	52.06	455.94
6.34	121.74	40.58	25.30	125.43	42.29	0.15	48.39	52.06	455.94
8.8	121.74	40.58	25.30	125.43	42.29	0.15	48.39	52.06	455.94
9.68	121.74	40.58	25.30	125.43	42.29	0.15	48.39	52.06	455.94
11.6	145.74	48.58	30.27	150.10	50.44	0.17	58.04	62.44	545.78
15.5	243.76	81.25	50.65	251.11	84.58	0.29	96.95	104.30	912.89
17.6	243.76	81.25	50.65	251.11	84.58	0.29	96.95	104.30	912.89
19.4	243.76	81.25	50.65	251.11	84.58	0.29	96.95	104.30	912.89
25.7	365.22	121.74	75.88	376.29	126.87	0.44	145.18	156.18	1367.80
26.4	365.22	121.74	75.88	376.29	126.87	0.44	145.18	156.18	1367.80
34.2	487.52	162.51	101.29	502.23	169.15	0.58	193.91	208.60	1825.79
35.2	487.52	162.51	101.29	502.23	169.15	0.58	193.91	208.60	1825.79
1.5 litre PET	543.95	181.32	113.91	565.42	203.80	0.63	207.88	223.63	2040.54

Table 22.9

Total energy required to pack 1000 non-returnable bottles in multipacks and shrinkwrap film

Container size fl. oz.	Electricity/MJ		Oil fuels/MJ			Other fuels/MJ			Total energy /MJ
	Fuel production and delivery energy	Energy content of fuel	Fuel production and delivery energy	Energy content of fuel	Feedstock energy	Fuel production and delivery energy	Energy content of fuel	Feedstock energy	
9.68	258.91	86.31	52.44	258.88	65.16	0.36	114.53	122.99	959.58
19.4	676.50	225.52	136.37	672.70	158.99	0.96	305.47	328.04	2504.55

22.5 MATERIALS FOR PACKAGING CANS

The most common methods for packaging full cans are:

(a) The use of Hicone with packing in shallow trays. This method is common for 9.68, 11.6 and 15.5 fl. oz. cans.

(b) Packing in shallow trays with the overall pack shrinkwrapped. This method is widely used for cans of capacity 9.68, 11.6, 15.5, 78, 98 and 136 fl. oz.

(c) Packaging in cartons. This method is widely used for 5.5 fl. oz. cans and for 11.6 fl. oz. cans for export.

(d) Packaging in multipacks with shrinkwrapping. This method is essentially used only for the 9.68 fl. oz. size although some 11.6 fl. oz. cans for carbonated soft drinks are packed in this way.

22.5.1 Cans in Hicone on shallow trays

This method of packaging is used for the three can sizes having the same external diameter. The usual arrangement is for cans to be Hiconed in groups of four prior to packing in shallow cardboard trays with 24 cans to a tray. This pack is sufficiently stable to be palletised. The average masses of materials associated with this method of packaging are given in Table 22.10. The energies associated with the provision of the board and Hicone are given in Table 5.2, hence, using the data of Table 22.10, the total energy associated with this pack is as shown in Table 22.11.

Table 22.10

Materials requirements for packaging 1000 cans with Hicone on trays

Container size fl. oz.	Mass of board in the tray kg
9.68	3.13
11.6	3.13
15.5	3.54

© *Crown Copyright 1981*

22.5.2 Cans in trays with shrinkwrap film

Small can sizes are packed with 24 cans to a tray, 78 and 98 fl. oz. cans with 6 cans to a tray and 136 fl. oz. cans with 4 to a tray. Typical average masses of packaging materials are shown in Table 22.12. The energies associated with the provisions of these materials are given in Table 5.2 so that the total energy of the materials will be as in Table 22.13.

Table 22.11

Total energy associated with the provision of Hicone and cardboard trays for packaging 1000 cans

Container size fl. oz.	Electricity/MJ			Oil fuels/MJ			Other fuels/MJ			Total energy /MJ
	Fuel production and delivery energy	Energy content of fuel	Fuel production and delivery energy	Energy content of fuel	Feedstock energy	Fuel production and delivery energy	Energy content of fuel	Feedstock energy	
9.68	119.03	39.68	23.81	118.49	47.04	0.16	52.05	56.00	456.26
11.6	119.03	39.68	23.81	118.49	47.04	0.16	52.05	56.00	456.26
15.5	130.53	43.51	25.97	129.00	47.04	0.18	58.87	63.33	498.43

Table 22.12

Masses of board (as trays) and shrinkwrap film required
to package 1000 cans

Can size fl. oz.	Mass of board kg	Mass of shrinkwrap film kg
5.5	2.91	0.83[†]
9.68	3.13	1.13
11.6	3.13	1.25
15.5	3.54	1.67
78	11.53	4.73
98	11.53	4.73[†]
136	17.3	7.10

[†]Denotes estimated values.

© *Crown Copyright 1981*

22.5.3 Cans in cartons
Some cans of all sizes are packed in cartons and the masses of board associated
with this form of packaging are given in Table 22.14. The energy associated with

Table 22.14

Mass of board as cartons to pack 1000
cans

Container size fl. oz.	Mass of board kg
5.5	5.42
9.68	12.50
11.6	12.50
15.5	16.70
78	28.33
98	28.33
136	42.50

© *Crown Copyright 1981*

the provision of this quantity of board has been obtained from Table 5.2 and the
total packaging energy is given in Table 22.15.

22.5.4 Cans in multipacks with shrinkwrap
At present the 9.68 fl. oz. can is the main can packed in this way. The materials
requirements per 1000 cans are 9.45 kg board and 0.98 kg of shrinkwrap film.

Table 22.13
Total energy required to pack 1000 cans in trays with shrinkwrap

Container size fl. oz.	Electricity/MJ		Oil fuels/MJ			Other fuels/MJ			Total energy /MJ
	Fuel production and delivery energy	Energy content of fuel	Fuel production and delivery energy	Energy content of fuel	Feedstock energy	Fuel production and delivery energy	Energy content of fuel	Feedstock energy	
5.5	121.74	40.58	25.30	125.43	42.29	0.15	48.39	52.06	455.94
9.68	142.41	47.47	30.07	149.44	57.57	0.16	52.05	56.00	535.17
11.6	148.21	49.41	31.52	156.80	63.69	0.16	52.05	56.00	557.84
15.5	180.01	60.00	38.75	193.03	85.09	0.18	58.87	63.33	679.26
78	552.02	184.01	117.62	585.27	240.99	0.58	191.74	206.27	2078.50
98	552.02	184.01	117.62	585.27	240.99	0.58	191.74	206.27	2078.50
136	828.41	276.14	176.53	878.35	361.75	0.87	287.70	309.50	3199.25

Table 22:15

Energy required to package 1000 cans in cartons

Container size fl. oz.	Electricity/MJ		Oil fuels/MJ			Other fuels/MJ			Total energy /MJ
	Fuel production and delivery energy	Energy content of fuel	Fuel production and delivery energy	Energy content of fuel	Feedstock energy	Fuel production and delivery energy	Energy content of fuel	Feedstock energy	
5.5	152.03	50.68	28.46	138.91	—	0.27	90.13	96.96	557.44
9.68	350.63	116.88	65.63	320.38	—	0.63	207.88	223.63	1285.66
11.6	350.63	116.88	65.63	320.38	—	0.63	207.88	223.63	1285.66
15.5	468.44	156.15	87.68	428.02	—	0.84	277.72	298.76	1717.61
78	794.66	264.89	148.73	726.10	—	1.42	471.13	506.82	2913.75
98	794.66	264.89	148.73	726.10	—	1.42	471.13	506.82	2913.75
136	1192.13	397.40	223.13	1089.28	—	2.13	706.78	760.33	4371.18

The energy for the provision of these materials is taken from Table 5.2 and Table 17.2 and is shown in Table 22.16.

Table 22.16

Total energy required to pack 1000 cans of capacity 9.68 fl. oz. in multipacks and shrinkwrap

Fuel type	Fuel production and delivery energy/MJ	Energy content of fuel /MJ	Feedstock energy /MJ	Total energy /MJ
Electricity	353.33	117.78	nil	471.11
Oil fuels	69.77	343.09	58.32	471.18
Other fuels	0.54	173.17	185.96	359.67
Totals/MJ	423.64	634.04	244.28	1301.96

© *Crown Copyright 1981*

22.6 PALLETS

To obtain an energy requirement for pallet use, the number of bottles carried during the lifetime of the pallet must be calculated. Pallet loadings vary from company to company and average values for each container type and size are given in Table 22.17 together with pallet use per 1000 containers delivered assuming a pallet lifetime of 10 trips.

The energy associated with the provision of pallets is given in Table 5.1. Hence, using the multipliers given in Table 22.17, the energies associated with pallet use can be readily calculated to give Table 22.18.

22.7 LABELS, GLUE AND CLOSURES

The energy associated with label production is given in Table 17.1. Wastage during labelling is approximately 1% so that the energy requirement for label provision is 1.01 × Table 17.1.

Labelling glue is typically consumed at the rate of 0.15 kg per 1000 bottles labelled. The energy associated with the provision of this input can be obtained from Table 17.3.

A variety of closures are used for sealing glass bottles. In general crown closures are used for the smaller bottle sizes (up to 20 fl. oz.) and aluminium closures are used on larger bottles. The use of crown closures is most widespread on beer bottles since containers capped in this way can be pasteurised. In addition, aluminium screw closures are invariably used where a reseal facility is required. The energy associated with the provision of closures of different types is given in Table 16.4.

Table 22.17
Average pallet loadings for beverage containers and pallet requirements

Container size fl. oz.	Container type	Average pallet loading (this is number of containers per pallet)	Number of pallets used in the delivery of 1000 containers
3.7	RB	3024	0.033
6.0	RB	1848	0.054
6.34	RB	1320	0.076
6.5	RB	1320	0.076
9.68	RB	1320	0.076
19.40	RB	924	0.108
25.0	RB	528	0.189
34.5	RB	528	0.189
35.2	RB	528	0.189
38.0	RB	528	0.189
39.1	RB	528	0.189
6.0	NRB	2016	0.050
6.34	NRB	2016	0.050
8.8	NRB	2016	0.050
9.68	NRB	2016	0.050
11.6	NRB	2016	0.050
15.5	NRB	1100	0.091
17.6	NRB	1100	0.091
19.4	NRB	1100	0.091
25.7	NRB	600	0.167
26.4	NRB	600	0.167
34.2	NRB	600	0.167
35.2	NRB	600	0.167
1.5 litre PET	NRB	480	0.209
5.5	C	3744	0.027
9.68	C	2640	0.038
11.6	C	2376	0.042
15.5	C	1944	0.051
78	C	360	0.278
98	C	336	0.298
136	C	222	0.450

Energy required to provide pellets for delivery of 1000 containers from filling plants

Container size fl. oz.	Container type	Electricity/MJ		Oil fuels/MJ			Other fuels/MJ			Total energy /MJ
		Fuel production and delivery energy	Energy content of fuel	Fuel production and delivery energy	Energy content of fuel	Feedstock energy	Fuel production and delivery energy	Energy content of fuel	Feedstock energy	
3.7	RB	10.62	3.54	1.23	6.01	neg	0.06	0.68	14.45	36.59
6.0	RB	17.39	5.79	2.01	9.83	neg	0.09	1.12	23.65	59.88
6.34	RB	24.47	8.16	2.83	13.84	neg	0.13	1.57	33.28	84.28
6.5	RB	24.47	8.16	2.83	13.84	neg	0.13	1.57	33.28	84.28
9.68	RB	24.47	8.16	2.83	13.84	neg	0.13	1.57	33.28	84.28
19.4	RB	34.77	11.59	4.03	19.67	neg	0.18	2.23	47.29	119.76
25.0	RB	60.85	20.28	7.05	34.42	neg	0.32	3.90	82.76	209.58
34.5	RB	60.85	20.28	7.05	34.42	neg	0.32	3.90	82.76	209.58
35.2	RB	60.85	20.28	7.05	34.42	neg	0.32	3.90	82.76	209.58
38.0	RB	60.85	20.28	7.05	34.42	neg	0.32	3.90	82.76	209.58
39.1	RB	60.85	20.28	7.05	34.42	neg	0.32	3.90	82.76	209.58
6.0	NRB	16.10	5.37	1.86	9.10	neg	0.09	1.03	21.90	55.45
6.34	NRB	16.10	5.37	1.86	9.10	neg	0.09	1.03	21.90	55.45
8.8	NRB	16.10	5.37	1.86	9.10	neg	0.09	1.03	21.90	55.45
9.68	NRB	16.10	5.37	1.86	9.10	neg	0.09	1.03	21.90	55.45
11.6	NRB	16.10	5.37	1.86	9.10	neg	0.09	1.03	21.90	55.45
15.5	NRB	29.30	9.77	3.39	16.57	neg	0.15	1.88	39.85	100.91
17.6	NRB	29.30	9.77	3.39	16.57	neg	0.15	1.88	39.85	100.91
19.4	NRB	29.30	9.77	3.39	16.57	neg	0.15	1.88	39.85	100.91
25.7	NRB	53.77	17.92	6.23	30.41	neg	0.28	3.45	73.13	185.19
26.4	NRB	53.77	17.92	6.23	30.41	neg	0.28	3.45	73.13	185.19
34.2	NRB	53.77	17.92	6.23	30.41	neg	0.28	3.45	73.13	185.19
35.2	NRB	53.77	17.92	6.23	30.41	neg	0.28	3.45	73.13	185.19
1.5 litre	NRB (PET)	67.29	22.43	7.79	38.06	neg	0.36	4.32	91.52	231.77
5.5	C	8.69	2.90	1.01	4.92	neg	0.05	0.56	11.82	29.95
9.68	C	12.23	4.08	1.42	6.92	neg	0.06	0.78	16.64	42.13
11.6	C	13.52	4.51	1.57	7.65	neg	0.07	0.87	18.39	46.58
15.5	C	16.42	5.47	1.90	9.29	neg	0.09	1.05	22.33	56.55
78	C	89.50	29.83	10.36	50.62	neg	0.47	5.74	121.74	308.26
98	C	95.94	31.98	11.11	54.26	neg	0.51	6.15	130.50	330.45
136	C	144.88	48.29	16.78	81.94	neg	0.77	9.29	197.06	499.01

CHAPTER 23

Total Energies for Filling and Packing

23.1 TOTAL ENERGIES FOR FILLING AND PACKING

It is clear from Chapter 22, that the number of different forms of outer packaging is very large. This is especially true for non-returnable containers and it is impractical to cover every conceivable variation. Furthermore new modes of packaging are constantly appearing so that those covered in Chapter 22 represent only a sample of the full range. Nevertheless the techniques illustrated, demonstrate how energy requirements can be calculated for the different types of packs.

In this Chapter the total energy required to fill and pack beverage containers is given in summary form. All of the data used is derived from Chapters 18 to 22. The results of these calculations are given in Tables 23.1 to 23.12. Note that these tables show only the total energy requirement as a function of contributing operation but the detailed fuel breakdown can be readily calculated from the data given in Chapters 18 to 22.

Table 23.1

Breakdown of the total filling energy for returnable bottles filled with beer and using crown closures. All values are in MJ/1000 bottles

Container size fl.oz.	6.34	9.68	19.4	39.1
Contribution				
filling and packing	1136.72	1314.45	2061.68	3434.23
CO_2, water, detergent, oil and grease	90.15	90.95	93.31	98.15
crates	123.40	164.97	328.55	536.55
pallets	84.28	84.28	119.76	209.58
labels	53.58	53.58	53.58	53.58
glue	11.58	11.58	11.58	11.58
crowns	221.87	221.87	221.87	221.87
Totals	1721.58	1941.68	2890.33	4565.54

© *Crown Copyright 1981*

Table 23.2

Breakdown of the total filling energy for returnable bottles filled with carbonated soft drinks using aluminium screw closures. All values are in MJ/1000 bottles

Container size/fl. oz.	3.7	6.0 6.5	9.68	25.0	34.5 35.2	38.0
Contribution						
filling and packing	678.02	829.03	851.57	1170.91	1205.24	1237.99
carbon dioxide, water, detergent, oil and grease	89.51	90.03	90.95	94.67	97.19	97.87
crates	56.84	117.85	164.97	328.55	328.55	536.55
pallets	36.59	59.88	84.28	209.58	209.58	209.58
labels	53.58	53.58	53.58	53.58	53.58	53.58
glue	11.58	11.58	11.58	11.58	11.58	11.58
closures	708.00	708.00	708.00	708.00	708.00	708.00
Totals	1634.12	1869.95	1964.93	2576.87	2613.72	2855.15

© *Crown Copyright 1981*

Table 23.3

Breakdown of the total filling energy for returnable bottles capped with crown or aluminium closures and filled with cider. All values are in MJ per 1000 bottles

Container size fl. oz.	3.33	6.0	9.68 (9.5)	39.1 (38.0)
Contribution				
filling and packing	678.02	829.03	1314.45	3434.23
carbon dioxide, water, detergent, oil and grease	89.51	90.03	90.95	98.15
crates	56.84	117.85	164.97	536.55
pallets	36.59	59.88	84.28	209.58
labels	53.58	53.58	53.58	53.58
glue	11.58	11.58	11.58	11.58
crowns	221.87	221.87	221.87	221.87
aluminium screw closures	708.00	708.00	708.00	708.00
Totals with crowns	1147.99	1383.82	1941.68	4565.54
Totals with aluminium screw closures	1634.12	1869.95	2427.81	5051.67

Table 23.4

Breakdown of the total filling energy for beer in non-returnable bottles packed in cartons. All values are in MJ per 1000 bottles

Container size fl.oz.	6.34	8.8	9.68	11.6	15.5	17.6	19.4	25.7	35.2
Contribution									
filling	727.50	912.25	1080.49	1078.29	1392.38	1424.42	1728.34	2164.74	2897.22
CO_2, water, oil and grease	44.87	45.47	45.67	46.15	47.11	47.63	48.03	49.39	51.91
pallets	55.45	55.45	55.45	55.45	100.91	100.91	100.91	185.19	185.19
labels	53.58	53.58	53.58	53.58	53.58	53.58	53.58	53.58	53.58
glue	11.58	11.58	11.58	11.58	11.58	11.58	11.58	11.58	11.58
cartons	843.38	1337.05	1337.05	1779.33	2427.26	2427.26	2427.26	2674.10	3754.06
crown closures	221.87	221.87	221.87	221.87	221.87	221.87	221.87	221.87	221.87
aluminium closures	708.00	708.00	708.00	708.00	708.00	708.00	708.00	708.00	708.00
Totals with crown closures	1958.23	2637.25	2805.69	3246.25	4254.69	4287.25	4591.57	5360.45	7175.41
Totals with aluminium screw closures	2444.36	3123.38	3291.82	3732.38	4740.82	4773.38	5077.70	5846.58	7661.54

Table 23.5

Breakdown of the total filling energy for carbonated soft drinks in non-returnable bottles packed in shrinkwrapped trays. All values in MJ/1000 bottles

Container size fl. oz.	6.34	8.8	9.68	11.6	15.5	17.6	19.4	25.7	35.2
Contribution									
filling	727.50	912.25	1080.49	1078.29	1392.38	1424.42	1728.34	2164.74	2897.22
CO_2, water, oil and grease	44.87	45.47	45.67	46.75	47.11	47.63	48.39	49.03	51.91
pallets	55.45	55.45	55.45	55.45	100.91	100.91	100.91	185.19	185.19
labels	53.58	53.58	53.58	53.58	53.58	53.58	53.58	53.58	53.58
glue	11.58	11.58	11.58	11.58	11.58	11.58	11.58	11.58	11.58
trays and shrinkwrap	455.94	455.94	455.94	545.78	912.89	912.89	912.89	1367.80	1825.79
crown closures	221.87	221.87	221.87	221.87	221.87	221.87	221.87	221.87	221.87
aluminium screw closures	708.00	708.00	708.00	708.00	708.00	708.00	708.00	708.00	708.00
Totals with crown closures	1570.79	1756.14	1924.58	2012.70	2740.32	2772.88	3077.20	4054.15	5247.14
Totals with aluminium screw closures	2056.92	2242.27	2410.71	2498.83	3226.45	3259.01	3563.33	4540.28	5733.27

© *Crown Copyright 1981*

Table 23.6
Breakdown of total filling energy for beer in non-returnable bottles packed in shrinkwrapped multipacks. Values are in MJ/1000 bottles

Container size/fl. oz.	9.68	9.68 wide mouth	19.4
Contribution			
Filling	1080.49	1080.49	1728.34
CO_2, water, oil and grease	45.67	45.67	48.03
Pallets	55.45	55.45	100.91
Labels	53.58	53.58	53.58
Glue	11.58	11.58	11.58
Multipacks and shrinkwrap	959.58	959.58	2504.55
Crown closures	221.87		221.87
Aluminium closures	708.00		708.00
Rip-cap closures		1043.09	
Totals with crown closures	2428.22		4668.86
Totals with aluminium closures	2914.35		5154.99
Total with rip-cap closures		3249.44	

Table 23.7

Breakdown of the total filling energy for carbonated soft drinks in non-returnable bottles packed in cartons. All values are in MJ per 1000 bottles

Container size/fl. oz.	6.0	8.8	17.6	26.4	34.2 35.2	1.5 litre (PET)
Contribution						
filling	386.68	730.64	910.96	963.55	1046.99	1440.83
CO_2, water, oil and grease	44.75	45.47	47.63	49.75	51.91	56.19
pallets	55.45	55.45	100.91	185.19	185.19	231.77
labels	53.58	53.58	53.58	53.58	53.58	53.58
glue	11.58	11.58	11.58	11.58	11.58	11.58
cartons	843.38	1337.05	2427.26	2674.10	3754.06	3890.81
crown closures	221.87	221.87	221.87	221.87	221.87	–
aluminium closures	708.00	708.00	708.00	708.00	708.00	708.00
Totals with crown closures	1617.29	2455.64	3773.79	4159.62	5325.18	–
Totals with aluminium screw closures	2103.42	2941.77	4259.92	4645.75	5811.31	6392.76

Table 23.8

Breakdown of the total filling energy for carbonated soft drinks in non-returnable bottles packed in shrinkwrapped trays. All values in MJ per 1000 bottles

Container size/fl. oz.	6.0	8.8	17.6	26.4	34.2	35.2	1.5 litre (PET)
Contribution							
filling	386.68	730.64	910.96	963.55	1046.99		1440.83
CO_2, water, oil and grease	44.75	45.47	47.63	49.75	51.91		56.19
pallets	55.45	55.45	100.91	185.19	185.19		231.77
labels	53.58	53.58	53.58	53.58	53.58		53.58
glue	11.58	11.58	11.58	11.58	11.58		11.58
trays and shrinkwrap	455.94	455.94	912.89	1367.80	1825.79		2040.54
crowns	221.87	221.87	221.87	221.87	221.87		–
aluminium closures	708.00	708.00	708.00	708.00	708.00		708.00
Totals with crown closures	1229.85	1574.53	2259.42	2853.32	3396.91		–
Totals with aluminium screw closures	1715.98	2060.66	2745.55	3339.45	3883.04		4542.49

© Crown Copyright 1981

Table 23.9

Breakdown of the total filling energy in MJ/1000 bottles for carbonated soft drinks in 8.8 fl. oz. wide mouth bottles using different packaging materials and closures

Contribution	Energy
filling	730.64
CO_2, water, oil and grease	45.47
pallets	55.45
labels	53.58
glue	11.58
cartons	1337.05
shrinkwrapped trays	455.94
aluminium rip-caps	1043.09
large aluminium screw closures	1156.60
Totals with:	
(a) trays and rip-caps	2395.75
(b) trays and large aluminium screws	2509.26
(c) cartons and rip-caps	3276.86
(d) cartons and large aluminium screws	3390.37

© *Crown Copyright 1981*

Table 23.10
Breakdown of the total filling energy for cider in non-returnable bottles. Values in MJ/1000 bottles

System	1	2	3	4
Contribution				
filling	2897.22	2897.22	2897.22	2897.22
carbon dioxide, water, oil				
and grease	51.91	51.91	51.91	51.91
pallets	185.19	185.19	185.19	185.19
labels	53.58	53.58	53.58	53.58
glue	11.58	11.58	11.58	11.58
cartons	3754.06	3754.06	—	—
trays and shrinkwrap	—	—	1825.79	1825.79
crown closures	221.87	—	221.87	—
screw closures	—	708.00	—	708.00
Total cartons and crowns	7175.41	—	—	—
Total cartons and screw closures	—	7661.54	—	—
Total trays and crowns	—	—	5247.14	—
Total trays and screw closures	—	—	—	5733.27

Table 23.11
Breakdown of the total filling and packing energy for beer in cans. MJ/1000 cans

Can size fl./oz.	5.5	9.68	11.6	15.5	78	98	136
Contribution							
filling	–	645.10	687.80	759.85	7428.55	8994.91	12085.68
CO_2, water, oil and grease	–	45.67	46.15	47.11	62.35	67.27	76.55
pallets	–	42.13	46.58	56.55	308.26	330.45	499.01
Hicone and trays	–	456.26	456.26	498.43	–	–	–
trays and shrinkwrap	–	535.17	557.84	679.26	2078.50	2078.50	3119.25
cartons	–	–	–	–	–	–	–
multipacks and shrinkwrap	–	1301.96	1285.66	–	–	–	–
TOTALS							
Hicone and trays	–	1189.16	1236.79	1361.94	–	–	–
Trays and shrinkwrap	–	1268.07	1338.37	1542.77	9877.66	11471.13	15780.49
Cartons	–	–	2066.19	–	–	–	–
Multipacks and shrinkwrap	–	2034.86	–	–	–	–	–

© Crown Copyright 1981

Table 23.12

Breakdown of the total filling energy for carbonated soft drinks in cans. Values in MJ/1000 cans

Can size/fl. oz.	5.5	11.6	15.5
Contribution			
filling	349.81	312.27	342.45
CO_2, water oil and grease	44.67	46.15	47.11
pallets	29.95	46.58	56.55
Hicone and trays	–	456.26	498.43
trays and shrinkwrap	455.94	557.84	679.26
cartons	557.44	1285.66	–
TOTALS			
Hicone and trays	–	861.26	944.54
Trays and shrinkwrap	880.37	962.84	1125.37
Cartons	981.87	1690.66	–

© *Crown Copyright 1981*

CHAPTER 24

Distribution of Full Containers

24.1 INTRODUCTION

Most packaged beer, cider and carbonated soft drinks is sold in the U.K. through supermarkets, off-licences, clubs, and public houses. Large volumes of carbonated soft drinks are sold through non-licensed premises such as confectioners and corner shops. As a broad generalisation, non-returnable containers are sold almost exclusively through off-licences and supermarkets whereas returnable bottles are sold predominantly through public houses, with a small but diminishing trade in off-licences and supermarkets. In the case of soft drinks, the non-licensed retail outlets also handle returnable containers in significant numbers.

Within this non-returnable sector, the trend in recent years has been strongly away from non-returnable bottles and towards cans although this situation may be modified to some extent in the future by the introduction, by the major glass manufacturers, of wide neck bottles with rip-cap closures.

The information presented in this Chapter is based on distribution information from five of the 'big six' brewers and from three major soft drinks fillers, together with a number of small independent manufacturers. No detailed distribution information was available from the cider manufacturers but their practices have been discussed in general terms with them. It is convenient initially to consider the distribution of the three types of beverage separately to highlight differences in practices.

24.2 DISTRIBUTION OF BEER

There is considerable long distance movement (trunking) of containers of beer within the U.K. because most filling plants handle only a restricted range of beers and container types and sizes whereas demand for the full range of container types and beers may exist across the whole country. Furthermore, the location of many filling plants is linked to the siting of the original independent breweries which now form part of larger brewing chains. In addition, economies of scale suggest that it is not sensible to fill a wide range of container types within all plants and in general, any filling plant will typically serve 10 to 15 major depots which are then responsible for retail distribution either directly or through sub-depots.

Many supermarkets sell beer under their own label so that both proprietary and own-label brands are shipped from some filling plants. Own-label beer is generally distributed initially to central supermarket warehouses by brewery transport or contract vehicles. These bulk warehouses act as centres for further distribution and the movement of containers from them is usually by the company's own transport with the beer now forming part of a mixed load with food and other goods.

Supermarkets also sell proprietary brands of beer and the way in which these reach the individual supermarket shops depends upon company marketing policy. Some supermarket chains distribute the proprietary brands in the same way as own-label beers. Others take advantage of the brewery distribution system. All transport is then owned or hired by the brewery and the containers follow the same route as those destined for sale in off-licences, public houses or other outlets. One or two supermarket chains use a mixture of both systems. Whichever of the two networks described above is considered, it can conveniently be divided into two stages; trunking from the filling plant to major brewery depots or supermarket warehouses, followed by local distribution to retail outlets.

24.3 THE TRUNKING OF BEER

Beer is trunked in the U.K. almost exclusively in 20 tonne payload articulated vehicles. Such vehicles carry up to 22 pallet loads of containers but the number of containers constituting a lorry load depends upon the container type and size and how it is packaged. Maximum lorry loadings vary slightly from company to company as shown in Table 24.1.

Data have been analysed for 11 complete distribution networks involving the movement of full containers to a total of 251 depots throughout the U.K. Because of the nature of the trade, the data are most comprehensive for the popular containers, for example the 9.68 fl. oz. returnable bottle and the 9.68 and 15.5 fl. oz. cans.

24.4 TRUNKING ENERGIES FOR BEER DELIVERY

24.4.1 Delivery of full containers

Data from companies is typically as shown in Table 24.2. In this example, six different containers are trunked to a total of 22 depots. The average number of vehicle-miles in trunking a load of full containers may be obtained by summing, for each container, the product of the one-way mileage and the number of loads delivered and then dividing this sum by the total number of loads. These weighted averages for all the data available are given in Tables 24.3 to 24.5 for the different types of container. They refer to the delivery leg of the journey only; the return journey is considered later. Where known, actual lorry loadings have been used in the calculation. In all other instances, the theoretical loadings given in Table 24.1 have been used.

Table 24.1

Number of containers constituting a full lorry load on a 20 tonne payload vehicle

Container size fl. oz.	Container type	Range of maximum loading
6.34	RB	26400-31104
9.68	RB	23040-28224
19.4	RB	14400-18480
39.1	RB	6235-7776
6.34	NRB	40320
9.68	NRB	no data
8.8 (0.25 1)	NRB	34320
11.6 (0.33 1)	NRB	23760
17.6 (0.50 1)	NRB	23320
35.2 (1.00 1)	NRB	9480
9.68	C	46080-60440
15.5	C	30600-38880
78	C	7200-7776
98	C	6720
136	C	4410

Note: This Table contains figures which are the theoretical maximum number of containers based on the weight of the vehicle when loaded. This ignores any practical limitation on the heights of loads arising from factors other than the vehicle.

© *Crown Copyright 1981*

Table 24.2
Typical filling plant distribution data (trunking)

Depot	One-way mileage	Loads of containers to each location on 20 tonne lorries					
		6.34 fl. oz. RB	9.68 fl. oz. RB	19.4 fl. oz. RB	39.1 fl. oz. RB	9.68 fl. oz. C	15.5 fl. oz. C
1	121	12	4	–	–	7	133
2	163	1	63	24	6	1	148
3	125	3	94	30	2	5	346
4	67	2	90	24	2	–	292
5	29	2	135	36	4	1	178
6	40	1	28	24	1	–	138
7	88	–	–	–	–	1	2
8	46	14	36	7	1	3	584
9	95	1	3	2	–	2	26
10	143	2	4	3	–	1	36
11	196	14	23	24	–	10	196
12	235	1	1	1	–	2	25
13	203	7	13	33	–	9	61
14	400	–	8	–	–	2	23
15	286	2	11	–	–	3	30
16	300	2	9	–	–	6	36
17	410	–	–	–	–	5	56
18	410	–	–	–	–	5	77
19	381	1	3	–	–	3	11
20	455	2	30	–	–	5	31
21	405	7	96	–	–	20	140
22	180	–	3	–	–	–	134
Totals		74	654	208	16	91	2703

Table 24.3
Trunking data for returnable bottles by container size (delivery only — no return journey is included)

Container size/fl. oz.	Total number of containers trunked	Number of depots served	Miles run per load	Miles run per 1000 containers
6.34	10,554,624	26	76.4	2.79
6.34	2,124,720	22	174.9	6.09
9.68	22,209,408	26	39.1	1.79
9.68	44,719,104	63	53.7	2.00
9.68	19,292,544	10	65.2	2.83
9.68	62,633,644	10	35.5	1.35
9.68	82,907,712	37	88.9	3.38
9.68	101,729,200	17	67.2	2.87
9.68	26,208,000	9	71.6	3.03
9.68	30,297,600	19	66.5	2.84
9.68	22,924,800	19	84.1	3.59
9.68	13,708,188	22	162.4	7.75
19.40	9,959,904	26	53.5	3.91
19.40	11,732,640	63	129.2	10.32
19.40	3,049,920	10	59.9	4.16
19.40	7,591,392	10	40.2	2.19
19.40	3,079,524	22	114.7	7.74
39.10	705,744	26	27.0	4.56
39.10	97,296	22	101.9	16.80

Table 24.4
Trunking data for non-returnable bottles by container size (delivery only – no return journey is included)

Container size/litre	Total number of containers trunked	Number of depots served	Miles run per load	Miles run per 1000 containers
0.25	3,877,125	37	86.6	2.59
0.25	1,645,800	9	89.3	2.60
0.25	12,660,000	19	106.4	3.06
0.25	2,278,000	19	39.2	1.14
0.33	2,853,600	37	109.9	4.66
0.5	2,772,900	26	110.7	5.00
1.0	1,246,778	26	74.0	8.22

Table 24.5

Trunking data for cans by container size (delivery only — no return journey is included)

Container size/fl. oz.	Total number of containers trunked	Number of depots served	Miles run per load	Miles run per 1000 containers
9.68	1,736,064	26	38.9	0.89
9.68	85,276,800	63	165.7	2.74
9.68	121,155,260	37	106.3	2.01
9.68	74,361,600	17	80.8	1.97
9.68	9,216,000	9	89.7	2.19
9.68	16,070,400	19	74.2	1.95
9.68	4,078,428	22	280.8	6.27
9.68	84,706,560	19	77.9	1.50
15.5	376,560	26	20.5	0.71
15.5	51,318,360	63	172.1	4.54
15.5	59,132,520	37	104.8	2.83
15.5	98,308,104	22	143.2	3.94
78	61,632	26	32.6	4.77
78	2,944,800	17	107.2	18.60
78	478,800	9	90.4	15.67
78	590,400	19	38.0	6.62
98	78,564	37	98.2	14.61
136	16,800	26	63.3	15.11

Calculation of the average energy requirements for trunking requires typical values for the vehicle miles run in the delivery of each type and size of container. These are shown in column 3 of Table 24.6 and have been calculated as the average of the delivery data in Tables 24.3 to 24.5 weighted by the number of containers delivered. Values have been deduced by interpolation for those containers for which data were not available. The average energy requirements for the delivery component of trunking can now be obtained by applying these values as multipliers to Table 3.1 to yield Table 24.7.

Table 24.6
Typical distribution data for trunking containers containing beer

Container size	Container type	Miles run per 1000 containers	Range of values for miles run per 1000 containers	Range of values for trunking energy per 1000 containers/MJ
6.34 fl. oz.	RB	3.34	2.79– 6.09	123.65–269.91
9.68 fl. oz.	RB	2.80	1.35– 7.75	59.83–343.48
19.40 fl. oz.	RB	6.02	2.19–10.32	97.06–457.38
39.10 fl. oz.	RB	6.04	4.56–16.80	202.10–744.58
0.25 litre	NRB	4.78	1.14– 3.06	50.52–135.62
0.33 litre	NRB	4.66	–	206.53
0.5 litre	NRB	5.00	–	221.60
1.0 litre	NRB	8.22	–	364.30
9.68 fl. oz.	C	2.09	0.89– 6.27	39.44–277.89
15.5 fl. oz.	C	3.77	0.71– 4.54	31.47–201.21
78 fl. oz.	C	16.31	4.77–18.60	211.41–824.35
98 fl. oz.	C	14.61	–	647.51
136 fl. oz.	C	15.11	–	669.68

Interpolated values[†]

6.34 fl. oz.	NRB	2.70	–	119.66
9.68 fl. oz.	NRB	3.30	–	146.26
19.4 fl. oz.	NRB	5.20	–	230.46

[†] In this and subsequent Tables, where precise operational data were not available for certain container sizes, values have been derived by interpolating from known data for other container sizes.

Table 24.7

Typical energies required to trunk 1000 full containers of beer to brewery depots or supermarket warehouses

	Container size	Container type	Electricity/MJ		Oil fuels/MJ			Other fuels/MJ			Total energy /MJ
			Fuel production and delivery energy	Energy content of fuel	Fuel production and delivery energy	Energy content of fuel	Feedstock energy	Fuel production and delivery energy	Energy content of fuel	Feedstock energy	
a	6.34 fl. oz.	RB	4.31	1.44	21.28	108.35	0.27	0.94	11.46	nil	148.05
b	9.68 fl. oz.	RB	3.61	1.20	17.84	90.83	0.22	0.78	9.60	nil	124.08
c	19.40 fl. oz.	RB	7.77	2.59	38.35	195.29	0.48	1.69	20.65	nil	266.82
d	39.10 fl. oz.	RB	7.79	2.60	38.47	195.94	0.48	1.69	20.72	nil	267.69
e	0.25 litre	NRB	3.51	1.17	17.33	88.24	0.22	0.76	9.33	nil	120.56
f	0.33 litre	NRB	6.01	2.00	29.68	151.17	0.37	1.30	15.98	nil	206.51
g	0.50 litre	NRB	6.45	2.15	31.85	162.20	0.40	1.40	17.15	nil	221.60
h	1.00 litre	NRB	10.60	3.53	52.36	266.66	0.66	2.30	28.19	nil	364.30
j	9.68 fl. oz.	C	2.70	0.90	13.31	67.80	0.17	0.60	7.17	nil	92.65
k	15.5 fl. oz.	C	4.86	1.62	24.01	122.30	0.30	1.06	12.93	nil	167.08
l	78 fl. oz.	C	21.04	7.01	103.89	529.10	1.30	4.57	55.94	nil	722.85
m	98 fl. oz.	C	18.85	6.28	93.07	473.95	1.17	4.09	50.11	nil	647.52
n	136 fl. oz.	C	19.49	6.50	96.25	490.17	1.21	4.23	51.83	nil	669.68
Interpolated values											
p	6.34 fl. oz.	NRB	3.48	1.16	17.20	87.59	0.22	0.76	9.26	nil	119.67
r	9.68 fl. oz.	NRB	4.26	1.42	21.02	107.05	0.26	0.92	11.32	nil	146.25
s	19.4 fl. oz.	NRB	6.71	2.24	33.12	168.69	0.42	1.46	17.84	nil	230.48

24.4.2 Return journeys

For return journeys, three variations in practices are of particular importance.

(a) Lorries returning to filling plants after delivery of returnable bottles carry mixed loads of empty bottles, pallets, etc. It is reasonable to assume that such lorries carry on average approximately half of the their maximum permitted loading and from [1] a half loaded 20 tonne payload articulated lorry attracts an energy which is typically only 0.86 of that of a fully loaded vehicle.

(b) Lorries returning from delivery of non-returnable containers will usually be empty. The vehicle energy when travelling unloaded is reduced by a factor of 0.7 (See Chapter 3).

(c) When contract hire lorries are used to deliver non-returnable containers, they would usually seek an alternative return load. In such circumstances, no energy will be attributable to the container system for this return load.

24.4.3 Total energies for beer trunking

For each of the cases described above the total energy associated with beer trunking can therefore be derived from Table 24.7 as follows.

(a) Total trunking energy for containers a, b, c or d
    = (entry a, b, c or d) + 0.86 (entry a, b, c or d)
    This data is shown in Table 24.8.

(b) Total trunking energy for containers e to s
    = (relevant entry e–s) + 0.7 (relevant entry e–s)
    This data is given in Table 24.9.

(c) Total trunking energy for containers e to s when no return journey is charged = relevant entry e to s in Table 24.7.

Table 24.8

Typical energies associated with trunking 1000 returnable bottles (delivery and return journeys)

Container size fl. oz.	Container type	Electricity/MJ		Oil fuels/MJ			Other fuels/MJ			Total energy /MJ
		Fuel production and delivery energy	Energy content of fuel	Fuel production and delivery energy	Energy content of fuel	Feedstock energy	Fuel production and delivery energy	Energy content of fuel	Feedstock energy	
6.34	RB	8.02	2.68	39.58	201.54	0.50	1.74	21.31	nil	275.37
9.68	RB	6.72	2.24	33.18	168.95	0.42	1.45	17.86	nil	230.82
19.4	RB	14.45	4.82	71.34	363.25	0.90	3.13	38.41	nil	496.30
39.1	RB	14.49	4.83	71.57	364.46	0.90	3.14	38.54	nil	497.93

© *Crown Copyright 1981*

Table 24.9

Typical total energies associated with trunking 1000 non-returnable beer containers (delivery and return journeys)

Container size	Container type	Electricity/MJ		Oil fuels/MJ			Other fuels/MJ			Total energy /MJ
		Fuel production and delivery energy	Energy content of fuel	Fuel production and delivery energy	Energy content of fuel	Feedstock energy	Fuel production and delivery energy	Energy content of fuel	Feedstock energy	
0.25 litre	NRB	5.96	1.99	29.46	150.01	0.38	1.30	15.86	nil	204.96
0.33 litre	NRB	10.20	3.40	50.46	257.00	0.65	2.23	27.16	nil	351.10
0.50 litre	NRB	10.95	3.65	54.15	275.75	0.70	2.40	29.15	nil	376.75
1.00 litre	NRB	18.00	6.00	89.02	453.34	1.15	3.94	47.92	nil	619.37
9.68 fl. oz.	C	4.58	1.53	22.63	115.26	0.29	1.02	12.19	nil	157.50
15.5 fl. oz.	C	8.25	2.75	40.82	207.92	0.53	1.81	21.98	nil	284.06
78 fl. oz.	C	35.72	11.90	176.63	899.50	2.28	7.83	95.08	nil	1228.94
98 fl. oz.	C	32.00	10.66	158.23	805.74	2.05	7.01	85.17	nil	1100.86
136 fl. oz.	C	33.09	11.03	163.64	833.32	2.12	7.25	88.09	nil	1138.54
Interpolated values										
6.34 fl. oz.	NRB	5.91	1.97	29.24	148.91	0.38	1.30	15.74	nil	203.45
9.68 fl. oz.	NRB	7.23	2.41	35.74	181.99	0.46	1.58	19.24	nil	248.65
19.4 fl. oz.	NRB	11.39	3.80	56.31	286.78	0.73	2.50	30.32	nil	391.83

24.5 RETAIL DISTRIBUTION OF BEER

Retail distribution is more complex than trunking because mixed loads predominate. For example, distribution from brewery depots, involves mixed loads of draught and packaged beer, wines, spirits and soft drinks. Distribution from supermarket warehouses will mix packaged beer with foods, soft drinks and other commodities. Furthermore, unlike trunking, retail distribution occurs in a wide variety of lorry sizes.

24.5.1 Retail distribution via the brewery network

In general, delivery lorries work with mixed loads, dropping off full containers at a number of points and returning to the depot with empty bottles, kegs and casks where appropriate. The lorry loading is therefore constantly changing throughout the journey so that a detailed analysis of the type applied to trunking, in which the outward and return journeys may be separated and load factors taken into account, cannot be used.

Seven brewery retail distribution networks have been analysed and these show that the average round trip delivery mileages vary as shown in Table 24.10. Although the classification of the delivery network into urban or mixed urban/rural can only be crude, there are detectable differences in the round-trip delivery mileage for the two types of network. The average round-trip mileage worked from a depot will naturally reflect the pattern of retail outlets served which itself will be a function of local geography. A smaller network is expected in urban areas compared with rural areas and there are also likely to be differences between the networks taking in city centre areas as compared with largely suburban areas. There are however insufficient data available to detect this latter variation. While noting these variations an average round trip delivery distance of 42 miles has been assumed.

Table 24.10

Round-trip mileages for the retail delivery of beer

Type of delivery area	Average round-trip mileage
urban	28.0
urban	36.5
urban	36.9
urban/rural	54.3
urban/rural	58.0
urban/rural	59.0
urban/rural	60.0

© *Crown Copyright 1981*

A variety of lorry sizes are employed in retail distribution and in the networks examined, lorries with payloads of 1.5, 3, 6, 7 and 13 tonne were used with the 6 and 7 tonne sizes predominating. To simplify the analysis a lorry size of 6 tonne payload has been taken as typical. One further assumption that has been made is that the volume occupied by a part load of a particular container type is the same as it would occupy in a lorry loaded with only that container. This is not strictly true since mixed loads will, in general, not pack together as efficiently as full loads of only one type. However, although this assumption will lead to a slight underestimate of the delivery energy, this can be set against the slight overestimate arising from the fact that once the vehicle starts replacing full containers with empties, its fuel consumption per unit load will increase.

Making these assumptions, the vehicle-miles involved in the retail delivery of beer are as shown in Table 24.11 and the associated energy requirements calculated from Tables 3.1 and 24.11 are as shown in Table 24.12.

Table 24.11

Vehicle-miles per 1000 containers for the retail delivery of beer by a 6 tonne payload lorry with a round-trip delivery distance of 42 miles

Container size	Container type	Miles run per 1000 containers
6.34 fl. oz.	RB	5.5
9.68 fl. oz.	RB	5.7
19.4 fl. oz.	RB	11.0
39.1 fl. oz.	RB	25.0
0.25 litre	NRB	5.0
0.33 litre	NRB	6.6
0.50 litre	NRB	7.2
1.00 litre	NRB	17.3
9.68 fl. oz.	C	2.8
15.5 fl. oz.	C	5.3
78 fl. oz.	C	22.5
98 fl. oz.	C	31.2
136 fl. oz.	C	36.1
Interpolated values		
6.34 fl. oz.	NRB	4.6
9.68 fl. oz.	NRB	5.5
19.4 fl. oz.	NRB	7.6

Table 24.12

Retail delivery energies via the brewery system for 1000 containers as a function of container type and size

Container size	Container type	Electricity/MJ		Oil fuels/MJ			Other fuels/MJ			Total energy /MJ
		Fuel production and delivery energy	Energy content of fuel	Fuel production and delivery energy	Energy content of fuel	Feedstock energy	Fuel production production energy	Energy content of fuel	Feedstock energy	
6.34 fl. oz.	RB	4.79	1.60	20.46	105.05	0.17	0.83	9.74	nil	142.64
9.68 fl. oz.	RB	4.96	1.65	21.20	108.87	0.17	0.86	10.09	nil	147.80
19.4 fl. oz.	RB	9.57	3.19	40.92	210.10	0.33	1.65	19.47	nil	285.23
39.1 fl. oz.	RB	21.75	7.25	93.00	477.50	0.75	3.75	44.25	nil	648.25
0.25 litre	NRB	4.35	1.45	18.60	95.50	0.15	0.75	8.85	nil	129.65
0.33 litre	NRB	5.74	1.91	24.55	126.06	0.20	0.99	11.68	nil	171.13
0.50 litre	NRB	6.26	2.09	26.78	137.52	0.22	1.08	12.74	nil	186.69
1.00 litre	NRB	15.05	5.02	64.36	330.43	0.52	2.60	30.62	nil	448.60
9.68 fl. oz.	C	2.44	0.81	10.42	53.48	0.08	0.42	4.96	nil	72.61
15.5 fl. oz.	C	4.61	1.54	19.72	101.23	0.16	0.80	9.38	nil	137.44
78 fl. oz.	C	19.58	6.53	83.70	429.75	0.68	3.38	39.83	nil	583.45
98 fl. oz.	C	27.14	9.05	116.06	595.92	0.94	4.68	55.22	nil	809.01
136 fl. oz.	C	31.41	10.47	134.29	689.51	1.08	5.42	63.90	nil	936.08
Interpolated values										
6.34 fl. oz.	NRB	4.00	1.33	17.11	87.86	0.14	0.69	8.14	nil	119.27
9.68 fl. oz.	NRB	4.79	1.60	20.46	105.05	0.17	0.83	9.74	nil	142.64
19.4 fl. oz.	NRB	6.61	2.20	28.27	145.16	0.23	1.14	13.45	nil	197.06

24.5.2 Retail distribution from supermarket warehouses

Beer is distributed in mixed loads with other commodities frequently in 20 tonne payload articulated vehicles. The available data indicate that supermarket warehouses typically supply 50 retail outlets. Analysing the movement of commodities to some 207 retail outlets produced an average round trip delivery distance of 98.25 miles, a factor of 2.3 greater than that for retail delivery from brewery depots. It is not surprising that this difference should emerge because of the very different nature of the two distribution systems. Hence the number of vehicle miles required to deliver 1000 containers of each type can be calculated (Table 24.13). Data are given for returnable bottles even though few such containers are currently handled by this system.

To convert these values to delivery energies, it is necessary to make some assumptions concerning the vehicle loading during delivery. Effectively, the

Table 24.13

Miles-run in the retail delivery of 1000 containers
of beer via the supermarket distribution system
with a round-trip delivery distance of 98.25 miles
and using 20 tonne payload vehicles

Container size	Container type	Miles run per 1000 containers delivered
6.34 fl. oz.	RB	3.45
9.68 fl. oz.	RB	3.85
19.4 fl. oz.	RB	6.95
39.1 fl. oz.	RB	15.90
0.25 litre	NRB	2.75
0.33 litre	NRB	4.15
0.50 litre	NRB	4.25
1.00 litre	NRB	10.40
9.68 fl. oz.	C	1.75
15.5 fl. oz.	C	2.70
78 fl. oz.	C	13.8
98 fl. oz.	C	14.6
136 fl. oz.	C	22.6
Interpolated values		
6.34 fl. oz.	NRB	2.30
9.68 fl. oz.	NRB	3.15
19.4 fl. oz.	NRB	5.45

© *Crown Copyright 1981*

vehicles will be full at the start of the delivery and apart from returnable bins or pallets, empty at the end. This differs from retail delivery via the brewery system where lorries pick up empties, casks and kegs. The energies for super-market retail deliveries have been calculated from Table 3.1 and 24.13 assuming that on average the vehicle is half loaded. This reduces the vehicle energy re-quirement by a factor of 0.86 [1]. These energies are shown in Table 24.14.

24.6 TRUNKING OF CARBONATED SOFT DRINKS

The distribution system for carbonated soft drinks from filler to retailer is more complex than that for beer. The carbonated soft drinks industry comprises a large number of small or medium sized fillers serving a restricted geographical area and a much smaller number of large regional and national organisations. This diversity within the industry means that there are almost as many distri-bution networks as there are members of the industry. In practical terms it means that whereas the majority of full containers of beer are trunked prior to retail distribution, in the carbonated soft drinks industry, the proportion of containers trunked is considerably smaller. As with beer, both proprietary and own label brands for supermarkets are produced.

Despite these complications, essentially two different types of delivery system can be identified for carbonated soft drinks: those in which containers are trunked to depots for further retail distribution and hence will possess both trunking and retail distribution energy contributions and those which possess only a local delivery component of energy. Both of these systems are considered here.

As with the long distance movement of beer, carbonated soft drinks are trunked in the U.K. almost exclusively on 20 tonne payload vehicles. The number of containers which constitute a lorry load is a function of container size and type and is shown in Table 24.15.

Data have been analysed for 19 complete distribution networks involving the movement of containers to 215 depots throughout the U.K. Using the same procedures as outlined for the trunking of beer, typical delivery vehicle-miles per 1000 containers can be derived to give the values shown in column 3 of Table 24.16. The energy associated with the delivery leg of trunking may be calculated from Table 3.1 using these values as multipliers and the resulting energy requirements are shown in Table 24.17.

As with beer trunking, there are three possible types of return journey (See Section 24.4.2). The total energies for cases a and b are derived as for beer making and form Tables 24.18 and 24.19. Values for case c where no return journey is charged are equal to the unmodified entries h to t in Table 24.17.

Table 24.14

Retail delivery energies for 1000 containers of beer via the supermarket distribution system

Container size	Container type	Electricity/MJ		Oil fuels/MJ			Other fuels/MJ			Total energy /MJ
		Fuel production and delivery energy	Energy content of fuel	Fuel production and delivery energy	Energy content of fuel	Feedstock energy	Fuel production and delivery energy	Energy content of fuel	Feedstock energy	
6.34 fl. oz.	RB	3.83	1.28	18.91	96.26	0.24	0.83	10.18	nil	131.53
9.68 fl. oz.	RB	4.27	1.42	21.10	107.42	0.27	0.92	11.36	nil	146.76
19.4 fl. oz.	RB	7.71	2.57	38.09	193.91	0.49	1.67	20.50	nil	264.94
39.10 fl. oz.	RB	17.65	5.88	87.13	443.61	1.11	3.82	46.91	nil	606.11
0.25 litre	NRB	3.05	1.02	15.07	76.73	0.19	0.66	8.11	nil	104.83
0.33 litre	NRB	4.61	1.54	22.74	115.79	0.29	1.00	12.24	nil	158.21
0.50 litre	NRB	4.72	1.57	23.29	118.58	0.30	1.02	12.54	nil	162.02
1.00 litre	NRB	11.54	3.85	56.99	290.16	0.73	2.50	30.68	nil	396.45
9.68 fl. oz.	C	1.94	0.65	9.59	48.83	0.12	0.42	5.16	nil	66.71
15.5 fl. oz.	C	3.00	1.00	14.80	75.33	0.19	0.65	7.97	nil	102.94
78 fl. oz.	C	15.32	5.11	75.62	385.02	0.97	3.31	40.71	nil	526.06
98 fl. oz.	C	16.21	5.40	80.01	407.34	1.02	3.50	43.07	nil	556.55
136 fl. oz.	C	25.09	8.36	123.85	630.54	1.58	5.42	66.67	nil	86.151
Interpolated values										
6.34 fl. oz.	NRB	2.55	0.85	12.60	64.17	0.16	0.55	6.79	nil	87.67
9.68 fl. oz.	NRB	3.50	1.17	17.26	87.89	0.22	0.76	9.29	nil	120.09
19.4 fl. oz.	NRB	6.05	2.02	29.87	152.06	0.38	1.31	16.08	nil	207.77

Table 24.15

Number of containers of carbonated soft drinks in a full trunking load

Container size fl. oz.	Container type	Number of containers which constitutes a full trunking load	
		Company 1	Company 2
3.7	RB	60480	–
6.0	RB	36960	–
6.5	RB	26400	–
9.6	RB	29568	–
25.0	RB	13200	10560
34.5	RB	10560	–
38.0	RB	9000	6240
6.0	NRB	–	61440
8.8	NRB	40320	42768
17.5	NRB	20160	–
26.4	NRB	14520	–
34.2	NRB	12000	–
35.2	NRB	10560	–
5.5	C	74880	92160
11.5	C	47520	48000
16.0	C	34000	38400
1.5 litre (PET)	NRB	10560	–

Distribution of Full Containers

Table 24.16

Typical values for miles-run per 1000 containers for trunking carbonated soft drinks

Container size fl. oz.	Container type	Typical values of miles-run per 1000 containers	Range of values of miles run per 1000 containers	Range of trunking energies in MJ per 1000 containers
3.7	RB	1.44	1.13–1.49	50.08– 66.04
6.0	RB	2.15	1.33–3.45	58.95–152.90
6.5	RB	2.22	1.83–3.18	81.11–140.94
9.6	RB	2.41	1.43–3.20	63.38–141.82
25.0	RB	4.63	2.68–9.05	118.78–401.10
34.5	RB	4.30	2.70–6.25	119.66–277.00
38.0	RB	7.58	7.25–7.69	321.32–340.82
6.0	NRB	no data	no data	no data
8.8	NRB	1.74	1.69–1.90	74.90– 84.21
17.6	NRB	3.23	2.43–8.07	107.70–357.66
26.4	NRB	3.18	3.18	140.94
34.2	NRB	5.25	5.25	232.68
35.2	NRB	6.86	4.86–7.39	215.40–327.52
5.5	C	1.77	1.77	78.45
11.5	C	1.89	0.95–2.62	42.10–116.12
15.5	C	no data	no data	no data

© *Crown Copyright 1981*

Table 24.17

Typical energies required to trunk 1000 containers of carbonated soft drinks to depots (delivery leg only)

Container size fl. oz.	Container type	Electricity/MJ		Oil fuels/MJ			Other fuels/MJ			Total energy /MJ
		Fuel production and delivery energy	Energy content of fuel	Fuel production and delivery energy	Energy content of fuel	Feedstock energy	Fuel production and delivery energy	Energy content of fuel	Feedstock energy	
a 3.7	RB	1.86	0.62	9.17	46.71	0.12	0.40	4.94	nil	63.82
b 6.0	RB	2.77	0.92	13.70	69.75	0.17	0.60	7.37	nil	95.28
c 6.5	RB	2.86	0.95	14.14	72.02	0.18	0.62	7.61	nil	98.38
d 9.6	RB	3.11	1.04	15.35	78.18	0.19	0.67	8.27	nil	106.81
e 25.0	RB	5.97	1.99	29.49	150.20	0.37	1.30	15.88	nil	205.20
f 34.5	RB	5.55	1.85	27.39	139.49	0.34	1.20	14.75	nil	190.57
g 38.0	RB	9.78	3.26	48.28	245.90	0.61	2.12	26.00	nil	335.95
h 6.0†	NRB†	1.86†	0.62†	9.17†	46.71†	0.12†	0.40†	4.94†	nil	63.82†
j 8.8 (¼ l)	NRB	2.24	0.75	11.08	56.45	0.14	0.49	5.97	nil	77.12
k 17.6 (½ l)	NRB	4.17	1.39	20.58	104.78	0.26	0.90	11.08	nil	143.16
l 26.4 (¾ l)	NRB	4.10	1.37	20.26	103.16	0.25	0.89	10.91	nil	140.94
m 34.2	NRB	6.77	2.26	33.44	170.31	0.42	1.47	18.01	nil	232.68
n 35.2 (1 l)	NRB	8.85	2.95	43.70	222.54	0.55	1.92	23.53	nil	304.04
p 5.5	C	2.28	0.76	11.27	57.42	0.14	0.50	6.07	nil	78.44
r 11.5	C	2.44	0.81	12.04	61.31	0.15	0.53	6.48	nil	83.76
s 15.5‡	C	4.86‡	1.62‡	24.01‡	122.30‡	0.30‡	1.06‡	12.93‡	nil	167.08‡
t 1.5 litre (PET	NRB	8.85	2.95	43.70	222.54	0.55	1.92	23.53	nil	304.04

† No data but assumed to be the same as 3.7 fl. oz. returnable bottle.
‡ No data but the value for the beer trunking has been used.

© *Crown Copyright 1981*

Table 24.18

Typical energies associated with trunking 1000 returnable bottles of carbonated soft drinks — outward and return journey

Container size fl. oz	Container type	Electricity/MJ		Oil fuels/MJ			Other fuels/MJ			Total energy /MJ
		Fuel production and delivery energy	Energy content of fuel	Fuel production and delivery energy	Energy content of fuel	Feedstock energy	Fuel production and delivery energy	Energy content of fuel	Feedstock energy	
3.7	RB	3.46	1.15	17.06	86.89	0.22	0.75	9.19	nil	118.72
6.0	RB	5.16	1.72	25.48	129.74	0.32	1.12	13.71	nil	177.25
6.5	RB	5.32	1.77	26.31	133.96	0.34	1.15	14.16	nil	183.01
9.6	RB	5.79	1.93	28.56	145.42	0.36	1.25	15.38	nil	198.69
25.0	RB	11.11	3.70	54.86	279.38	0.69	2.41	29.54	nil	381.69
34.5	RB	10.32	3.44	50.95	259.46	0.64	2.23	27.44	nil	354.48
38.0	RB	18.19	6.06	89.82	457.38	1.14	3.94	48.36	nil	624.89

Table 24.19

Typical total energies associated with trunking 1000 non-returnable containers of carbonated soft drinks – outward and return journey

Container size fl. oz.	Container type	Electricity/MJ		Oil fules/MJ			Other fuels/MJ			Total energy /MJ
		Fuel production of delivery energy	Energy content of fuel	Fuel production and delivery energy	Energy content of fuel	Feedstock energy	Fuel production and delivery energy	Energy content of fuel	Feedstock energy	
6.0	NRB	3.16	1.05	15.59	79.41	0.21	0.69	8.40	nil	108.51
8.8 (¼ l)	NRB	3.81	1.27	18.84	95.97	0.24	0.84	10.15	nil	131.12
17.6 (½ l)	NRB	7.08	2.36	34.99	178.13	0.45	1.55	18.83	nil	243.39
26.4 (¾ l)	NRB	6.96	2.32	34.44	175.38	0.44	1.53	18.54	nil	239.61
34.2	NRB	11.50	3.84	56.86	289.54	0.74	2.52	30.61	nil	395.61
35.2 (1 l) 1.5 litre (PET)	NRB NRB	15.02	5.01	74.30	378.33	0.96	3.29	39.99	nil	516.90
5.5	C	3.87	1.29	19.16	97.62	0.25	0.85	10.32	nil	133.36
11.5	C	4.14	1.38	20.47	104.23	0.26	0.91	11.02	nil	142.41
15.5	C	8.25	2.75	40.82	207.92	0.53	1.81	21.98	nil	284.06

24.7 RETAIL DELIVERY OF CARBONATED SOFT DRINKS

24.7.1 Distribution from small fillers and soft drinks depots

In general, lorries work with mixed loads on a circular route, dropping off full containers at a number of points and returning with a full or part-load of empty bottles and cases. The nature of the return load differs from operation to operation and depends upon the proportion of the company's trade in returnable bottles. Analysis of 7 such retail networks shows average round trip delivery mileages between 20 and 100 miles. The data obtained for the retail distribution of carbonated soft drinks were less detailed than those for beer and it is not possible to comment on the differences in the delivery mileages.

The range of round-trip distances observed in practice makes it reasonable to assume the same average mileage (42 miles) as was chosen for beer deliveries. This makes further sense in that some soft drinks are distributed through the brewery network. Lorries in the range 6 to 9 tonne (payload) are the most widely used and again 6 tonne lorries have been chosen as typical. This assumption may understate future operations because a number of distributors are moving to 8 and 10 tonne capacities for off-trade deliveries.

Assuming once again that the space occupied by a part load is proportionally the same as in a bulk load, and following the same procedure as for beer, the average vehicle-miles run in delivering 1000 containers are as shown in Table 24.20. The associated energy requirements are given in Table 24.21.

24.7.2 Distribution from supermarket warehouses

Soft drinks distributed from supermarket warehouses are treated in exactly the same way as for beer, i.e. as mixed loads on 20 tonne articulated vehicles assuming a round trip delivery distance of 98.25 miles. The number of vehicle-miles per 1000 containers delivered is given in Table 24.22 and the associated energy requirement is as shown in Table 24.23.

24.8 DISTRIBUTION OF CIDER

No direct information relating to the distribution of cider was available but from discussion with cider manufacturers, it is apparent that cider distribution most closely parallels that of carbonated soft drinks in that the major manufacturers trunk bottles to depots for subsequent local distribution whereas the small manufacturers operate on a local basis only. The situation is complicated slightly by the fact that at least one major manufacturer combined both trunking and retail distribution stages using 20 tonne vehicles. However, the volume of cider sold, compared with beer and carbonated soft drinks is small. Hence the use of data already computed for carbonated soft drinks is unlikely to introduce any serious errors into the overall system energies.

Table 24.20

Miles-run per 1000 containers for deliveries of carbonated soft drinks by 6 tonne (payload) lorry with a round trip delivery distance of 42 miles

Container size fl. oz	Container type	Miles run per 1000 containers
3.7	RB	2.5
6.0	RB	3.7
6.5	RB	5.7
9.6	RB	5.7
25.0	RB	11.7
34.5	RB	21.8
38.0	RB	21.8
6.0	NRB	2.5
8.8 (¼ litre)	NRB	5.0
17.6 (½ litre)	NRB	7.2
26.4 (¾ litre)	NRB	8.5
34.2	NRB	17.3
35.2 (1 litre)	NRB	17.3
5.5	C	1.9
11.5	C	2.9
15.5	C	5.3
1.5 litre PET	NRB	17.3

© *Crown Copyright 1981*

Table 24.21

Retail delivery energies for 1000 containers of carbonated soft drinks via the soft drinks distribution network

Container size fl. oz.	Container type	Electricity/MJ		Oil fuels/MJ			Other fuels/MJ			Total energy /MJ
		Fuel production and delivery energy	Energy content of fuel	Fuel production and delivery energy	Energy content of fuel	Feedstock energy	Fuel production and delivery energy	Energy content of fuel	Feedstock energy	
3.7	RB	2.18	0.73	9.30	47.75	0.08	0.38	4.43	nil	64.85
6.0	RB	3.22	1.07	13.76	70.67	0.11	0.56	6.55	nil	95.94
6.5	RB	4.96	1.65	21.20	108.87	0.17	0.86	10.09	nil	147.80
9.6	RB	4.96	1.65	21.20	108.87	0.17	0.86	10.09	nil	147.80
25.0	RB	10.18	3.39	43.52	223.47	0.35	1.76	20.71	nil	303.38
35.2	RB	18.97	6.32	81.10	416.38	0.65	3.27	38.59	nil	565.28
38.0	RB	18.97	6.32	81.10	416.38	0.65	3.27	38.59	nil	565.28
6.0	NRB	2.18	0.73	9.30	47.75	0.08	0.38	4.43	nil	64.85
8.8 (¼ l)	NRB	4.35	1.45	18.60	95.50	0.15	0.75	8.85	nil	129.65
17.6 (½ l)	NRB	6.26	2.09	26.78	137.52	0.22	1.08	12.74	nil	186.69
26.4 (¾ l)	NRB	7.40	2.47	31.62	162.35	0.26	1.28	15.05	nil	220.43
34.2	NRB	15.05	5.02	64.36	330.43	0.52	2.60	30.62	nil	448.60
35.2 (1 l) / 1.5 litre PET }	NRB	15.05	5.02	64.36	330.43	0.52	2.60	30.62	nil	448.60
5.5	C	1.65	0.55	7.07	36.29	0.06	0.29	3.36	nil	49.27
11.5	C	2.52	0.84	10.79	55.39	0.09	0.44	5.13	nil	75.20
15.5	C	4.61	1.54	19.72	101.23	0.16	0.08	9.38	nil	137.44

Table 24.22

Vehicle miles run in the retail delivery of full containers of carbonated soft drinks via the supermarket network with a round trip delivery distance of 98.25 miles using 20 tonne payload vehicles

Container size fl. oz.	Container type	Miles-run per 1000 containers
3.7	RB	1.62
6.0	RB	2.65
6.5	RB	3.72
9.6	RB	3.32
25.0	RB	7.42
34.5	RB	9.28
38.0	RB	11.10
6.0	NRB	1.62
8.8 (0.25 litre)	NRB	2.30
17.6 (0.5 litre)	NRB	4.75
26.4 (0.75 litre)	NRB	6.75
34.2	NRB	8.15
35.2 (1 litre)	NRB	9.25
5.5	C	1.06
11.5	C	2.05
15.5	C	2.55
1.5 litre PET	NRB	9.25

Table 24.23

Retail delivery energy for 1000 containers of carbonated soft drinks via the supermarket system

Container size fl. oz.	Container type	Electricity/MJ		Oil fuels/MJ			Other fuels/MJ			Total energy /MJ
		Fuel production and delivery energy	Energy content of fuel	Fuel production and delivery energy	Energy content of fuel	Feedstock energy	Fuel production and delivery energy	Energy content of fuel	Feedstock energy	
3.7	RB	1.80	0.60	8.88	45.20	0.11	0.39	4.78	nil	61.76
6.0	RB	2.94	0.98	14.52	73.94	0.19	0.64	7.82	nil	101.03
6.5	RB	4.13	1.38	20.39	103.79	0.26	0.89	10.97	nil	141.81
9.6	RB	3.69	1.23	18.19	92.63	0.23	0.80	9.79	nil	126.56
25.0	RB	8.24	2.75	40.66	207.02	0.52	1.78	21.89	nil	282.86
34.5	RB	10.30	3.43	50.85	258.91	0.65	2.23	27.38	nil	353.75
38.0	RB	12.32	4.11	60.83	309.69	0.78	2.66	32.75	nil	423.14
6.0	NRB	1.80	0.60	8.88	45.20	0.11	0.39	4.78	nil	61.76
8.8 (¼ l)	NRB	2.55	0.85	12.60	64.17	0.16	0.55	6.79	nil	87.67
17.6 (½ l)	NRB	5.27	1.76	26.03	132.53	0.33	1.14	14.01	nil	181.07
26.4 (¾ l)	NRB	7.49	2.50	36.99	188.33	0.47	1.62	19.91	nil	257.31
34.2	NRB	9.05	3.02	44.62	227.39	0.57	1.96	24.04	nil	310.65
35.2 (1 l) } 1.5 litre PET	**NRB**	10.27	3.42	50.69	258.08	0.65	2.22	27.29	nil	352.62
5.5	C	1.18	0.39	5.81	29.57	0.07	0.25	3.13	nil	40.40
11.5	C	2.28	0.76	11.23	57.20	0.14	0.49	6.05	nil	78.15
15.5	C	2.83	0.94	13.97	71.15	0.18	0.61	7.52	nil	97.20

© *Crown Copyright 1981*

CHAPTER 25

Retail Sale

25.1 INTRODUCTION

The final link between production of packaged beverages and consumer is the retail outlet. This Chapter examines the energy required to operate retail outlets which sell directly to the consumer. Export sales are not considered

Information relating to beverage sales through particular types of outlet is virtually non-existent. However, from data presented earlier and from discussions within the relevant trade, it is possible to paint a broad picture for each beverage type.

As a broad generalisation, returnable bottles are sold through public houses (either as on- or off-sales) and, in much smaller quantities, through specialist off-licences. On the other hand essentially all the beer packaged in non-returnable containers is destined for home consumption and is sold through supermarkets and specialist off-licences. Approximately 25% of the beer packed in returnable bottles is also drunk at home although the main volume is sold across the counter in licensed premises. Thus in total, approximately 55% of all packaged beer is consumed at home. Hence for beer there are three types of retail outlet of interest; public houses, off-licences and supermarkets.

The carbonated soft drinks trade is superficially similar to the beer trade in that some 43% of the total volume produced is packaged in non-returnable containers and 57% is in returnable bottles. However, the total volume of carbonated soft drinks packed in non-returnable containers plus approximately 72% of that packed in returnable bottles is consumed at home. This implies that approximately 72% of total production is consumed at home and only 28% is consumed on the premises. The pattern of sales by container type through different types of retail outlet is however less well defined for carbonated soft drinks than for beers. Most outlets handle most types of containers with some noticeable exceptions. Supermarkets essentially handle non-returnable bottles whereas public houses and door-to-door sales use predominantly returnable bottles.

Information relating to cider sales is less complete than for beer and carbonated soft drinks but the major volume of packaged cider is still sold in returnable 39.1 fl. oz. bottles (flagons) through public houses and specialist off-licences. Small returnable bottles are mostly sold across the bar for consumption on licensed premises and cider in large non-returnable bottles goes for home consumption through specialist off-licences and supermarkets.

This description of beverage sales is intended only as a broad picture but it highlights the fact that direct retail selling occurs through three main types of outlet — public houses, off-licences and supermarkets. Door-to-door sales incur a delivery energy similar to that for retail delivery. Sales through small shops are assumed to incur an energy similar to that arising from sales through supermarkets where beverages form only a small part of the total sales. Such an assumption will probably result in an overestimate of the energy for sales from small shops because supermarkets are usually lit and heated to a higher level than the average small shop.

25.2 RETAIL SALES IN PUBLIC HOUSES

Public houses exist to sell liquor so the energy used to heat and light their sales and storage areas, including any off-licence areas, must be attributable to all commodities sold. In addition to liquor, many public houses also sell confectionary, cigarettes and sundry other items and some also include a restaurant. Such sales must also be taken into account when apportioning energy.

We have examined in detail the operation of 15 public houses. Such a small number cannot be regarded as a satisfactory statistical sample of the many thousands of such establishments in the U.K. but by including houses in city centres, suburbs, estates and villages it is thought by the industry to give a good indication of the spread of operational characteristics. Reliable information has been obtained for the total energy used to heat and light these establishments for a 12 month period, together with data, in terms of both cash and number of units sold of draught beer, bottled beers by container type, canned beer, soft drinks, wines, spirits and sundries. In addition, data were made available for the floor area of the establishment and the proportion of this area used for sales and storage.

Two main problems arise when apportioning the energy used in a public house. The first is the problem of deciding how to split the total energy consumption of the premises between residential accommodation and that devoted to the sale and storage of saleable commodities. The second problem arises from the need to apportion this latter energy between the range of commodities and services provided from the sales area.

In this analysis, bars, cellars, restaurants, and off-licences areas are regarded as sales areas and the rest of the establishment as a private dwelling. The total energy used by public houses has been apportioned between residential and sales areas on the basis of relative floor area.

This procedure may underestimate energy use in the sales area because the fuel used to heat and light these areas per square metre may be greater than in domestic accommodation. This arises because of display lighting in the bars and the greater heat losses from these areas as customers enter and leave the premises. Furthermore, during opening hours, when staff are in the bars, lighting and heating in residential areas may be lower than when these areas are occupied. On the other hand, although restaurants and bars may consume more energy per unit floor area than residential accommodation, cellars and other storage areas are expected to use far less fuel, as will any off-licence areas where heating does not have to be such as to provide a comfortable temperature for sitting.

When the energy used in the sales areas has been estimated, it must be apportioned between sales of all goods before an average energy per container can be calculated. This has been done on the basis of the relative cash turnover. This parameter is unsatisfactory in many respects but alternative methods are equally arbitrary. The energy associated with the sales of beer, cider and carbonated soft drinks has then been divided between the total number of containers sold, to obtain an average retail energy per container. Because the mix of container sizes and types varies from house to house and also, to some extent, from season to season, it is not meaningful to calculate a different energy for each container size and type.

Table 25.1 shows a typical breakdown of total cash turnover between the different types of sales. In houses with a large food trade, the first category will obviously grow at the expense of the others. In all of the houses examined however, packaged beer, cider and carbonated soft drinks accounted for 12 to 16% of the total cash turnover. This percentage, calculated individually for each public house was used to apportion both electricity and thermal fuels consumption. Energies for retail sale in public houses per 1000 containers are given in Table 25.2. This shows that the energy per 1000 containers sold, lies in the range 912.23 to 2188.05 MJ. There is no apparent correlation between energy requirement and the location of the public house in terms of city centre, suburb, etc. The arithmetic mean of the data in Table 25.2 is shown in Table 25.3 and this value has been used later to evaluate system energies.

Table 25.1

Typical breakdown of total cash turnover in public houses

Commodity	% of total
Nuts, crisps, tobacco, food	10
Cordials, wines and spirits	11
Packaged beer, cider and carbonated soft drinks	14
Draught beer	65
Total	100

© *Crown Copyright 1981*

Table 25.2

Energy required to sell 1000 containers of beer, cider or carbonated soft drinks in all types of public house

Public house code	Electricity/MJ		Oil fuels/MJ			Other fuels/MJ			Total energy /MJ
	Fuel production and delivery energy	Energy content of fuel	Fuel production and delivery energy	Energy content of fuel	Feedstock energy	Fuel production and delivery energy	Energy content of fuel	Feedstock energy	
A	732.45	244.15	209.16	1002.29	—	—	—	—	2188.05
B	471.54	157.18	121.69	583.12	—	—	—	—	1333.53
C	665.70	221.90	—	—	—	66.41	807.67	—	1761.68
D	676.08	225.36	—	—	—	42.66	518.76	—	1462.86
E	334.14	111.38	—	—	—	35.46	431.25	—	912.23
F	312.78	104.26	86.31	413.59	—	—	—	—	916.94
G	582.54	194.18	—	—	—	27.57	335.30	—	1139.59
H	710.76	236.92	—	—	—	51.07	621.04	—	1619.79
I	493.56	164.52	—	—	—	31.13	378.53	—	1067.74
J	360.60	120.20	87.09	417.31	—	—	—	—	985.20
K	1313.82	437.94	—	—	—	0.05	0.55	—	1752.36
L	675.42	225.14	—	—	—	8.42	100.17	—	1008.97
M	837.75	279.25	—	—	—	3.81	46.39	—	1167.20
N	1109.82	369.94	—	—	—	—	—	—	1479.76
P	768.54	256.18	—	—	—	—	—	—	1024.72

Table 25.3

Average energy requirement for the retail sale of 1000 containers in a public house

Fuel type	Fuel production and delivery energy/MJ	Energy content of fuel /MJ	Feedstock energy /MJ	Total energy /MJ
Electricity	669.69	223.23	nil	892.92
Oil fuels	33.62	161.09	nil	194.71
Other fuels	17.76	215.98	nil	233.74
Totals/MJ	721.07	600.30	nil	1321.37

© Crown Copyright 1981

No satisfactory distinction can be made between containers of different sizes but the average volume per container sold within the sample of houses was 0.31 litre.

25.3 RETAIL SALE FROM SUPERMARKETS

Supermarkets are stores with floor areas typically in the range 5,000 to 20,000 square feet. Many of these stores have licences to sell alcoholic beverages and general policy is to sell all carbonated beverages in non-returnable containers only. This eliminates the need to handle and store empty bottles which tends to be labour intensive and expensive on space.

Many stores sell alcoholic beverages across the counter from separate off-licence sections but increasingly, where permitted by the licensing authorities, part of the normal self-service display area carries such commodities alongside soft drinks. In both cases the fraction of the total sales area given over to the sale of beverages is typically 100 to 600 square feet.

The energy to heat and light this area of the supermarket is chargeable to all containers sold from the area. Because beer, wines, spirits, cider and carbonated soft drinks all tend to share the same selling space, it is not possible to partition energy in any satisfactory way between container types. Hence an average energy per container has been calculated, applicable to all containers sold from this area of the store.

The calculations have been performed by partitioning the total energy requirement of a supermarket between the beverage and non-beverage sales areas on the basis of floor area. The energy per container sold has then been calculated by dividing the energy of the beverage sales area by the total container throughput from that area. This energy is shown in Table 25.4.

Table 25.4

Energy required to sell 1000 containers of carbonated beverages of all types in a supermarket

Fuel type	Fuel production and delivery energy/MJ	Energy content of fuel /MJ	Feedstock energy /MJ	Total energy /MJ
Electricity	602.73	200.91	nil	803.64
Oil fuels	nil	nil	nil	nil
Other fuels	4.79	58.26	nil	63.05
Totals/MJ	607.52	259.17	nil	866.69

© *Crown Copyright 1981*

Taken over all sales, the average container size sold from supermarkets was 0.49 litre. Considered separately, the average container size for beer was 0.32 litre, for cider 0.83 litre and for carbonated soft drinks, 0.65 litre.

25.4 RETAIL SALES IN OFF-LICENCES

Specialist off-licences are substantially of two types; lock-up shops or shops with associated accommodation. The sample examined included both types. In most instances, companies record fuel consumption in the shop separately from that used in any living accommodation, thus removing the need to carry out any mathematical apportionment of the total fuel use. Within the shop, energy consumption can be apportioned in two ways; in terms of cash turnover or in terms of total shelf area devoted to sales of carbonated beverages of different types. Both methods have been used and the results are shown in Table 25.5 for a number of off-licences.

The two methods of apportionment yield quite different results because carbonated beverages are low value products compared with wines and spirits. Consequently although they may typically occupy some 30 to 40% of the shelf area, they typically only account for 10 to 30% of the cash turnover. In this instance, apportionment on the basis of relative shelf area occupied is thought to be the most appropriate technique and Table 25.6 gives the average value for this type of retailing.

The values derived are very close to those obtained for sales through supermarkets. This is not unexpected since the layout and operation of a specialist off-licence is similar to the drinks area of a supermarket. A further similarity lies in the average volume per container. For an off-licence, it is 0.47 litre whereas for the supermarket it is 0.49 litre. It is also significant that the energy requirements are independent of whether the shop is of the lock-up type or whether there is associated living accommodation.

Table 25.5

Energy required to sell all types of carbonated beverage (1000 containers) in specialist off-licences

Off-licence code	Type of premises	Apportioned as % of cash sales			Apportioned as % shop area heated			Average volume per container sold (litre)
		Electricity/MJ			Electricity/MJ			
		Fuel production and delivery energy	Energy content of fuel	Total energy /MJ	Fuel production and delivery energy	Energy content of fuel	Total energy /MJ	
A	(1)	693.90	231.30	925.20	817.89	272.63	1090.52	0.50
B	(1)	837.42	279.14	1116.56	1495.47	498.49	1993.96	0.56
C	(2)	215.58	71.86	287.44	552.96	184.32	737.28	0.45
D	(1)	920.58	306.86	1227.44	1022.97	340.99	1363.96	0.44
E	(1)	784.50	261.50	1046.00	1352.70	450.90	1803.60	0.56
F	(2)	318.27	106.09	424.36	424.32	141.44	565.76	0.50
G	(1)	158.76	52.92	211.68	364.08	121.36	485.44	0.44
H	(1)	384.15	128.05	512.20	749.40	249.80	999.20	0.51
I	(2)	119.22	39.74	158.96	297.87	99.29	397.16	0.40
J	(2)	99.57	33.19	132.76	224.43	74.81	299.24	0.44
K	(1)	146.88	48.96	195.84	369.57	123.19	492.76	0.40
L	(2)	236.94	78.98	315.92	540.75	180.25	721.00	0.44

Notes: (1) denotes shops with associated accommodation.
(2) denotes lock-up shops.

Table 25.6

Average energy required to sell 1000 containers of carbonated beverage of any type through a specialist off-licence

Fuel type	Fuel production and delivery energy/MJ	Energy content of fuel /MJ	Feedstock energy /MJ	Total energy /MJ
Electricity	684.36	228.12	nil	912.48
Oil fuels	nil	nil	nil	nil
Other fuels	nil	nil	nil	nil
Totals/MJ	684.36	228.12	nil	912.48

© *Crown Copyright 1981*

CHAPTER 26

Consumer Energy

26.1 INTRODUCTION

The beverage systems considered in this report result in the delivery of containers of carbonated beverages to a number of different types of outlet for retail sale to the public. Although each type of outlet operates in a different way, one factor common to all is that the consumer must make a journey either to consume the beverage on the premises, as in the case of the public house, or to purchase drinks for home consumption. It is arguable therefore that some proportion of the energy expended by the consumer in making the journey to and from the point of sale should be charged to the beverage containers purchased.

Much time can be spent fruitlessly arguing whether this consumer energy should or should not be included in the overall system energy, but it is important to separate potential arguments into two groups; those which are based on the definition of the system function and those which are based on the difficulty in arriving at a value for consumer energy.

If the definition of the system function requires the container plus beverage to be supplied to the consumer, then the system must include all operations needed to achieve this objective. It is insufficient to devise a system which delivers the container to a retail outlet but does not complete the distribution to the point of consumption. In this respect, it is important to note that it is logically of no consequence whether the container is taken to the consumer or the consumer is brought to the container.

Systems are only comparable if their functions are identical. If therefore final delivery to the consumer is omitted from systems involving sales through shops and supermarkets as well as public houses, then these systems cannot be compared with systems which do include the final delivery step, for example door-to-door sales.

In the case of returnable bottles sold through public houses for consumption on the premises, the return efficiency of this container system relies heavily upon the fact that the container does not pass out of the control of the retailer.

To achieve this constraint, the customer is required to expend the energy needed to avail himself of the material product of the system. The system therefore requires this energy use before it will function and the energy should therefore be included as part of the total system energy.

One further system property that should be considered is the influence of the delivery method on energy. Delivery energies tend to be lower for bulk distribution; trunking bulk palletised loads in 20 tonne lorries for example is energetically more favourable than retail delivery using part mixed loads over the same distance. We may therefore reasonably assume that the final delivery by vehicle from the retail outlet to the consumer will be energetically unfavourable in comparison with retail distribution to the outlet because of the very low load that is in general carried. In the worst possible case, a consumer may use his car to 'deliver' a single bottle or can to his home. To exclude this final distribution energy because it is expected to be 'high' can hardly be regarded as a serious argument; there is little point in attempting to minimise the production system energy by improving the efficiency of the operations prior to the consumer, if the consumer is to be allowed a free rein on his energy expenditure. The absurdity of such arguments is clear in that if the consumer energy is to be excluded, the total system energy could be significantly reduced by insisting that the consumer should collect his containers direct from the filling plant thus eliminating trunking, retail distribution and retail selling.

We would maintain therefore that the consumer energy should logically be included as part of the overall beverage system energy. At the same time however, it is recognised that there are difficulties inherent in attempting to calculate a value for such an energy contribution. These difficulties arise from variations in the mode of transport used by the consumer, the distance travelled and the nature of the total 'load' carried per journey.

26.2 ESTIMATING CONSUMER ENERGY

The consumer may walk, cycle or use public or private transport and the effect of these different modes of transport on energy requirements is illustrated in Table 26.1. Walking and cycling by definition attract zero energy [1]. The energy associated with other forms of transport has been calculated solely on the basis of the fuel requirement and takes no account of factors such as lubricating oil, tyres, etc.

The important feature of Table 26.1 is the very wide spectrum of energies that arise. The precise mixture of travelling modes is not known although a recent survey [60] suggests that of visitors to two shopping precincts, between 60% and 70% came by car and up to 30% walked with public transport coming a very poor third. No information is available for customers of public houses.

The distance travelled by the consumer is equally uncertain. Examination of the locations of retail outlets in relation to population centres for a number of towns in the U.K. shows that for most urban dwellers there is a public house or off-licence within half a mile. This will not be true for rural dwellers nor can it be assumed that urban dwellers will patronise the nearest retail outlet.

The load carried on a journey is equally uncertain. For visits to supermarkets, beverages will usually form only a small part of the total goods purchased. Equally, for visits to public houses by car, more than one passenger may be carried and the number of containers of beverage consumed will be variable. Table 26.1 shows how the total fuel energy will vary if each vehicle trip is responsible for the carrying or consuming of four containers.

It is clear then that no typical value can be assigned to this component of energy until some extensive research into the behaviour of consumers is carried out. Such a survey is outside the scope of the present work and so a notional value for consumer energy has been calculated based on the assumption that a round-trip of one mile is used to collect or consume four containers and that the average car performance is 30 m.p.g. The corresponding energy per 1000 containers is therefore as given in Table 26.2.

Table 26.1

Total fuel energy required to convey passengers 1 mile in vehicles of different types

Vehicle type	Energy per passenger-mile MJ[†]	Energy per 1000 containers[‡] MJ
Private car carrying 1 passenger with a performance of 25 m.p.g.	7.8	1950
Private car carrying 1 passenger with a performance of 40 m.p.g.	4.9	1225
Private car carrying 2 passengers with a performance of 25 m.p.g.	3.9	975
Private car carrying 2 passengers with a performance of 40 m.p.g.	2.4	600
31 seater bus half full	0.9	225
86 seater double deck bus half full	0.52	130
31 seater bus full	0.47	118
86 seater double deck bus full	0.26	65
Walking/cycling	0	0

Notes: † Bus data from reference [16].
‡ Based on carrying a single 'four pack' from a shop or consuming 4 containers of beverage in a public house.

© *Crown Copyright 1981*

Table 26.2

Fuel energy required per 1000 containers to collect a four-pack of beverages from a retail outlet by private car.

Fuel type	Fuel production and delivery energy/MJ	Energy content of fuel /MJ	Feedstock energy /MJ	Total energy /MJ
Electricity	nil	nil	nil	nil
Oil fuels	259.58	1362.75	nil	1622.33
Other fuels	nil	nil	nil	nil
Totals/MJ	259.58	1362.75	nil	1622.33

CHAPTER 27

Disposal

27.1 SOURCE OF INFORMATION

The mass of refuse handled by collection and disposal authorities is well documented [69] but no equivalent information is available for fuel use in these operations. This information was obtained directly from selected Councils by questionnaire. A sample was chosen to cover the range of areas, populations and population densities exhibited by authorities. In addition, in selecting the sample it was recognised that in England, District Councils are responsible for refuse collection and County Councils are responsible for disposal whereas in Scotland and Wales, District Councils are responsible for both of these functions. For authorities with this dual responsibility, it is often difficult to separate the requirements of the two functions with any accuracy. Separate energy requirements for collection and disposal were therefore obtained by deriving most of the data from English Councils. Additional data were obtained from some Scottish Councils to examine the effect of the combined function on energy requirements.

Outline information was obtained on the sizes of authorities from published data [69] and questionnaires were sent to 110 Councils in England and Scotland. Completed questionnaires were returned by 75% of all District Councils approached and by 41% of County Councils. Some returns were incomplete and some contained errors which could not be satisfactorily resolved so that the results presented here are based on returns from 35 District Councils and 14 County Councils. Information which was incomplete or was noticeably inaccurate in some respects was not used in the numerical calculations but was used as supporting evidence for any conclusions derived from the calculations.

27.2 AVERAGE REFUSE COLLECTION ENERGY REQUIREMENTS

The energy associated with refuse collection arises from three components; that due to direct consumption of fuel by collection vehicles, that used to provide ancilliary services for collection vehicles and that used to provide dustbins and/or paper and plastic sacks for refuse storage by the householder.

27.2.1 Collection vehicle fuel requirements

Aggregated over all collection authorities in the sample, the fuel and materials requirements for the collection of domestic refuse are as shown in Table 27.1, where diesel and petrol refer to the fuels used directly by the collection vehicles. When converted to energy requirements using the data of Table 2.2, these fuel requirements provide the first row of Table 27.2.

Table 27.1
Summary of data for the collection of domestic refuse based on returns from 35 collection authorities in England and Scotland

	Whole sample per year	Per tonne refuse collected
Total population	5,411,106	–
Total refuse collected/tonne	1,449,251	–
Diesel used/gallon	1,114,980	0.7693
Petrol used/gallon	27,010	0.0186
Plastic sacks used/number	35,002,000	24.15
Paper sacks used/number	1,140,600	0.79

27.2.2 Provision of ancillary services

Few data were directly available for the ancillary services provided for the collection vehicles. However, it has been shown [1] that the fuel used directly by general service road vehicles contributes only approximately 67% of the total energy required. The remainder arises from vehicle construction and maintenance. Because of the more arduous conditions under which refuse collection vehicles operate, it might be expected that these additional contributions to the total energy requirement would be greater than for general service vehicles. However, in the absence of specific data for refuse collection, it has been assumed that their performance is similar to that for 20 tonne payload general service vehicles and the energy associated with the ancillary services has been calculated by reference to earlier data [1]. This energy is shown as the second row of Table 27.2. Note that these transport energy requirements refer to all operations up to the point at which the disposal authority takes over, that is landfill site, incinerator, etc.

27.2.3 Provision of refuse storage facilities at the consumer

Most domestic waste is stored in dustbins which are tipped directly into the collection vehicle. Estimating the energy associated with the provision of dustbins is however difficult because of variations in sizes, types and lifetimes.

Table 27.2

Energy associated with the collection and disposal of 1 tonne of domestic refuse in Great Britain based on average aggregated data for the sample of authorities used in this work

Contribution	Electricity/MJ		Oil fuels/MJ			Other fuels/MJ			Total energy /MJ
	Fuel production and delivery energy	Energy content of fuel	Fuel production and delivery energy	Energy content of fuel	Feedstock energy	Fuel production and delivery energy	Energy content of fuel	Feedstock energy	
Fuel for collection vehicles	–	–	26.35	159.31	–	–	–	–	185.66
Ancillary services for collection vehicles	8.48	2.69	8.88	46.09	0.50	3.49	21.25	–	91.38
Provision of:									
(a) dustbins	5.65	2.16	0.79	3.80	–	2.60	16.52	–	31.52
(b) plastic sacks	90.47	30.18	17.37	87.21	52.30	–	–	–	277.53
(c) paper sacks	2.76	0.92	0.57	2.82	–	0.01	2.38	2.56	12.02
Total collection energy	107.36	35.95	53.96	299.23	52.80	6.10	40.15	2.56	598.11
Disposal energy	27.08	9.03	8.30	42.11	–	–	–	–	86.52
Totals/MJ	134.44	44.98	62.26	341.34	52.80	6.10	40.15	2.56	684.63

An estimate can however be made as follows. Suppose that typically a galvanized dustbin weighing 6 kg lasts for 10 years and serves a household of three people. From Table 27.1, the average annual rate of refuse generation per person in the sample is 0.27 tonne. Hence during the estimated lifetime of the dustbin the total mass of refuse held is 8.1 tonne. If the energy required to produce galvanized steel products is 50 mJ/kg [1], the energy per tonne of refuse associated with the use of the dustbin is 37 MJ. A more accurate calculation leads to Table 27.3. From the sample data, however, approximately 14.6% of the total mass of refuse is regularly collected in sacks. Therefore assuming dustbins to be used only in those areas where sacks are not issued regularly, Table 27.3 should be reduced by a factor (1-0.146) = 0.854 to give a reasonable estimate of the contribution made to average collection energies by using dustbins. This reduced value is shown as the entry in Table 27.2.

Table 27.3

Estimated average energy per tonne of refuse associated with the use of a galvanized steel dustbin

Fuel type	Fuel production and delivery energy/MJ	Energy content of fuel /MJ	Total energy /MJ
Electricity	6.62	2.53	9.15
Oil fuels	0.92	4.45	5.37
Other fuels	3.04	19.34	22.38
Totals/MJ	10.58	26.32	36.90

Sacks are increasingly used in refuse collection as a way of improving conditions for the workforce and as a temporary measure in holiday periods when regular collections are disrupted. For the sample as a whole, the average requirement is 24.15 plastic sacks and 0.79 paper sacks per tonne of refuse collected. The precise mass of sacks depends upon design but typical values are 0.085 kg for low density polyethylene sacks and 0.179 kg for paper sacks.

The gross energy required to produce 1 kg of low density polyethylene film from virgin raw materials is 189 MJ/kg (Table 5.2) so that the gross energy associated with the provision of 24.15 polyethylene sacks is 388 MJ, assuming that the energy required to produce the sack from film is negligible. However, a significant proportion of the plastic sacks used in refuse collection are made from recycled polyethylene and the energy associated with the provision of sacks must therefore be reduced to take account of this. A correction can be made by halving the energy associated with the production of the polyethylene resin thus accommodating the two uses that are made of it. This reduces the gross energy required to provide plastic sacks for refuse collection to 277.5 MJ per tonne of refuse collected; the detailed fuel breakdown is shown in Table 27.2.

The gross energy required to produce 1 kg of paper sacks is 85 MJ (Table 5.2). Hence the energy associated with the provision of 0.79 paper sacks per tonne of refuse collected is 12 MJ/tonne. A detailed calculation leads to the fuel breakdown shown in Table 27.2.

The total energy associated with the collection of 1 tonne of domestic refuse is therefore the sum of the above components and is also shown in Table 27.2.

27.3 AVERAGE ENERGY REQUIREMENTS FOR REFUSE DISPOSAL

The aggregated data for refuse disposal from the sample are given in Table 27.4. The corresponding energy requirements are shown in Table 27.2. In general, plant used on disposal sites requires less maintenance than vehicles using public roads and no allowance has been made for maintenance energy in Table 27.2. It is important to note that the average disposal energy shown in Table 27.2 is a composite of the different energies used in landfill operations, incinerators, pulverisers and transfer stations, employing the mix of methods used within the sample.

Table 27.4

Aggregated data for the disposal of domestic refuse based on information supplied by 14 disposal authorities in England and Scotland

	Total per annum	Requirements per tonne of refuse
Total population	10,335,965	—
Total area (acres)	9,540,042	—
Total refuse (tonne)	4,140,004	—
Diesel use/gallon	621,900	0.1502
Petrol use/gallon	3,800	0.0009
Gas oil use/gallon	389,681	0.0941
LPG use/kg	1,452	0.0004
Electricity use/kWh	10,380,305	2.5073

27.4 TOTAL ENERGY TO COLLECT AND DISPOSE OF WASTE

When collection and disposal energy requirements are summed, the total, shown in Table 27.2 represents a typical value for current practices within the U.K. Of the gross energy requirement of 684 MJ per tonne of refuse handled, the most striking feature is that over 42% arises from the use of paper and plastic sacks. Since only 14.6% of the total mass of refuse in the sample was regularly collected in this way, it is clear that if this practice were to become widespread, the associated energy would be considerably increased. If *all* refuse were collected in plastic sacks then, from Table 27.2, the gross energy requirement for collection and disposal would be of the order of 2242 MJ per tonne or between 3 and 4 times the present energy requirement. The equivalent value for the total collection in paper sacks is even higher at 2879 MJ/tonne.

CHAPTER 28

Total Energy Requirements of Beverage Container Systems

28.1 DEFINING THE BEVERAGE CONTAINER SYSTEM

The number of different possible beverage container systems is very large and when calculating their energy and materials requirements, it is important to define the system clearly if misinterpretation of the results is to be avoided. In general systems can be defined by identifying *eight* system properties. These are:

28.1.1 Container specification
This requires definition of

  (a) container type (i.e. glass bottle, metal can, etc.)
  (b) container size (i.e. volume capacity)
  (c) empty container mass in the case of bottles,
  (d) container composition in the case of cans and PET bottles (e.g. three-piece tinplate can, PET bottle with polyethylene base)
  (e) whether the container is returnable or non-returnable.

28.1.2 Beverage type
The use of the container for beer, cider or carbonated soft drinks requires definition because this affects the filling energy.

28.1.3 Empty container delivery
This requires definition of the type of packaging materials used because this affects not only the energy and materials associated with the production of the package itself but also influences lorry loadings and hence transport energy.

28.1.4 Type of closure
For bottles a distinction must be made between crown closures, standard aluminium closures, large aluminium screw closures and rip cap closures.

28.1.5 Outer packaging
The method of packaging filled containers is a significant contributor to energy and materials requirements. Practices vary widely so that the form of outer packaging must be closely identified.

28.1.6 Trunking
The most significant variable affecting the energy associated with trunking (apart from container size) is the nature of the return load. In general, returnable bottle systems use trunking vehicles to return empty bottles and crates whereas non-returnable bottle systems usually involve empty return loads.

28.1.7 Retail distribution
The essential distinction is between brewery distribution systems which involve multiple dropping points whilst collecting returnable items, and supermarket systems in which only one or two dropping points are made with negligible return loads.

28.1.8 Retail sale
It is assumed that supermarket sales require the same energy as sales through off-licences so that the fundamental distinction is between supermarkets and public houses.

28.2 CALCULATION OF ENERGY REQUIREMENTS
Energy requirements for the overall systems have been calculated by summing the contributions from the following nine sub-systems:

    (1) Container manufacture
    (2) Delivery of empty containers – vehicle energy
    (3) Delivery of empty containers – packaging energy
    (4) Filling and packing the container
    (5) Trunking
    (6) Retail distribution
    (7) Retail sale
    (8) Consumer use
    (9) Disposal.

These nine contributions have been divided into three groups:

Group 1
This consists of filling and packing the container, trunking and retail distribution (that is items 4, 5 and 6 above). Each of these components is independent of whether the container is returnable or non-returnable and so is independent of trippage.

Group 2

This group consists of container manufacture, packing and delivery of empty containers and disposal (that is items 1, 2, 3 and 9 above). All of these components depend upon whether the container is returnable or non-returnable and hence are influenced by trippage.

Group 3

This group of contributions comprises retail sale and consumer use. Both should arguably be included in the total system energy but are often excluded. They are separated here to allow their influence to the overall total to be easily identified.

Using the results presented in the earlier chapters of this book the total system energies may be readily calculated by summing the relevant energy requirements for each of these contributions. These overall energy requirements are given for a selection of systems in Appendix 1.

28.3 MATERIALS REQUIREMENTS OF A SYSTEM

The materials requirements of any system are the materials inputs passing across the system boundary. As with energy requirements, the absolute masses comprising these materials flows depend upon system throughput, a dependence which can be eliminated by normalising with respect to the output of the system. For beverage container systems it is usually more convenient to normalise with respect to an internal flow (see section 4.1) and the number of containers delivered to the consumer is usually chosen.

Whilst this procedure may appear simple, its direct application to extended systems such as those described here can lead to lists of materials that are unmanageable and of little direct use. Moreover, if some production systems are traced back to raw materials from the ground, the quantities become so small as to appear negligible (per container delivered).

A further problem is that of identifying the precise nature of some inputs. This can arise from the use of complex mixtures such as lining compounds and solder fluxes as well as from the existence of competing routes to the same finished product as for example the production of polyethylene from crude oil or natural gas. Since the aim here is to consider the comparative efficiency with which materials are used materials requirements have been expressed in terms of the list shown in Table 28.1 — some of which are intermediates.

Note that some synthetic materials are easily referred back to recognisable raw materials; soda ash requirements for example have been converted to sodium chloride and limestone requirements. Hence in terms of the inputs listed in Table 28.1, the major inputs to the beverage system have been calculated as shown in Tables 28.2 and 28.3.

Table 28.1
Materials inputs to beverage systems

Minerals	Chemical intermediates	Stock products	Complex products or products of variable composition
Bauxite (1)	Carbon dioxide*	Epoxy adhesive*	Board (4)
Calcium sulphate	Methanol*	Lead	Cullet (6)
Feldspar (2)	Petroleum coke*	Manganese	Detergent
Fluorspar	Sulphuric acid	Polyethylene*	Flux (solder)*
Iron chromite	Toluene*	PET*	Glue
Iron ore (3)		Polypropylene*	Ink*
Limestone		Selenium	Lacquer*
Sand		Tin	Pallets (5)
Sodium chloride			Paper (4)
			Lining compound*
			Steel scrap (6)
			Varnish*
			Wood (7)

Inputs marked with an asterisk (*) are based on oil products.
Notes: (1) Actual requirements depend upon ore grade; calculations have been based on an assumed 51% alumina content.
(2) Includes nepheline syenite.
(3) Based on the average ore grade available to British Steel.
(4) Not referred back to wood because of the use of recycled pulp in the manufacturing of many low grade packaging qualities.
(5) Expressed in the requirements as a *number* of pallets because of the wide variation in masses used.
(6) Scrap originating outside the glass factory or outside the steel production system.
(7) Used as wooden packaging material; does not include the wood in pallets.

Table 28.2

Materials required to produce 1 kg of saleable glass as containers

Input	Mass/kg
Sand	0.757
Limestone	0.446
Sodium chloride	0.342
Feldspar minerals	0.260
Bauxite	0.028
Foreign cullet	0.007
Calcium sulphate	0.007
Sulphuric acid	0.002
Iron chromite	0.0008
Selenium	0.00001

Table 28.3

Materials required to produce 1 kg of primary aluminium metal

Input	Mass/kg
Bauxite	5.838
Sodium chloride	0.274
Limestone	1.165
Fluorspar	1.715
Sulphuric acid	2.160
Petroleum coke*	0.463

Notes: (1) Based on the requirements of U.K. production given in Section 7.
(2) The bauxite requirement will depend upon ore grade; here the values of Table 7.1 have been used.
(3) Soda ash has been referred to sodium chloride and limestone requirements using [62].
(4) Item marked with an asterisk also appears in the energy table.

The materials requirements for tinplate manufacture are slightly more difficult to calculate. In all of the flow diagrams presented earlier, to accommodate recycling, tinplate production has been separated into a hot metal (pig iron) feed, for which the materials requirements are as shown in Table 28.4 and a tinplate output for which the additional materials requirements are as in Table 28.5. The production routes to tin-free steel and steel strapping do not possess this complication and the materials requirements for their production are given in Tables 28.6 and 28.7 respectively.

Table 28.4

Materials required to produce 1 kg of pig iron

Input	Mass/kg
Iron ore	2.4976
Limestone	0.2237
Manganese	0.0199

© *Crown Copyright 1981*

Table 28.5

Additional additives to produce 1 kg of tinplate from pig iron

Input	Mass/kg
Scrap	0.4388
Limestone	0.2365
Tin	0.0114

© *Crown Copyright 1981*

Table 28.6

Materials required to produce 1 kg of tin-free steel from ore

Input	Mass/kg
Iron ore	2.020
Steel scrap	0.436
Limestone	0.312
Manganese	0.016

© *Crown Copyright 1981*

Table 28.7
Materials required to produce 1 kg of
steel strapping from ore

Input	Mass/kg
Iron ore	2.020
Limestone	0.375
Steel scrap	0.360
Manganese	0.016

© *Crown Copyright 1981*

It is important to note that materials requirements are not independent of the system energy requirements. For materials which may be used alternatively as fuels, (oil, coal, gas or wood), an entry will appear in both the energy table (as feedstock) and the materials table.

28.4 ENERGY AND MATERIALS REQUIREMENTS OF BEVERAGE CONTAINER SYSTEMS

Using the procedures outlined above, the energy and materials requirements of some 46 different beverage container systems have been evaluated in Appendix 1, taking into account some of the more common variations occurring in practice.

CHAPTER 29

Interpretation of Results

29.1 INTRODUCTION

The primary aim of this book is the calculation of the total energy and materials requirements of beverage container systems for the delivery of beer, cider, and carbonated soft drinks in the U.K. The data in the earlier chapters and the summary tables in Appendix 1 satisfy this aim and provide for the first time a detailed, quantitative description of these systems.

The use of this information is however of some concern. Anyone who has worked in this area will be aware of the conflicting arguments, both commercial and ecological, which arise from the same, apparently uncontroversial facts. With the amount of information presented here, it would be impossible to provide a comprehensive discussion of the results in the available space. This chapter therefore considers how the results might be used and attempts to highlight some of the problems that can arise.

In the following sections, data has been drawn from the systems described in Appendix 1 and the system numbers refer to those in the Appendix.

29.2 CRITERIA FOR COMPARISONS

The efficiencies with which container systems use energy or materials can be compared using a number of different criteria. Hitherto, comparisons have almost invariably been made on the basis of gross energy requirements since this is a measure of the total energy resource that must be extracted from the ground in order to sustain the system. However, this is not the only basis of comparison and in some instances may not be the most important. For example, because of the political sensitivity of oil supplies the dependence of a system upon oil fuels may be of greater significance. Using the detailed breakdowns by fuel types given in Appendix 1, container systems may be compared using a number of different criteria and by way of illustration Table 29.1 shows such comparisons for the 9.68 fl. oz. non-returnable glass bottle and the three-piece tinplate can for beer packaging (Systems 17 and 18 in Appendix 1).

If the more efficient system is regarded as the one with the lower energy requirement, then clearly from Table 29.1 the advantage can lie with either container depending upon which energy criterion is chosen. In general, it is unusual to find two comparable container systems in which the advantage lies solely with one container in all respects.

Table 29.1

Comparison of the energy requirements of 9.68 fl. oz. (0.28 litre) non-returnable glass bottles and cans for beer packaging using different criteria. All values are in MJ/1000 containers and all data have been taken from Appendix 1

Comparison parameter	Non-returnable glass bottle	Three-piece tinplate can
Gross energy requirement	9934	8757
Total direct energy (1)	6568	5308
Total electricity (2)	2818	3244
Total oil (excluding feedstock) (2)	5120	3369
Total other fuel (excluding feedstock) (2)	1747	1776
Oil feedstock	96	276
Other feedstock	153	92
Total feedstock	249	368

Notes: (1) Sum of the energy content of all fuels consumed.
       (2) Sum of fuel production, delivery and energy content.

One other feature of Table 29.1 which is of great potential importance is the total feedstock requirement of the system. At present, there is virtually no post consumer waste recycling practised in the U.K. and so the total feedstock energy requirement like the fuel energy requirement, represents a net consumption of energy. However, one form of recycling which may be introduced at some time in the future is the recovery of refuse derived fuel from post-consumer waste. A proportion of the feedstock energy can then be recovered from the system in this form. Because the data in this report calculates the system feedstock energy requirement it may be used to estimate the energy potentially available from such recovery schemes.

In the following sections, most of the illustrative examples have been drawn from a comparison of the gross energy requirement of systems. However, as indicated above other comparison criteria could be used in which case different conclusions could be drawn.

29.3 GRAPHICAL COMPARISON OF SYSTEM ENERGIES

A convenient method of illustrating energy requirements of any system is the use of cumulative energy. Fig. 29.1 for example, shows how the gross energy requirement increases as the number of containers delivered increases, in non-returnable container systems for beer packaging. Note that with increasing container capacity there is a progressive increase in the gross energy requirement for any one container type.

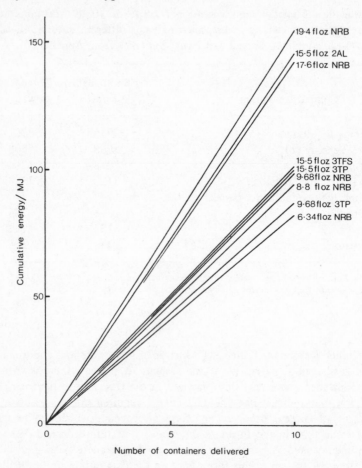

Fig. 29.1 — Cumulative energy requirements to deliver beer in non-returnable containers. (Abbreviations: NRB = non-returnable glass bottle, 3TP = three-piece tinplate can, 3TFS = three-piece tin-free steel can, 2AL = two-piece aluminium can.)

For returnable bottle systems, the graphical representation is complicated by trippage. For example, the gross energy requirement for the 9.68 fl. oz.

returnable glass bottle system for beer (System 14 in Appendix 1) is summarised by the expression $(5265 + 7390/t)$ MJ/1000 containers where t is trippage. Graphically this will appear as shown in Fig. 29.2 where cumulative energy is plotted as a function of the number of trips (that is the number of times the

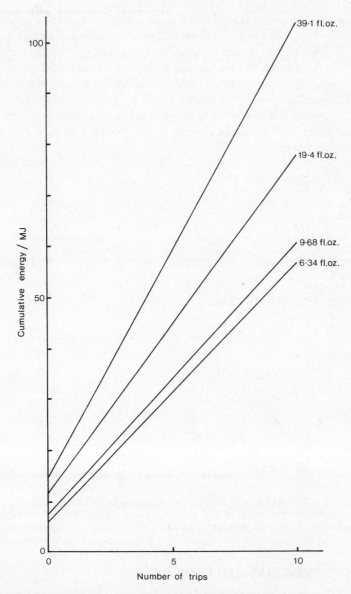

Fig. 29.2 – Cumulative energy requirements for returnable glass bottle systems for beer packaging.

container is delivered to the consumer). The important difference between the returnable and non-returnable systems (Figs. 29.2 and 29.1 respectively) is that the cumulative energy requirement graphs for returnable bottles do not pass through the origin of the graph.

The main advantage of this graphical representation, is that it enables returnable and non-returnable systems to be readily compared. Fig. 29.3 for example shows the situation for three different container systems. Comparing the returnable and non-returnable glass bottles systems, it is clear that the former must achieve an average trippage of at least 1.5 to be the more energy efficient. Similarly, when compared with the metal can system, the returnable glass bottle must achieve a trippage of 2.1. It is therefore a relatively simple task to determine the minimum trippage which a returnable container must achieve in order that the system energy be lower than that of a comparable non-returnable system.

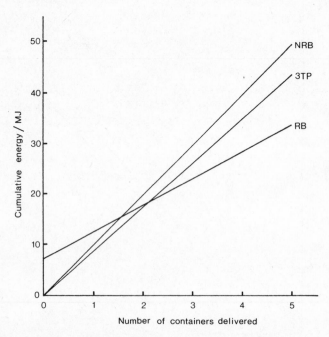

Fig. 29.3 – Comparison of gross energy requirements for 9.68 fl. oz. container systems for beer packaging. (RB = returnable glass bottle, NRB = non-returnable glass bottle, 3TP = three-piece tinplate can).

29.4 VARIATIONS IN SYSTEM ENERGIES

The above discussion indicates the relative simplicity with which system energies may be compared when systems involved are precisely defined. However if the criteria discussed in Chapter 28 are not satisfied it is impossible to specify a single valued system energy and a range of values must be stated.

In addition, it must be recognised that many of the energies quoted in Appendix 1 are average or 'typical' values. In practice, there may be considerable variation about this average, which must also be taken into account when specifying acceptable ranges. The possible sources of such variations are considered below.

29.4.1 Accuracy of data
The calculations presented in this report are subject to three potential sources of error.

(a) Reliability of the information supplied.
Most of the information used was supplied by the industries involved. While every attempt was made to ensure that the information is accurate it was impossible to check the validity of all the data supplied. However, no gross errors have been detected.

(b) Accuracy of data supplied.
Physical quantities of materials and fuels used by a factory are usually determined from deliveries and periodic stock-checking. There is however, a time delay between receipt of raw materials and fuels and their use in the production of finished saleable products. This inevitably leads to a mis-match between input materials and output products. By choosing a sufficiently long time period over which to calculate average requirements, the influence of such differences may be minimised. For most of the organisations examined, data were obtained for a 12 month period.

(c) Accuracy of the calculations.
Throughout the work we have attempted to avoid introducing errors into the data during the calculations by retaining an apparently unnecessary number of significant figures to prevent accumulation of rounding errors.

It is difficult to establish accurately overall variations in the information supplied but it is generally thought by most of the companies who provided data to be approximately 5%.

29.4.2 Variations in the mass of glass bottles
The energies associated with the production of glass bottles are directly proportional to the bottle mass. However, the mass of a bottle of specified capacity often varies significantly. From the available data, the maximum variations in bottle masses are as shown in Table 29.2.

Table 29.2
Variations in masses for glass bottles

Container size and type			Mass used in Appendix 1 /g	Range of masses available/g		Range as % of mass in Section 29	
fl. oz.	litre			min.	max.	min.	max.
6.34	0.18	RB beer	240	200	240	−17%	0
9.68	0.28	RB beer	300	300	380	0	+27%
19.4	0.55	RB beer	470	455	485	−3%	+ 3%
39.1	1.11	RB beer	675	630	790	− 7%	+17%
6.34	0.18	NRB beer	150	150	160	0	+ 7%
8.8	0.25	NRB beer	190	175	215	− 8%	+13%
9.68	0.28	NRB beer	200	200	250	0	+25%
17.6	0.50	NRB beer	330	330	350	0	+ 6%
19.4	0.55	NRB beer	365	365	400	0	+ 9%
35.2	1.00	NRB beer	600	545	740	− 9%	+33%
3.33	0.10	RB cider	165	165	235	0	+42%
6.0	0.17	RB cider	240	200	240	−17%	0
9.5	0.27	RB cider	300	300	375	0	+25%
38.0	1.08	RB cider	675	625	790	− 7%	+17%
35.2	1.00	NRB cider	600	545	740	− 9%	+23%
3.7	0.11	RB CSD	170	155	170	− 9%	0
6.0	0.17	RB CSD	300	275	335	− 8%	+12%
9.68	0.28	RB CSD	355	310	370	−13%	+ 4%
25.0	0.71	RB CSD	580	535	680	− 8%	+17%
35.2	1.00	RB CSD	705	650	740	− 8%	+ 5%
38.0	1.08	RB CSD	850	655	850	−23%	0
6.0	0.17	NRB CSD	170	170	200	0	+18%
8.8	0.25	NRB CSD	210	190	210	−10%	0
17.6	0.50	NRB CSD	350	320	350	− 9%	0
26.4	0.75	NRB CSD	500	500	660	0	+32%
35.2	1.00	NRB CSD	630	630	750	0	+19%

Abbreviations: RB = returnable bottle; NRB = non-returnable bottle;
CSD = carbonated soft drink.

29.4.3 Variations in glass factory energy

The energy required by U.K. glass factories in 1977 varied by ± 19% about the average value (Chapter 10). This variation in bottle production energy is significant and must be taken into account.

29.4.4 Variations in empty bottle delivery energy

The main glass bottle systems described in Appendix 1 involve the delivery of empty bottles in bulk palletised packs. However, although bulk palletising dominates, a variety of other packaging methods may be employed. The variations in vehicle energy with changes in pack type are negligible (Chapter 13) and

Table 29.3

Energy requirements for different types of packs for the delivery of empty glass bottles expressed as a fraction of that used for bulk palletising

Bottle size		Energy of pack/Energy of bulk pack		
fl. oz.	litre	Alternative packs for delivery only	Alternative packs used as outer packaging	Customers' crates
3.7	0.11	1.91	0.42	0.18
5.5	0.16	1.89	0.44	0.19
6.0	0.17	1.88	0.45	0.19
6.34	0.18	1.87	0.45	0.19
6.5	0.19	1.87	0.45	0.19
8.8	0.25	1.85	0.46	0.20
9.68	0.28	1.85	0.47	0.20
11.6	0.33	1.84	0.47	0.20
15.5	0.44	1.83	0.48	0.22
17.6	0.50	1.83	0.48	0.21
19.4	0.55	1.82	0.49	0.21
25.0	0.71	1.82	0.48	0.21
25.7	0.73	1.82	0.48	0.20
26.4	0.75	1.82	0.49	0.21
34.5	0.98	1.83	0.48	0.20
35.2	1.00	1.83	0.48	0.20
38.0	1.08	1.83	0.48	0.20
39.1	1.11	1.84	0.47	0.20
78	2.22	1.83	0.48	0.20

hence it is necessary only to consider changes arising directly from the provision of the different types of packaging materials. The most convenient way of highlighting the relative energy requirements of the different packaging modes is shown in Table 29.3. The use of alternative packaging methods in which the packaging materials for empty bottle delivery are subsequently used as the outer packaging for the filled containers, can lead to more than a 50% reduction in this component of energy. Similarly, the use of customer crates to deliver empty returnable glass bottles can produce a reduction of 80% in this contribution.

29.4.5 Variations in outer packaging for full glass bottles
There is currently no serious competitor to the use of crates for returnable glass bottles. However for non-returnable glass bottles, three forms of outer packaging are in use; shrinkwrapped trays, multipacks and cartons each with a different energy requirement. These variations must therefore be considered where used.

29.4.6 Variations in metal quality
A significant proportion of the total energy for can systems arises from the production of the primary metal. One problem experienced by all can and closure manufacturers is the quality of the coil received and up to 10% of any batch may be rejected. The overall system energy will consequently vary with the proportion rejected. An average rejection rate of 5% has been used in the calculations but if actual variations are considered then a 5% variation about this mean value is expected.

29.4.7 Variations in can making energy
Based on the operations examined in this report the maximum variation in the energy requirements of factories producing can bodies and ends is ± 22%.

29.4.8 Variations in outer packaging for filled cans
The energies associated with the different forms of outer packaging used with metal cans are shown in Table 29.4. The most commonly used pack for small beer cans is Hicone on shrinkwrapped trays whereas for the larger sizes, the commonest pack is shrinkwrapped trays. For carbonated soft drinks however, Hicone is seldom used.

29.4.9 Variations in PET bottle production
When the variations in bottle mass and base type are considered, there is an 8% variation in the bottle making energy (Chapter 12).

Table 29.4

Total energies in MJ/can associated with the provision of different forms of outer packaging for filled metal cans

| Can size | | Energy associated with outer packaging (MJ/can) | | | |
fl. oz.	litre	Hicone and trays	Shrinkwrapped trays	Cartons	Multipacks
5.5	0.16	–	0.46	0.56	–
9.68	0.18	0.46	0.54	1.29	1.30
11.6	0.33	0.46	0.56	1.29	–
15.5	0.44	0.50	0.68	1.72	–
78	2.22	–	2.08	2.91	–
98	2.78	–	2.08	2.91	–
136	3.86	–	3.12	4.37	–

© *Crown Copyright 1981*

29.4.10 Variations in outer packaging for filled PET bottles
The main system has been evaluated using cartons as the outer packaging.However using shrinkwrapped trays there is a potential energy reduction of the order of 1.85 MJ/bottle delivered.

29.4.11 Overall effects of variations
The combined effect of the variations discussed above has been calculated for each of the systems studied and this data is presented in Appendix 1. Note that these ranges *must* be used when systems are specified only by container type and beverage filled.

29.4.12 Implications of variations in system energies
When an allowance is made for system energy variations in the comparison of different container systems the results are frequently quite different from those obtained using precisely defined average systems. To illustrate this, consider once again the 9.68 fl. oz. returnable glass bottle and the three-piece tinplate can systems for the delivery of beer. The cumulative energy associated with these systems is shown in Fig. 29.4. Two lines now exist for each container representing the minimum and maximum values of the system energy. As can be seen there are four cross-over points but the two of special significance are those labelled A and B. Point A is the intersection of the lines for the least energy intensive bottle system and the most energy intensive can system. Point B refers to the reverse situation. In practice the energy associated with the

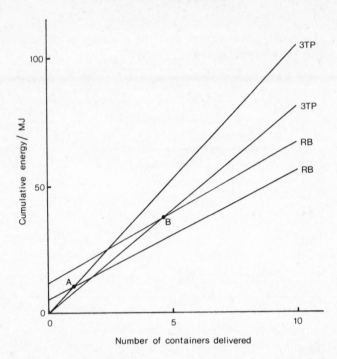

Fig. 29.4 – Comparison of 9.68 fl. oz. systems for beer packaging employing the returnable glass bottle (RB) and the three-piece tinplate can (3TP) making an allowance for system variations.

production and use of these containers will lie between these two extremes. Therefore the trippage which must be achieved by the returnable bottle to ensure that it is more energy efficient than the can system will lie between 1 and 4.6. To ensure that the returnable bottle system is *always* the more energy efficient, it must achieve a minimum trippage of 4.6. Compare this with the value of 2.1 calculated from the closely specified systems of Fig. 29.3. Similar comparisons may be carried out for the other container types using the various comparison criteria outlined earlier.

29.5 THE EFFECT OF CONTAINER VOLUME

For systems employing the same type of container, the system energy requirement increases with container volume. This is particularly well illustrated in Fig. 29.5 for non-returnable glass bottles used to package carbonated soft drinks but other systems exhibit a similar property. Consequently when the energy requirements of two container systems employing containers of different capacity are compared, the container of greater capacity will exhibit the higher

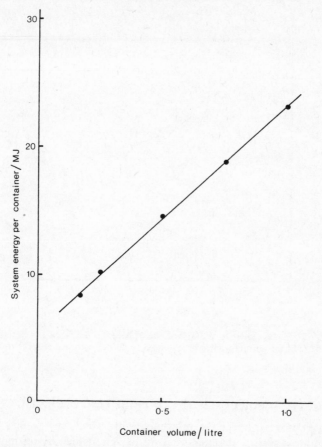

Fig. 29.5 – System energy requirement as a function of container size for non-returnable glass bottles used in the delivery of carbonated soft drinks.

system energy per container. In these circumstances it is important to recognise that the property being compared is not only dependent upon the container type but also upon the container volume.

An alternative way of considering the effect of volume is to calculate the energy required to deliver unit volume of beverage. This is shown in Fig. 29.6 for the data of Fig. 29.5 and illustrates the general conclusion for all container systems, that the greater the size of the container employed, the lower is the energy required to deliver unit volume. Hence it follows that when containers of different sizes are compared, although the smaller container size will possess the lower energy requirement *per container* delivered, it will also exhibit the greater energy *per unit volume* delivered. Great care is therefore needed when drawing conclusions from comparisons of different sized containers.

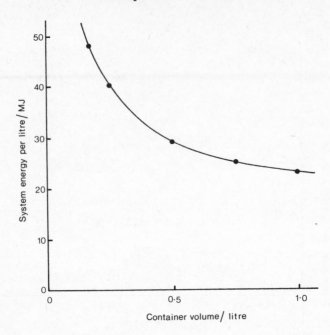

Fig. 29.6 — System energy per unit volume of beverage delivered as a function of container size for non-returnable bottles used to deliver carbonated soft drinks.

29.6 RELATIVE IMPORTANCE OF CONTRIBUTING OPERATIONS

In seeking improvements in the efficiency with which energy and materials are used within a system, it is important to be able to identify the most significant contributors to the total energy or materials requirement.

To illustrate this point, the detailed breakdown of energy requirement by operation given in Appendix 1 has been used to construct Fig. 29.7 which shows the influence of different groups of operations in the non-returnable glass bottle systems for beer delivery.

In this instance it is clear that the production and delivery of the empty container dominates the total system energy requirement for all container sizes, a factor which has led glass manufacturers to lightweight non-returnable bottles. One further interesting feature of Fig. 29.7 is the energy associated with retail sale and consumer use. Although the values used for these contributions in the Tables in Appendix 1 are only approximate, Fig. 29.7 shows that at small container sizes, this contribution may be more significant than filling and outer packaging.

For returnable containers, this type of argument is complicated by the influence of trippage. To illustrate this, Fig. 29.8 shows the relative importance

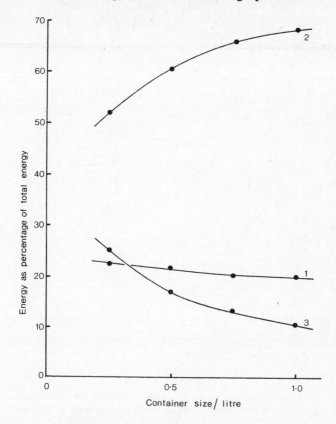

Fig. 29.7 – Relative significance of different groups of operations in the energy required to deliver carbonated soft drinks in non-returnable glass bottles; 1: filling, trunking and retail delivery. 2: container manufacture, packing, delivery and final disposal. 3: retail sale and consumer use.

of the different contributions to the total system energy as a function of trip-page for the 9.68 fl. oz. returnable beer bottle system. The most striking feature of Fig. 29.8 is the rapid decrease of the container production and delivery energy with increasing trippage. This underlines the very great importance that should be attached to achieving high trippage rates with returnable bottles. It should also be noted that by the time the trippage has reached about 4, the energy associated with container manufacture and delivery is less than the energy associated with either of the other groups of operation. It is also clear that lightweighting of returnable glass bottles is a less important feature than achieving high trippages.

Once again Fig. 29.8 illustrates the dominant nature of the retail and consumer energy. It is always higher than the filling energy and, at relatively

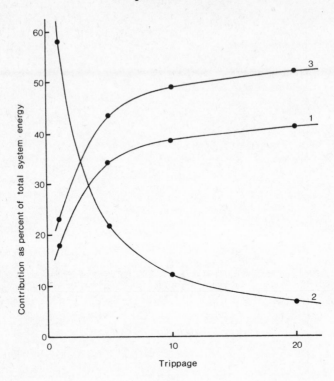

Fig. 29.8 – Relative significance of different groups of operations in the energy required to deliver beer in returnable glass bottles of capacity 9.68 fl. oz. (0.28 litre) as a function of trippage. 1: filling, packing, trunking and retail delivery. 2: container manufacture, packing, delivery and disposal. 3: consumer use and retail sale.

low trippages becomes more important than even the container production energy. This is a common feature of most beverage container systems and its closer examination is clearly an urgent necessity.

29.7 OVERALL COMPARISONS

Some idea of the relative energy requirements of all of the different systems for which data are given in Appendix 1 can be gained by plotting energy requirement as a function of container size. This is shown in Fig. 29.9 for the gross energy requirement per container delivered. Note that similar plots could be drawn for individual fuel types and also for system energy per unit volume of beverage delivered. Note also that possible variations have been omitted for clarity, but if systems are not precisely defined then these variations must be included.

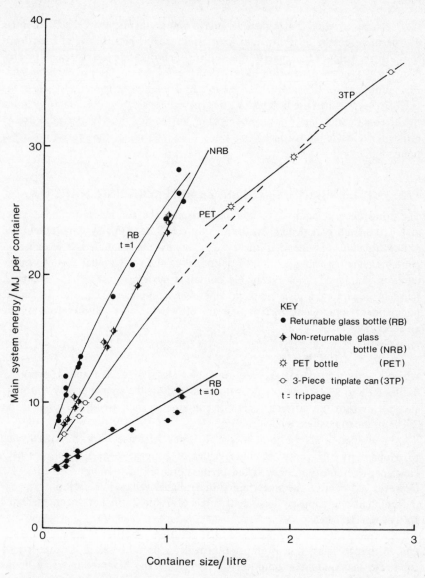

Fig. 29.9 – Gross system energy requirement per container as a function of container size for a number of the main systems listed in Appendix 1.

Diagrams such as Fig. 29.9 are useful in providing, at a glance, an overview of the energy requirements of the whole range of beverage containers and are particularly useful in predictive work. For example, if the line for PET bottles is extrapolated to lower sizes, it intersects the non-returnable glass bottle line at approximately 0.8 litre. This suggests that if no significant changes occur in the two systems to alter the respective energy requirements, then the PET bottle will be more energy efficient than the non-returnable glass bottle at sizes down to 0.8 litre. Below this capacity non-returnable bottles are the more energy efficient. This type of conclusion has particular relevance at the present time since PET containers of 0.5 and 0.75litre capacity are undergoing market trials in the U.K. However, when predictions of this type are considered, it is vital to examine the effects of such factors as lightweighting bottles and introducing more energy efficient forms of outer packaging since these can affect the precise crossover point.

29.8 TOTAL ENERGY USED IN BEVERAGE PACKAGING IN THE U.K.

The total energy required to manufacture and use containers for beer, cider and carbonated soft drinks can be readily calculated. For systems employing non-returnable containers, if the system energy per container is E and if N containers are used in a 12 month period, then the total annual energy requirement is simply NE. For returnable container systems however, the energy E must be separated into a trippage dependent component, E_d, and a trippage independent component, E_i. The delivery of N containers will therefore require an energy

$$N(E_i + E_d/t)$$

where t is trippage. Using the data of Tables 18.2, 19.2, 20.1 and Appendix 1, Tables 29.5 to 29.10 have been constructed to show the annual energy consumption for each of the different types of container system used in 1976/77. Table 29.11 summarises these data.

To evaluate a precise total energy, it is of course necessary to assign a value to trippage. If sufficiently accurate values of trippage were available for each container size and type, they could be incorporated into Tables 29.8 to 29.10. However, in view of the uncertainty in available values [63, 64] the trippage independent and trippage dependent totals of Table 29.11 have been evaluated for two special cases.

In case 1, it has been assumed that returnable bottles make only a single trip. Although false, this assumption provides a useful baseline because it indicates the maximum consumption of energy by the beverage industry in the worst possible case. The energy requirement corresponding to this condition is shown in Table 29.12.

Table 29.5

Calculation of the total energy arising from trippage independent operations in the delivery of beer in 1977 in the U.K. All values are in units of millions of MJ

Container size and type fl. oz. (litre)	Electricity		Oil fuels			Other fuels			Total energy /MJ
	Fuel production and delivery energy	Energy content of fuel	Fuel production and delivery energy	Energy content of fuel	Feedstock energy	Fuel production and delivery energy	Energy content of fuel	Feedstock energy	
6.34 (0.18) RB	220	74	84	427	11	9	90	8	923
9.68 (0.28) RB	3567	1191	1404	7131	222	146	1582	126	15369
19.4 (0.55) RB	265	89	102	516	24	10	114	10	1130
39.1 (1.11) RB	16	5	7	34	2	1	8	1	74
6.34 (0.18) NRB	35	12	14	70	2	2	24	3	162
8.8 (0.25) NRB	22	7	9	44	1	2	16	2	103
9.68 (0.28) NRB	25	8	10	50	1	2	19	2	117
17.6 (0.5) NRB	25	9	10	51	1	2	22	2	122
35.2 (1.00) NRB	25	8	10	50	1	2	24	3	123
9.68 (0.28) Can	1437	559	346	1750	167	148	1089	56	5552
11.6 (0.33) Can	170	66	40	205	21	18	132	7	659
15.5 (0.44) Can	1636	630	405	2037	188	201	1447	71	6615
78 (2.22) Can	81	29	25	122	6	18	126	6	413
98 (2.78) Can	46	16	13	62	3	10	68	3	221
136 (3.86) Can	10	4	2	12	1	2	14	1	46
Totals	7580	2707	2481	12561	651	573	4775	301	31629

Table 29.6

Calculation of the total energy arising from trippage independent operations in the delivery of carbonated soft drinks in 1977 in the U.K. All values in units of millions of MJ

Container type and size fl. oz. (litre)		Electricity		Oil fuels			Other fuels			Total energy /MJ
		Fuel production and delivery energy	Energy content of fuel	Fuel production and delivery energy	Energy content of fuel	Feedstock energy	Fuel production and delivery energy	Energy content of fuel	Feedstock energy	
3.7	(0.11) RB	878	294	347	1774	30	30	307	21	3681
6.0	(0.17) RB	903	302	382	1944	50	32	328	28	3969
9.68	(0.28) RB	38	13	29	84	3	1	14	2	184
25.0	(0.71) RB	768	272	299	1518	117	18	224	48	3264
38.0	(1.08) RB	637	225	268	1363	123	16	189	38	2859
6.0	(0.17) NRB	220	73	96	479	8	21	210	8	1115
8.8	(0.25) NRB	1049	362	394	1971	79	67	697	70	4689
9.68	(0.28) NRB	50	17	20	102	2	4	38	4	237
17.6	(0.5) NRB	362	125	142	708	27	27	289	31	1711
26.4	(0.75) NRB	105	36	41	205	8	9	92	10	506
35.2	(1.00) NRB	560	190	227	1125	43	50	520	59	2774
5.5	(0.16) Can	92	35	22	114	10	8	61	4	346
11.6	(0.33) Can	3365	1308	798	4043	423	364	2607	133	13041
15.5	(0.44) Can	38	15	9	47	5	5	33	2	154
Totals		9065	3267	3074	15477	928	652	5609	458	38530

© *Crown Copyright 1981*

Table 29.7

Calculation of the total energy arising from trippage independent operations in the delivery of cider in 1977 in the U.K. All values are in units of millions of MJ

Container type and size fl. oz. (litre)		Electricity		Oil fuels			Other fuels			Total energy /MJ
		Fuel production and delivery energy	Energy content of fuel	Fuel production and delivery energy	Energy content of fuel	Feedstock energy	Fuel production and delivery energy	Energy content of fuel	Feedstock energy	
3.3 (0.1)	RB	194	65	77	392	7	7	68	5	815
9.5 (0.27)	RB	27	9	10	53	2	1	12	1	115
38.0 (1.08)	RB	106	37	42	211	15	4	49	5	469
35.2 (1.0)	NRB	261	88	101	502	19	22	235	25	1253
9.68 (0.28)	Can	33	13	8	40	4	4	26	1	129
Totals		621	212	238	1198	47	38	390	37	2781

© Crown Copyright 1981

Table 29.8

Calculation of the total energy arising from trippage dependent operations in the delivery of beer in 1977 in the U.K. All values are in units of millions of MJ

Container size and type fl. oz. (litre)		Electricity		Oil fuels			Other fuels			Total energy /MJ
		Fuel production and delivery energy	Energy content of fuel	Fuel production and delivery energy	Energy content of fuel	Feedstock energy	Fuel production and delivery energy	Energy content of fuel	Feedstock energy	
6.34 (0.18)	RB	187	62	82	408	4	27	265	10	1045
9.68 (0.28)	RB	3794	1255	1681	8246	88	531	5347	204	21146
19.4 (0.55)	RB	347	115	155	758	9	50	491	19	1944
39.1 (1.11)	RB	26	9	11	56	1	4	36	2	145
Totals		4354	1441	1929	9468	102	612	6139	235	24280

© Crown Copyright 1981

Table 29.9

Calculation of the total energy arising from trippage dependent operations in the delivery of carbonated soft drinks in 1977 in the U.K. All values are in units of millions of MJ

Container type and size fl. oz. (litre)	Electricity		Oil fuels			Other fuels			Total energy /MJ
	Fuel production and delivery energy	Energy content of fuel	Fuel production and delivery energy	Energy content of fuel	Feedstock energy	Fuel production and delivery energy	Energy content of fuel	Feedstock energy	
3.7 (0.11) RB	623	210	272	1341	15	87	872	37	3457
6.0 (0.17) RB	1096	363	491	2405	19	156	1579	46	6155
9.68 (0.28) RB	54	18	25	120	1	8	78	3	307
25.0 (0.71) RB	1320	442	598	2910	34	192	1866	74	7436
38.0 (1.08) RB	1510	503	677	3319	38	216	2128	85	8476
Totals	4603	1536	2063	10095	107	659	6523	245	25831

© Crown Copyright 1981

Table 29.10

Calculation of the total energy arising from trippage dependent operations in the delivery of cider in 1977 in the U.K. All values are in millions of MJ

Container size and type fl. oz. (litre)	Electricity		Oil fuels			Other fuels			Total energy /MJ
	Fuel production and delivery energy	Energy content of fuel	Fuel production and delivery energy	Energy content of fuel	Feedstock energy	Fuel production and delivery energy	Energy content of fuel	Feedstock energy	
3.3 (0.1) RB	137	46	60	296	3	19	192	8	761
9.5 (0.27) RB	29	9	13	62	1	4	40	2	160
38.0 (1.08) RB	154	51	68	332	5	21	212	11	854
Totals	320	106	141	690	9	44	444	21	1775

© Crown Copyright 1981

Table 29.11

Summary of the total energies arising from the delivery of beer, cider and carbonated soft drinks in 1977 in the U.K. All values are in units of millions of MJ

Contribution	Electricity		Oil fuels			Other fuels			Total energy /MJ
	Fuel production and delivery energy	Energy content of fuel	Fuel production and delivery energy	Energy content of fuel	Feedstock energy	Fuel production and delivery energy	Energy content of fuel	Feedstock energy	
Trippage independent									
Beer systems	7580	2707	2481	12561	651	573	4775	301	31629
Soft drink systems	9065	3267	3074	15477	928	652	5609	458	38530
Cider systems	621	212	238	1198	47	38	390	37	2781
Totals	17266	6186	5793	29236	1626	1263	10774	796	72940
Trippage dependent									
Beer systems	4354	1441	1929	9468	102	612	6139	235	24280
Soft drink systems	4603	1536	2063	10095	107	659	6523	245	25831
Cider systems	320	106	141	690	9	44	444	21	1775
Totals	9277	3083	4133	20253	218	1315	13106	501	51886

Table 29.12

Total energy used by the beverage industry in 1976/77 in the U.K. in packaging beer, cider and carbonated soft drinks, assuming that all returnable containers have a trippage of 1. All values are in millions of MJ

Fuel type	Fuel production and delivery energy	Energy content of fuel	Feedstock energy	Total energy
Electricity	26543	9269	–	35812
Oil fuels	9926	49489	1844	61259
Other fuels	2578	23880	1297	27755
Totals	39047	82638	3141	124,826

© *Crown Copyright 1981*

In case 2, an average trippage of 10 has been assumed for all returnable containers. This again is a somewhat artifical assumption since it is known that trippage varies with container size and type of beverage packed. However, when compared with case 1, it does provide a useful guide to the influence of trippage variations on energy demand and probably gives a reasonable description of current practices. These values are shown in Table 29.13.

Table 29.13

Total energy used by the beverage industry in 1976/77 in the U.K. in packaging beer, cider and carbonated soft drinks, assuming that all returnable containers have a trippage of 10. All values are in millions of MJ

Fuel type	Fuel production and delivery energy	Energy content of fuel	Feedstock energy	Total energy
Electricity	18194	6494	–	24688
Oil fuels	6206	31261	1648	39115
Other fuels	1395	12085	846	14326
Totals	25795	49840	2494	78129

© *Crown Copyright 1981*

Their significance in relation to the total national demand for fuels can be considered using [68]. The total consumption of energy as primary fuels in the U.K. in 1977 was 8.829×10^{12} MJ. In the worst case described by Table 29.12, the beverage container systems examined in this report required an aggregate of 0.125×10^{12} MJ which represents 1.4% of national consumption. The more realistic values of Table 29.13, however, represent 0.8% of national consumption.

The net total electricity supplied to all consumers in 1977 was 242.5×10^9 kWh or 873×10^9 MJ. In the worst case (Table 29.12) the beverage container systems consumed some 9×10^9 MJ, approximately 1% of total net consumption. This falls to 0.7% when the values of Table 29.13 are used.

In 1977, the total inland delivery of petroleum products was 82.8×10^6 tonne or 3.7×10^{12} MJ of which 9.7×10^6 tonne $(0.4 \times 10^{12}$ MJ) was feedstock and the remaining 73.1×10^6 tonne $(3.3 \times 10^{12}$ MJ) was used as fuel. Oil feedstock use by the beverage container industries was 0.002×10^{12} MJ or 0.5% of national consumption. Direct use of oil fuels in these industries was 0.049×10^{12} MJ or 1.5% in the worst case (Table 29.12) and 0.031×10^{12} MJ (0.9%) for $t = 10$ (Table 29.13).

The use of 'other fuels' is more difficult to ascertain because of the mixed nature of this energy contribution. However, the total energy supplied as coal or gas to consumers in 1977 was 4.808×10^{12} MJ. From Tables 29.12 and 29.13, the beverage container industry would use 0.023×10^{12} MJ (0.5%) in the worst case or 0.012×10^{12} MJ (0.25%) in the more reasonable case.

References

[1] Boustead, I. and Hancock, G. F., *Handbook of Industrial Energy Analysis*, Horwood/Wiley, 1979.

[2] Smith, H., *Proceedings of the World Energy Conference*, 1969, 18E.

[3] Markus, T. and Slesser, M., *Energy Policy*, March 1977, p. 76.

[4] Chapman, P. F., Energy costs of producing copper and aluminium from primary sources, *Open University Energy Research Group Report No. 1,* 1973.

[5] Odum, E. P., *Fundamentals of Ecology*, Saunders & Co. (Philadelphia), 1971.

[6] British Petroleum Company, *Our Industry – Petroleum*, London 1971.

[7] Sundstrom, G., Investigation of the energy requirements from raw materials to garbage treatment for four Swedish beer packaging alternatives, *Report for Rigello Pak AB*, 1973.

[8] Chapman, P. F., Leach, G. and Slessser, M., *Energy Policy*, September 1974, p. 231.

[9] Boustead, I. and Hancock, G. F., The Glass Industry, *Report to The Glass Manufacturers Federation*, London 1975.

[10] Leach, G. and Slesser, M., *Energy equivalent of network inputs in food production*, University of Strathclyde, 1973.

[11] Makhijani, A. B. and Lichtenberg, A. J. *Environment*, 1972, **14**, No. 5, p. 10.

[12] Thornton, D. S. and Williams, D. I. T., *Ironmaking and Steelmaking Quarterly*, 1975, No. 4, p. 241.

[13] Barnes, R. S., *Ironmaking and Steelmaking Quarterly*, 1975, No. 4, p. 271.

[14] *T.R.R.L. Supplementary Report No. 251*, Future Transport Fuels, 1976.

[15] Mortimer, N. D., *Energy analysis of burner reactor power systems*, Ph.D. Thesis, The Open University, 1978.

[16] Commercial Motor, *Tables of Operating Costs*, I.P.C. Business Press, 1978.

[17] Mortimer, N. D., The energy cost of road and rail freight in the U.K. in 1968, *Open University Energy Research Group Report No. 4,* 1974.

[18] Leach, G., *Energy and Food Production,* International Institute for Environment and Development, London 1975.

[19] Everall, P. F., The effect of road traffic conditions on fuel consumption, *TRRL Report LR226,* 1968. (See also [18]).

[20] Tuininga, E. J., Energy analysis of transportation systems, *Proceedings of the 9th International TNO Conference,* Rotterdam 1976.

[21] Samples, D. K., *Energy and the automobile,* Seminar at the Institute of Sciences and Technology, University of Michigan, Chicago, 23 August 1974.

[22] Boustead, I. and Hancock, G. F., *The Hepsleve System,* Report to the Hepworth Iron Company, Stocksbridge, Sheffield, 1977.

[23] Clemow, C. J., *Understanding energy systems.* Paper presented to the joint meeting of the Operational Research Society and the Institute of Fuel, London, April 1975.

[24] British Rail, Private communication.

[25] Mortimer, N. D., Gross energy requirements of marine transport, *Open University Energy Research Group Report No. 7,* 1974.

[26] Boustead, I. and Hancock, G. F., *Merolite and other containers,* Report to Imperial Chemical Industries Ltd. (1976).

[27] Glass Manufacturers Federation, London. Report of the Glass Industry Liaison Working Party with Government Departments. *The Glass Container and the Environmental Debate.* Undated but issued 1974.

[28] Stamper, J. W., *Mineral Facts and Problems 1970,* – Aluminium. U.S. Bureau of Mines (1970), p. 437.

[29] Maddox, K. P., *Mineral Industries Bulletin,* Colorado School of Mines, **18**, July, 1975.

[30] U.S. Environmental Protection Agency, Resource and process analysis of nine beverage container alternatives. *Report No. EPA/530/SW-19c,* 1974.

[31] Smith, F. R. A., *Metals and Materials,* March 1974, p. 182.

[32] Altenpohl, D., Daugherty, T. S. and Blum, W. *Proceedings 9th International TNO Conference,* Rotterdam 1976.

[33] Lath, P. and Menon, P. M., *Trans. Ind. Inst. Metals,* December 1969, p. 39.

[34] Peters, F. A., Johnson, P. W. and Kirby, R. C., A cost estimate of the Bayer process for producing alumina. *U.S. Bureau of Mines Report No. 6730,* 1965.

[35] U.K. Aluminium Federation, *Energy Usage Survey,* Private communication, 1976.

[36] Ryan, W. (Ed.), *Non-ferrous extractive metallurgy in the United Kingdom.* Institute of Mining and Metallurgy, 1968.

[37] Shreve, R. N., *Chemical Process Industries*, 3rd ed., McGraw Hill, 1967.

[38] Edison Technical Services Report, *The Plastics Beer Bottle*, 1974.

[39] Imperial Chemical Industries Ltd., *The competitiveness of LDPE, PP and PVC after the 1973 oil crisis – the ICI view in 1978*, I.C.I. Plastics Division, Welwyn Garden City, 1978.

[40] Newton, G. E. H., *Proc. Conference on Fuel and Productivity*, November 1963. Paper entitled, Fuel and productivity in the pulp and paper industry.

[41] Wahlman, M. *Pulp and Paper International*, November 1974, p. 60.

[42] Pehlke, R. D., *Unit processes of extractive metallurgy*, Elsevier, New York, 1975.

[43] Bravard, J. C., Flora, H. B. and Portal, C., Energy expenditures associated with the production and recycle of metals. *Oak Ridge National Laboratory Report, ORNL.NSF.EP24*, 1972.

[44] Atkins, P. R., Recycling can cut energy demand dramatically, *Engineering Mining Journal*, 1973.

[45] Sharp, D., *Ironmaking and Steelmaking Quarterly*, No. 4, 1975, p. 320.

[46] Feltoe, F. J., Energy requirements and the problems of energy management in a newly commissioned steelworks, *J. Inst. Fuel*, February 1965, p. 66.

[47] Rubber and Plastics Research Association, *An initial study of the energy content of products and energy savings in their manufacture*, RAPRA, 1975.

[48] Waller, R. F., Primary energy requirements for the production of iron and steel in the U.K., *Open University Research Group Report No. 11*, 1976.

[49] Barnes, R. S., *Ironmaking and Steelmaking Quarterly*, No. 4, 1975, p. 271.

[50] Thornton, D. S. and Williams, D. I. T., *Ironmaking and Steelmaking Quarterly*, No. 4, 1975, p. 241.

[51] Wright, P. A., *The extractive metallurgy of tin*, Elsevier, London (1966).

[52] U.S. Department of Commerce, Energy uses patterns in metallurgical and non-metallic mineral processing, *Report No. PB-245759*, 1975.

[53] Shreve, R. N. and Brink, J. A., *Chemical process industries*, McGraw Hill, (1977).

[54] Pick, H. J. and Becker, P. E., *Applied Energy*, 1, 31, 1975.

[55] Gartner, E. M. and Smith, M. A., *Energy Policy*, June 1976, p. 144.

[56] Hinde, P. T. and Probert, S. D., *Applied Energy*, 2, 17, 1976.

[57] Kellogg, H. H., Energy use and conservation in the metals industry, *Met. Soc. AIME*, 1975, p. 87.

[58] Open University, *Zinc – A case study*, ISBN 0-335-04189-2, 1975.

[59] The Electricity Council, *Statement of Accounts and Statistics 1977-78*, HMSO, 1978.

[60] *Glass View No. 13,* Glass Manufacturers Federation, London, Autumn 1978.

[61] *Statutory Instrument No. 1532,* HMSO, 1962.

[62] Lowenheim, F. A. and Moran, M. K., *Industrial Chemicals,* Wiley Interscience, New York, 1975.

[63] Weymes, E., *Trippage rates of carbonated soft drinks,* A report prepared for the National Association of Soft Drinks Manufacturers and the Glass Manufacturers Federation, Cranfield School of Management, July 1978.

[64] National Association of Soft Drinks Manufacturers, *Trippage Study,* September 1976.

[65] Lightfoot, C. A., *Energy uses in glassworks,* Paper read to the Institute of Glass Technology, 1978.

[66] British Glass Industry Research Association, *Energy audit of six glass container factories,* 1978.

[67] Grauer, H. and Ladue, S. E., *Ceramic Bulletin,* 1978, 57, No. 11, 1061.

[68] *United Kingdom Energy Statistics 1979,* Department of Energy, HMSO, 1979.

[69] *Municipal Yearbook 1979,* Municipal Journal Ltd., London.

Appendix 1 -

Energy and materials requirements of selected beverage container systems

This appendix gives the energy and materials requirements for some 46 different beverage container systems. In addition to the main system, defined using the parameters outlined in Chapter 28, a number of the more common variations in it are also considered. For details of the procedure employed in the data sheets and the nature of the breakdowns, refer to Chapter 28. The data sheets also give the range of system energies which should be used when the system is defined only by container type and beverage filled.

System 1   375

System 1

3.33 fl. oz. (0.100 litre) RETURNABLE GLASS BOTTLE

CIDER

Bottle mass = 0.165 kg

Main system

Empty container delivery	: Bulk palletised
Closure employed	: Crown
Outer packaging	: Returnable crates
Trunking (return journey)	: Carrying empty bottles, crates, etc.
Retail distribution	: Brewery system
Retail sale	: Public houses

Variations considered

Empty container delivery	: Customer crates
Retail sale	: Off-licences

SUMMARY OF TOTAL SYSTEM ENERGIES AS A FUNCTION OF TRIPPAGE (t)

All values are in MJ/1000 containers

Main system
  (i) excluding retail sale and consumer use: Energy $= 1330 + 4095/t$
  (ii) including retail sale and consumer use: Energy $= 4275 + 4095/t$

Alternative systems
(a) Empty bottles packed in customer crates, sales through public houses
  (i) excluding retail sale and consumer use: Energy $= 1330 + 3885/t$
  (ii) including retail sale and consumer use: Energy $= 4275 + 3885/t$

(b) Empty bottles packed in customer crates, sales through off-licences
  (i) excluding retail sale and consumer use: Energy $= 1330 + 3885/t$
  (ii) including retail sale and consumer use: Energy $= 3865 + 3885/t$

(c) Empty bottles bulk palletised, sales through off-licences
  (i) excluding retail sale and consumer use: Energy $= 1330 + 4095/t$
  (ii) including retail sale and consumer use: Energy $= 3865 + 4095/t$

SYSTEM ENERGY VARIATIONS IN MJ/container – see Chapter 29

Minimum	$4.06 + 3.03/t$
Maximum	$4.49 + 6.89/t$
Average	$4.28 + 4.10/t$

SYSTEM ENERGY REQUIREMENTS PER 1000 CONTAINERS

Contribution	Electricity/MJ		Oil fuels/MJ			Other fuels/MJ			Total energy /MJ
	Fuel production and delivery energy	Energy content of fuel	Fuel production and delivery energy	Energy content of fuel	Feedstock energy	Fuel production and delivery energy	Energy content of fuel	Feedstock energy	
Main system									
Filling and packing	345	116	83	402	35	16	127	24	1148
Trunking	3	1	17	87	–	1	9	–	118
Retail delivery	2	1	9	48	–	–	4	–	64
Total 1	350	118	109	537	35	17	140	24	1330
Container manufacture	650	220	290	1420	–	100	980	–	3660
Container delivery	2	1	9	48	–	1	5	–	66
Container packaging	70	23	13	66	17	–	26	43	258
Disposal	21	7	10	56	8	2	7	–	111
Total 2	743	251	322	1590	25	103	1018	43	4095
Retail sale	670	223	34	161	–	18	216	–	1322
Consumer use	–	–	260	1363	–	–	–	–	1623
Total 3	670	223	294	1524	–	18	216	–	2945
Total 1 + 2	1093	369	431	2127	60	120	1158	67	5425
Total 1 + 2 + 3	1763	592	725	3651	60	138	1374	67	8370
Alternative systems									
(a) Customer crates used for empty bottles									
Revised Total 1 + 2 + 3	1707	574	714	3593	43	138	1349	42	8160
(b) Retail sales through off-licences									
Revised Total 1 + 2 + 3	1777	597	691	3490	60	120	1158	67	7960

MATERIALS REQUIREMENTS IN kg/1000 containers FOR THE MAIN SYSTEM

(t = trippage)

Bauxite	4.62/t	Iron chromite	0.132/t	PET	–
Board	1.50/t	Iron ore	6.971	Polypropylene	0.40
Calcium sulphate	1.16/t	Lacquer	0.039	Sand	124.91/t
Carbon dioxide	0.27	Lead	–	Selenium	0.002/t
Cullet	1.16/t	Limestone	1.287 + 73.59/t	Sodium chloride	56.43/t
Detergent	1.85	Lining compound	0.214	Steel scrap	1.231
Epoxy adhesive	–	Manganese	0.056	Sulphuric acid	0.33/t
Feldspar	42.90/t	Methanol	0.003	Tin	0.032
Fluorspar	–	Pallets/number	0.033 + 0.038/t	Toluene	0.004
Flux (solder)	–	Paper	0.456	Varnish	0.023
Glue	0.15	Petroleum coke	–	Wood	–
Ink	0.033	Polyethylene	0.33/t		

System 2

3.7 fl. oz. (0.11 litre) RETURNABLE GLASS BOTTLE

CARBONATED SOFT DRINKS

Bottle mass = 0.170 kg

Main system

Empty container delivery   : Bulk palletised
Closure employed   : Crown
Outer packaging   : Returnable crates
Trunking (return journey) : Carrying empty bottles, etc.
Retail distribution   : Soft drinks system
Retail sale   : Public houses

Variations considered

Empty container delivery   : Customer crates
Retail sale   : Off-licences

SUMMARY OF TOTAL SYSTEM ENERGIES AS A FUNCTION OF TRIPPAGE (t)

All values in MJ/1000 containers

Main system
    (i) excluding retail sale and consumer use: Energy = $1330 + 4229/t$
    (ii) including retail sale and consumer use: Energy = $4275 + 4229/t$

Alternative systems
(a) Empty bottles packed in customer crates, sales through public houses.
    (i) excluding retail sale and consumer use: Energy = $1330 + 4019/t$
    (ii) including retail sale and consumer use: Energy = $4275 + 4019/t$

(b) Empty bottles packed in customer crates, sales through off-licences.
    (i) excluding retail sale and consumer use: Energy = $1330 + 4019/t$
    (ii) including retail sale and consumer use: Energy = $3865 + 4019/t$

(c) Empty bottles bulk palletised, sales through off-licences.
    (i) excluding retail sale and consumer use: Energy = $1330 + 4229/t$
    (ii) including retail sale and consumer use: Energy = $3865 + 4229/t$

SYSTEM ENERGY VARIATIONS IN MJ/container – see Chapter 29
Minimum $4.06 + 2.81/t$
Maximum $4.49 + 5.43/t$
Average $4.28 + 4.23/t$

SYSTEM ENERGY REQUIREMENTS PER 1000 CONTAINERS

Contribution	Electricity/MJ		Oil fuels/MJ			Other fuels/MJ			Total energy /MJ
	Fuel production and delivery energy	Energy content of fuel	Fuel production and delivery energy	Energy content of fuel	Feedstock energy	Fuel production and delivery energy	Energy content of fuel	Feedstock energy	
Main system									
Filling and packing	345	116	83	402	35	16	127	24	1148
Trunking	3	1	17	87	–	1	9	–	118
Retail delivery	2	1	9	48	–	–	4	–	64
Total 1	350	118	109	537	35	17	140	24	1330
Container manufacture	670	220	300	1470	–	100	1010	–	3770
Container delivery	3	1	12	62	–	1	7	–	86
Container packaging	70	23	13	66	17	–	26	43	258
Disposal	22	7	10	58	9	2	7	–	115
Total 2	765	251	335	1656	26	103	1050	43	4229
Retail sale	670	223	34	161	–	18	216	–	1322
Consumer use	–	–	260	1363	–	–	–	–	1623
Total 3	670	223	294	1524	–	18	216	–	2945
Total 1 + 2	1115	369	444	2193	61	120	1190	67	5559
Total 1 + 2 + 3	1785	592	738	3717	61	138	1406	67	8504
Alternative systems									
(a) Using customer crates to deliver empty bottles									
Revised Total 2	709	233	324	1598	9	103	1025	18	4019
(b) Retail sale through off-licences									
Revised Total 3	684	228	260	1363	–	–	–	–	2535

Appendix 1

System 2

MATERIALS REQUIREMENTS IN kg/1000 containers FOR THE MAIN SYSTEM

Bauxite	4.62/t	Iron chromite	0.132/t	PET	–
Board	1.5/t	Iron ore	6.971	Polypropylene	0.40
Calcium sulphate	1.155/t	Lacquer	0.039	Sand	124.91/t
Carbon dioxide	0.27	Lead	–	Selenium	0.002/t
Cullet	1.155/t	Limestone	1.287 + 73.59/t	Sodium chloride	56.43/t
Detergent	1.85	Lining compound	0.214	Steel scrap	1.231
Epoxy adhesive	–	Manganese	0.056	Sulphuric acid	0.33/t
Feldspar	42.9/t	Methanol	0.003	Tin	0.032
Fluorspar	–	Pallets/number	0.033 + 0.038/t	Toluene	0.004
Flux (solder)	–	Paper	0.456	Varnish	0.023
Glue	0.15	Petroleum coke	–	Wood	–
Ink	0.033	Polyethylene	0.33/t		

5.5 fl. oz. (0.16 litre) METAL CAN

CARBONATED SOFT DRINKS

Can type: Three-piece tinplate

Main system
Outer packaging   : Shrinkwrapped trays
Trunking (return journey) : Empty return load
Retail distribution   : Supermarket system
Retail sale   : Supermarket/Off-licence

Variations considered
Outer packaging   : Cartons

SUMMARY OF TOTAL SYSTEM ENERGIES

All values in MJ/1000 containers

Main system
  (i) excluding retail sale and consumer use: Energy = 4717
  (ii) including retail sale and consumer use: Energy = 7252

Alternative systems
Full cans packed in cartons
  (i) excluding retail sale and consumer use: Energy = 4821
  (ii) including retail sale and consumer use: Energy = 7356

SYSTEM ENERGY VARIATIONS IN MJ/container — see Chapter 29

Minimum   6.68
Maximum   7.96
Average   7.25

SYSTEM ENERGY REQUIREMENTS PER 1000 CONTAINERS

Contribution	Electricity/MJ		Oil fuels/MJ			Other fuels/MJ			Total energy /MJ
	Fuel production and delivery energy	Energy content of fuel	Fuel production and delivery energy	Energy content of fuel	Feedstock energy	Fuel production and delivery energy	Energy content of fuel	Feedstock energy	
Main system									
Filling and packing	315	105	49	239	42	1	64	64	879
Trunking	4	1	19	98	–	1	10	–	133
Retail delivery	1	–	6	30	–	–	3	–	40
Total 1	320	106	74	367	42	2	77	64	1052
Container manufacture	942	403	134	659	179	135	1071	1	3524
Container delivery	1	–	5	24	–	1	3	–	33
Container packaging	32	11	2	12	–	1	17	12	87
Disposal	4	1	2	11	2	–	1	–	21
Total 2	979	415	143	706	181	136	1092	13	3665
Retail sale	684	228	–	–	–	–	–	–	912
Consumer use	–	–	260	1363	–	–	–	–	1623
Total 3	684	228	260	1363	–	–	–	–	2535
Total 1 + 2	1299	521	217	1073	223	138	1169	77	4717
Total 1 + 2 + 3	1983	749	477	2436	223	138	1169	77	7252
Alternative systems									
(a) Full cans packed in cartons Revised Total 1 + 2 + 3	2014	759	480	2450	181	139	1211	122	7356

MATERIALS REQUIREMENTS IN kg/1000 containers FOR THE MAIN SYSTEM

Material	Value	Material	Value	Material	Value
Bauxite	22.161	Iron chromite	—	PET	—
Board	3.09	Iron ore	88.568	Polypropylene	—
Calcium sulphate	—	Lacquer	0.275	Sand	—
Carbon dioxide	0.42	Lead	—	Selenium	—
Cullet	—	Limestone	20.066	Sodium chloride	1.04
Detergent	—	Lining compound	0.247	Steel scrap	14.307
Epoxy adhesive	—	Manganese	0.705	Sulphuric acid	8.199
Feldspar	—	Methanol	—	Tin	0.791
Fluorspar	6.51	Pallets/number	0.37	Toluene	—
Flux (solder)	0.10	Paper	0.075	Varnish	0.107
Glue	—	Petroleum coke	1.758	Wood	0.078
Ink	0.053	Polyethylene	0.83		

System 4

6.0 fl. oz. (0.17 litre) RETURNABLE GLASS BOTTLE

CIDER

Bottle mass = 0.240 kg

Main system
Empty container delivery   : Bulk palletised
Closure employed   : Crown
Outer packaging   : Returnable crates
Trunking (return journey)   : Carrying empty bottles, crates, etc.
Retail distribution   : Brewery system
Retail sale   : Public houses

Variations considered
Empty container delivery   : Customer crates
Retail sale   : Off-licences

SUMMARY OF TOTAL SYSTEM ENERGIES AS A FUNCTION OF TRIPPAGE (t)

All values in MJ/1000 containers

Main system
  (i) excluding retail sale and consumer use: Energy $= 1658 + 5873/t$
  (ii) including retail sale and consumer use: Energy $= 4603 + 5873/t$

Alternative systems
(a) Empty bottles packed in customer crates, sales through public houses.
  (i) excluding retail sale and consumer use: Energy $= 1658 + 5617/t$
  (ii) including retail sale and consumer use: Energy $= 4603 + 5617/t$

(b) Empty bottles packed in customer crates, sales through off-licences.
  (i) excluding retail sale and consumer use: Energy $= 1658 + 5617/t$
  (ii) including retail sale and consumer use: Energy $= 4193 + 5617/t$

(c) Empty bottles bulk palletised, sales through off-licences.
  (i) excluding retail sale and consumer use: Energy $= 1658 + 5873/t$
  (ii) including retail sale and consumer use: Energy $= 4193 + 5873/t$

SYSTEM ENERGY VARIATIONS IN MJ/container – see Chapter 29
Minimum $4.37 + 3.53/t$
Maximum $4.83 + 7.51/t$
Average $4.60 + 5.87/t$

SYSTEM ENERGY REQUIREMENTS PER 1000 CONTAINERS

Contribution	Electricity/MJ		Oil fuels/MJ			Other fuels/MJ			Total energy /MJ
	Fuel production and delivery energy	Energy content of fuel	Fuel production and delivery energy	Energy content of fuel	Feedstock energy	Fuel production and delivery energy	Energy content of fuel	Feedstock energy	
Main system									
Filling and packing	369	124	110	530	58	17	143	33	1384
Trunking	5	2	25	130	–	1	14	–	177
Retail delivery	3	1	14	71	–	1	7	–	97
Total 1	377	127	149	731	58	19	164	33	1658
Container manufacture	940	310	420	2070	–	150	1420	–	5310
Container delivery	3	1	12	62	–	1	7	–	86
Container packaging	86	29	16	80	22	–	30	53	316
Disposal	31	10	14	82	12	2	10	–	161
Total 2	1060	350	462	2294	34	153	1467	53	5873
Retail sale	670	223	34	161	–	18	216	–	1322
Consumer use	–	–	260	1363	–	–	–	–	1623
Total 3	670	223	294	1524	–	18	216	–	2945
Total 1 + 2	1437	477	611	3025	92	172	1631	86	7531
Total 1 + 2 + 3	2107	700	905	4549	92	190	1847	86	10476
Alternative systems									
(a) Using customer crates to deliver empty bottles									
Revised Total 2	991	327	448	2224	12	153	1438	24	5617
(b) Retail sale through off-licences									
Revised Total 3	684	228	260	1363	–	–	–	–	2535

Appendix 1

System 4

MATERIALS REQUIREMENTS IN kg/1000 containers FOR THE MAIN SYSTEM

Bauxite	6.72/t	Iron chromite	0.192/t	PET	—
Board	1.75/t	Iron ore	6.971	Polypropylene	0.83
Calcium sulphate	1.68/t	Lacquer	0.039	Sand	181.68/t
Carbon dioxide	0.45	Lead	—	Selenium	0.024/t
Cullet	1.68/t	Limestone	1.287 + 107.04/t	Sodium chloride	82.08/t
Detergent	1.85	Lining compound	0.214	Steel scrap	1.231
Epoxy adhesive	—	Manganese	0.056	Sulphuric acid	0.48/t
Feldspar	62.4/t	Methanol	0.003	Tin	0.032
Fluorspar	—	Pallets/number	0.054 + 0.049/t	Toluene	0.004
Flux (solder)	—	Paper	0.456	Varnish	0.023
Glue	0.15	Petroleum coke	—	Wood	—
Ink	0.033	Polyethylene	0.43/t		

6.0 fl. oz. (0.17 litre) RETURNABLE GLASS BOTTLE

CARBONATED SOFT DRINKS

Bottle mass = 0.300 kg

Main system

Empty container delivery	: Bulk palletised
Closure employed	: Crown
Outer packaging	: Returnable crates
Trunking (return journey)	: Carrying empty bottles, crates, etc.
Retail distribution	: Brewery system
Retail sale	: Public houses

Variations considered

Empty container delivery	: Customer crates
Retail sale	: Off-licences

SUMMARY OF TOTAL SYSTEM ENERGIES AS A FUNCTION OF TRIPPAGE (t)

All values in MJ/1000 containers

Main system

(i) excluding retail sale and consumer use: Energy $= 1658 + 7285/t$
(ii) including retail sale and consumer use: Energy $= 4603 + 7285/t$

Alternative systems

(a) Empty bottles packed in customer crates, sales through public houses.
  (i) excluding retail sale and consumer use: Energy $= 1658 + 7029/t$
  (ii) including retail sale and consumer use: Energy $= 4603 + 7029/t$

(b) Empty bottles packed in customer crates, sales through off-licences.
  (i) excluding retail sale and consumer use: Energy $= 1658 + 7029/t$
  (ii) including retail sale and consumer use: Energy $= 4193 + 7029/t$

(c) Empty bottles packed as bulk palletised loads, retail sales through off-licences.
  (i) excluding retail sale and consumer use: Energy $= 1658 + 7285/t$
  (ii) including retail sale and consumer use: Energy $= 4193 + 7285/t$

SYSTEM ENERGY VARIATIONS IN MJ/container − see Chapter 29

Minimum	$4.37 + 4.98/t$
Maximum	$4.83 + 10.11/t$
Average	$4.60 + 7.28/t$

SYSTEM ENERGY REQUIREMENTS PER 1000 CONTAINERS

Contribution	Electricity/MJ		Oil fuels/MJ			Other fuels/MJ			Total energy /MJ
	Fuel production and delivery energy	Energy content of fuel	Fuel production and delivery energy	Energy content of fuel	Feedstock energy	Fuel production and delivery energy	Energy content of fuel	Feedstock energy	
Main system									
Filling and packing	369	124	110	530	58	17	143	33	1384
Trunking	5	2	25	130	–	1	14	–	177
Retail delivery	3	1	14	71	–	1	7	–	97
Total 1	377	127	149	731	58	19	164	33	1658
Container manufacture	1180	390	530	2590	–	180	1790	–	6660
Container delivery	3	1	16	79	–	1	8	–	108
Container packaging	86	29	16	80	22	–	30	53	316
Disposal	39	12	18	102	15	3	12	–	201
Total 2	1308	432	580	2851	37	184	1840	53	7285
Retail sale	670	223	34	161	–	18	216	–	1322
Consumer use	–	–	260	1363	–	–	–	–	1623
Total 3	670	223	294	1524	–	18	216	–	2945
Total 1 + 2	1685	559	729	3582	95	203	2004	86	8943
Total 1 + 2 + 3	2355	782	1023	5106	95	221	2220	86	11888
Alternative systems									
(a) Empty bottles delivered in customer crates									
Revised Total 2	1239	409	566	2781	15	184	1811	24	7029
(b) Retail sales through off-licences									
Revised Total 3	684	228	260	1363	–	–	–	–	2535

MATERIALS REQUIREMENTS IN kg/1000 containers FOR THE MAIN SYSTEM

Bauxite	6.72/t	Iron chromite	0.192/t	PET	—
Board	1.75/t	Iron ore	6.971	Polypropylene	0.83
Calcium sulphate	1.68/t	Lacquer	0.039	Sand	181.68/t
Carbon dioxide	0.45	Lead	—	Selenium	0.024/t
Cullet	1.68/t	Limestone	1.287 + 107.04/t	Sodium chloride	82.08/t
Detergent	1.85	Lining compound	0.214	Steel scrap	1.231
Epoxy adhesive	—	Manganese	0.056	Sulphuric acid	0.48/t
Feldspar	62.4/t	Methanol	0.003	Tin	0.032
Fluorspar	—	Pallets/number	0.054 + 0.049/t	Toluene	0.004
Flux (solder)	—	Paper	0.456	Varnish	0.023
Glue	0.15	Petroleum coke	—	Wood	—
Ink	0.033	Polyethylene	0.43/t		

6.0 fl. oz. (0.17 litre) NON-RETURNABLE GLASS BOTTLE

CARBONATED SOFT DRINKS

Bottle mass = 0.170 kg

Main system
Empty container delivery : Bulk palletised
Closure employed   : Crown
Outer packaging   : Shrinkwrapped trays
Trunking   : Empty return load
Retail distribution   : Supermarket system
Retail sale   : Supermarkets/Off-licences

Variations considered
Empty container delivery : Alternative packaging
Outer packaging   : Cartons

SUMMARY OF TOTAL SYSTEM ENERGIES

All values in MJ/1000 containers

Main system
    (i) excluding retail sale and consumer use: Energy = 5697
   (ii) including retail sale and consumer use: Energy = 8232.

Alternative systems
(a) Empty bottles bulk palletised, full bottles in cartons.
    (i) excluding retail sale and consumer use: Energy = 6084
   (ii) including retail sale and consumer use: Energy = 8619

(b) Empty bottles in alternative packaging, full bottles in shrinkwrapped trays.
    (i) excluding retail sale and consumer use: Energy = 5972
   (ii) including retail sale and consumer use: Energy = 8507

(c) Empty bottles in alternative packaging, full bottles in cartons.
    (i) excluding retail sale and consumer use: Energy = 6359
   (ii) including retail sale and consumer use: Energy = 8894

SYSTEM ENERGY VARIATIONS IN MJ/container – see Chapter 29
Minimum   6.97
Maximum   10.39
Average   8.23

SYSTEM ENERGY REQUIREMENTS PER 1000 CONTAINERS

Contribution	Electricity/MJ		Oil Fuels/MJ			Other fuels/MJ			Total energy /MJ
	Fuel production and delivery energy	Energy content of fuel	Fuel production and delivery energy	Energy content of fuel	Feedstock energy	Fuel production and delivery energy	Energy content of fuel	Feedstock energy	
Main system									
Filling and packing	382	129	71	344	56	14	151	83	1230
Trunking	3	1	16	79	–	1	8	–	108
Retail delivery	2	1	9	45	–	–	5	–	62
Total 1	387	131	96	468	56	15	164	83	1400
Container manufacture	670	220	300	1470	–	100	1010	–	3770
Container delivery	3	1	14	70	–	1	7	–	96
Container packaging	86	29	16	80	22	–	30	53	316
Disposal	22	7	10	58	9	2	7	–	115
Total 2	781	257	340	1678	31	103	1054	53	4297
Retail sale	684	228	–	–	–	–	–	–	912
Consumer use	–	–	260	1363	–	–	–	–	1623
Total 3	684	228	260	1363	–	–	–	–	2535
Total 1 + 2	1168	388	436	2146	87	118	1218	136	5697
Total 1 + 2 + 3	1852	616	696	3509	87	118	1218	136	8232
Alternative systems									
(a) Empty bottles bulk palletised, full bottles in cartons									
Revised Total 1 + 2 + 3	1960	652	713	3594	45	118	1306	231	8619
(b) Empty bottles in alternative packaging, full bottles in shrink-wrapped trays									
Revised Total 1 + 2 + 3	1927	641	710	3577	87	118	1262	185	8507
(c) Empty bottles in alternative packaging, full bottles in cartons									
Revised Total 1 + 2 + 3	2035	677	727	3662	45	118	1350	280	8894

System 6

MATERIALS REQUIREMENTS IN kg/1000 containers FOR THE MAIN SYSTEM

Bauxite	12.98	Iron chromite	0.136	PET	—
Board	3.66	Iron ore	—	Polypropylene	—
Calcium sulphate	1.19	Lacquer	0.193	Sand	128.69
Carbon dioxide	0.45	Lead	—	Selenium	0.002
Cullet	1.19	Limestone	77.46	Sodium chloride	58.526
Detergent	—	Lining compound	0.523	Steel scrap	—
Epoxy adhesive	—	Manganese	—	Sulphuric acid	3.381
Feldspar	44.2	Methanol	0.003	Tin	—
Fluorspar	2.415	Pallets/number	0.099	Toluene	0.004
Flux (solder)	—	Paper	0.456	Varnish	0.193
Glue	0.15	Petroleum coke	0.652	Wood	—
Ink	0.031	Polyethylene	1.26		

6.34 fl. oz. (0.18 litre) RETURNABLE GLASS BOTTLE

BEER

Bottle mass = 0.240 kg

Main system
Empty container delivery : Bulk palletised
Closure employed   : Crown
Outer packaging   : Returnable crates
Trunking (return journey) : Carrying empty bottles, crates, etc.
Retail distribution   : Brewery system
Retail sale   : Public houses

Variations considered
Empty container delivery : Customer crates

SUMMARY OF TOTAL SYSTEM ENERGIES AS A FUNCTION OF TRIPPAGE (t)

All values in MJ/1000 containers

Main system
 (i) excluding retail sale and consumer use: Energy $= 2142 + 5877/t$
 (ii) including retail sale and consumer use: Energy $= 5087 + 5877/t$

Alternative systems
Empty bottles packed in customer crates.
 (i) excluding retail sale and consumer use: Energy $= 2142 + 5621/t$
 (ii) including retail sale and consumer use: Energy $= 5087 + 5621/t$

SYSTEM VARIATIONS IN MJ/container — see Chapter 29
Minimum $4.83 + 3.53/t$
Maximum $5.34 + 7.52/t$
Average $5.09 + 5.88/t$

SYSTEM ENERGY REQUIREMENTS PER 1000 CONTAINERS

Contribution	Electricity/MJ		Oil fuels/MJ			Other fuels/MJ			Total energy /MJ
	Fuel production and delivery energy	Energy content of fuel	Fuel production and delivery energy	Energy content of fuel	Feedstock energy	Fuel production and delivery energy	Energy content of fuel	Feedstock energy	
Main system									
Filling and packing	532	179	108	522	60	26	252	43	1722
Trunking	8	3	40	202	1	2	21	–	277
Retail delivery	5	2	20	105	–	1	10	–	143
Total 1	545	184	168	829	61	29	283	43	2142
Container manufacture	940	310	420	2070	–	150	1420	–	5310
Container delivery	3	1	13	64	–	1	7	–	89
Container packaging	86	29	16	81	22	–	30	53	317
Disposal	31	10	14	82	12	2	10	–	161
Total 2	1060	350	463	2297	34	153	1467	53	5877
Retail sale	670	223	34	161	–	18	216	–	1322
Consumer use	–	–	260	1363	–	–	–	–	1623
Total 3	670	223	294	1524	–	18	216	–	2945
Total 1 + 2	1605	534	631	3126	95	182	1750	96	8019
Total 1 + 2 + 3	2275	757	925	4650	95	200	1966	96	10964
Alternative systems									
(a) Empty bottles packed in customer crates									
Revised Total 2	992	327	449	2226	12	153	1438	24	5621

MATERIALS REQUIREMENTS IN kg/1000 containers FOR THE MAIN SYSTEM

Bauxite	6.72/t	Iron chromite	0.192/t	PET	—
Board	1.75/t	Iron ore	6.971	Polypropylene	0.87
Calcium sulphate	1.68/t	Lacquer	0.039	Sand	181.68/t
Carbon dioxide	0.48	Lead	—	Selenium	0.024/t
Cullet	1.68/t	Limestone	1.287 + 107.04/t	Sodium chloride	82.08/t
Detergent	1.85	Lining compound	0.214	Steel scrap	1.231
Epoxy adhesive	—	Manganese	0.056	Sulphuric acid	0.48/t
Feldspar	62.4/t	Methanol	0.003	Tin	0.032
Fluorspar	—	Pallets/number	0.076 + 0.050/t	Toluene	0.004
Flux (solder)	—	Paper	0.456	Varnish	0.023
Glue	0.15	Petroleum coke	—	Wood	—
Ink	0.033	Polyethylene	0.44/t		

6.34 fl. oz. (0.18 litre) NON-RETURNABLE GLASS BOTTLE

BEER

Bottle mass = 0.150 kg

Main system

Empty container delivery	: Bulk palletised
Closure type	: Crown
Outer packaging	: Shrinkwrapped trays
Trunking (return load)	: Empty load
Retail distribution	: Supermarket system
Retail sale	: Supermarket/Off-licence

Variations considered

Closure type	: Standard aluminium screw
Outer packaging	: Cartons

SUMMARY OF SYSTEM ENERGIES

All values in MJ/1000 containers

Main system
- (i) excluding retail sale and consumer use: Energy = 5710
- (ii) including retail sale and consumer use: Energy = 8245

Alternative systems
- (a) Using aluminium screw closures and shrinkwrapped trays.
  - (i) excluding retail sale and consumer use: Energy = 6198
  - (ii) including retail sale and consumer use: Energy = 8733

- (b) Using crown closures and packaging in cartons.
  - (i) excluding retail sale and consumer use: Energy = 6099
  - (ii) including retail sale and consumer use: Energy = 8634

- (c) Using aluminium screw closures and packaging in cartons.
  - (i) excluding retail sale and consumer use: Energy = 6584
  - (ii) including retail sale and consumer use: Energy = 9119

SYSTEM ENERGY VARIATIONS IN MJ/container — see Chapter 29

Minimum	7.07
Maximum	9.85
Average	8.25

SYSTEM ENERGY REQUIREMENTS PER 1000 CONTAINERS

Contribution	Electricity/MJ		Oil fuels/MJ			Other fuels/MJ			Total energy /MJ
	Fuel production and delivery energy	Energy content of fuel	Fuel production and delivery energy	Energy content of fuel	Feedstock energy	Fuel production and delivery energy	Energy content of fuel	Feedstock energy	
Main system									
Filling and packing	418	141	102	495	56	22	253	83	1570
Trunking	6	2	29	149	–	1	16	–	203
Retail delivery	3	1	13	64	–	1	7	–	89
Total 1	427	144	144	708	56	24	276	83	1862
Container manufacture	590	200	270	1290	–	90	890	–	3330
Container delivery	3	1	14	72	–	1	8	–	99
Container packaging	86	29	16	81	22	–	30	53	317
Disposal	20	6	9	51	8	2	6	–	102
Total 2	699	236	309	1494	30	93	934	53	3848
Retail sale	684	228	–	–	–	–	–	–	912
Consumer use	–	–	260	1363	–	–	–	–	1623
Total 3	684	228	260	1363	–	–	–	–	2535
Total 1 + 2	1126	380	453	2202	86	117	1210	136	5710
Total 1 + 2 + 3	1810	608	713	3565	86	117	1210	136	8245
Alternative systems									
(a) Using aluminium screw closures and shrinkwrapped trays Revised Total 1 + 2 + 3	2005	703	734	3670	171	111	1203	136	8733
(b) Using crown closures and packaging in cartons Revised Total 1 + 2 + 3	1918	644	731	3650	44	118	1298	231	8634
(c) Using aluminium screw closures and packaging in cartons Revised Total 1 + 2 + 3	2113	739	752	3754	128	111	1291	231	9119

System 8

MATERIALS REQUIREMENTS IN kg/1000 containers FOR THE MAIN SYSTEM

Material	Value	Material	Value	Material	Value
Bauxite	4.20	Iron chromite	0.12	PET	—
Board	4.66	Iron ore	6.971	Polypropylene	—
Calcium sulphate	1.05	Lacquer	0.039	Sand	113.55
Carbon dioxide	0.48	Lead	—	Selenium	0.002
Cullet	1.05	Limestone	68.187	Sodium chloride	51.3
Detergent	—	Lining compound	0.214	Steel scrap	1.231
Epoxy adhesive	—	Manganese	0.056	Sulphuric acid	0.3
Feldspar	39.0	Methanol	0.003	Tin	0.032
Fluorspar	—	Pallets/number	0.100	Toluene	0.004
Flux (solder)	—	Paper	0.456	Varnish	0.023
Glue	0.15	Petroleum coke	—	Wood	—
Ink	0.033	Polyethylene	1.27		

8.8 fl. oz. (0.25 litre) NON-RETURNABLE GLASS BOTTLE

BEER

Bottle mass = 0.190 kg

Main system
Empty container delivery : Bulk palletised
Closure employed : Crown
Outer packaging : Shrinkwrapped trays
Trunking (return journey) : Empty load
Retail distribution : Supermarket system
Retail sale : Supermarket/Off-licence

Variations considered
Closures employed : Standard aluminium screw
Outer packaging : Cartons

SUMMARY OF TOTAL SYSTEM ENERGIES

All values in MJ/1000 containers

Main system
   (i) excluding retail sale and consumer use: Energy = 6919
  (ii) including retail sale and consumer use: Energy = 9454

Alternative systems
(a) Using aluminium screw closures and shrinkwrapped trays.
   (i) excluding retail sale and consumer use: Energy = 7407
  (ii) including retail sale and consumer use: Energy = 9942

(b) Using crown closures and packaging in cartons.
   (i) excluding retail sale and consumer use: Energy = 7801
  (ii) including retail sale and consumer use: Energy = 10336

(c) Using aluminium screw closures and packaging in cartons.
   (i) excluding retail sale and consumer use: Energy = 8286
  (ii) including retail sale and consumer use: Energy = 10821

SYSTEM ENERGY VARIATIONS IN MJ/container – see Chapter 29
Minimum   7.70
Maximum   11.68
Average   9.45

SYSTEM ENERGY REQUIREMENTS PER 1000 CONTAINERS

Contribution	Electricity/MJ		Oil fuels/MJ			Other fuels/MJ			Total energy /MJ
	Fuel production and delivery energy	Energy content of fuel	Fuel production and delivery energy	Feedstock energy	Energy content of fuel	Fuel production and delivery energy	Energy content of fuel	Feedstock energy	
Main system									
Filling and packing	469	158	117	56	565	25	283	83	1756
Trunking	6	2	29	–	150	1	16	–	204
Retail delivery	3	1	15	–	77	1	8	–	105
Total 1	478	161	161	56	792	27	307	83	2065
Container manufacture	740	240	340	–	1640	120	1130	–	4210
Container delivery	4	1	18	–	91	2	10	–	126
Container packaging	106	35	20	28	99	–	36	65	389
Disposal	25	8	11	10	65	2	8	–	129
Total 2	875	284	389	38	1895	124	1184	65	4854
Retail sale	684	228	–	–	–	–	–	–	912
Consumer use	–	–	260	–	1363	–	–	–	1623
Total 3	684	228	260	–	1363	–	–	–	2535
Total 1 + 2	1353	445	550	94	2687	151	1491	148	6919
Total 1 + 2 + 3	2037	673	810	94	4050	151	1491	148	9454
Alternative system									
(a) Using aluminium screw closures and shrinkwrapped trays Revised Total 1 + 2 + 3	2232	768	831	179	4155	144	1484	149	9942
(b) Using crown closures and packaging in cartons Revised Total 1 + 2 + 3	2280	754	853	52	4258	151	1659	329	10336
(c) Using aluminium screw closures and packaging in cartons Revised Total 1 + 2 + 3	2475	849	874	136	4362	145	1651	329	10821

MATERIALS REQUIREMENTS IN kg/1000 containers FOR THE MAIN SYSTEM

Bauxite	5.32	Iron chromite	0.152	PET	—
Board	5.01	Iron ore	6.971	Polypropylene	—
Calcium sulphate	1.33	Lacquer	0.039	Sand	143.83
Carbon dioxide	0.67	Lead	—	Selenium	0.002
Cullet	1.33	Limestone	86.027	Sodium chloride	64.98
Detergent	—	Lining compound	0.214	Steel scrap	1.231
Epoxy adhesive	—	Manganese	0.056	Sulphuric acid	0.38
Feldspar	49.4	Methanol	0.003	Tin	0.032
Fluorspar	—	Pallets/number	0.113	Toluene	0.004
Flux (solder)	—	Paper	0.456	Varnish	0.023
Glue	0.15	Petroleum coke	—	Wood	—
Ink	0.033	Polyethylene	1.38		

System 10

8.8 fl. oz. (0.25 litre) NON-RETURNABLE GLASS BOTTLE

CARBONATED SOFT DRINKS

Bottle mass = 0.176 kg

Main system
Empty bottle delivery   : Bulk palletised
Closure employed   : Rip cap
Outer packaging   : Shrinkwrapped trays
Trunking (return journey)   : Empty return load
Retail distribution   : Supermarket system
Retail sale   : Supermarkets/Off-licences

Variations considered
Empty bottle delivery   : Alternative packaging
Outer packaging   : Cartons

SUMMARY OF TOTAL SYSTEM ENERGIES

All values in MJ/1000 containers

Main system
   (i) excluding retail sale and consumer use: Energy = 7150
   (ii) including retail sale and consumer use: Energy = 9685

Alternative systems
(a) Empty bottles bulk palletised, full bottles in cartons.
   (i) excluding retail sale and consumer use: Energy = 8030
   (ii) including retail sale and consumer use: Energy = 10565

(b) Empty bottles in alternative packaging, full bottles in shrinkwrapped trays.
   (i) excluding retail sale and consumer use: Energy = 7481
   (ii) including retail sale and consumer use: Energy = 10016

(c) Empty bottle in alternative packaging, full bottles in cartons.
   (i) excluding retail sale and consumer use: Energy = 8361
   (ii) including retail sale and consumer use: Energy = 10896

SYSTEM ENERGY VARIATIONS IN MJ/container – see Chapter 29
Minimum   7.79
Maximum   11.38
Average   9.69

SYSTEM ENERGY REQUIREMENTS PER 1000 CONTAINERS

Contribution	Electricity/MJ		Oil fuels/MJ			Other fuels/MJ			Total energy /MJ
	Fuel production and delivery energy	Energy content of fuel	Fuel production and delivery energy	Energy content of fuel	Feedstock energy	Fuel production and delivery energy	Energy content of fuel	Feedstock energy	
Main system									
Filling and packing	729	290	151	736	187	13	207	84	2397
Trunking	4	1	19	96	–	1	10	–	131
Retail delivery	3	1	13	64	–	1	7	–	89
Total 1	736	292	183	896	187	15	224	84	2617
Container manufacture	690	230	310	1520	–	110	1040	–	3900
Container delivery	4	1	18	91	–	1	10	–	125
Container packaging	106	35	20	99	28	–	36	65	389
Disposal	23	7	11	60	9	2	7	–	119
Total 2	823	273	359	1770	37	113	1093	65	4533
Retail sale	684	228	–	–	–	–	–	–	912
Consumer use	–	–	260	1363	–	–	–	–	1623
Total 3	684	228	260	1363	–	–	–	–	2535
Total 1 + 2	1559	565	542	2666	224	128	1317	149	7150
Total 1 + 2 + 3	2243	793	802	4029	224	128	1317	149	9685
Alternative systems									
(a) Empty bottles bulk palletised, full bottles in cartons									
Revised Total 1 + 2 + 3	2486	874	845	4237	181	128	1484	330	10565
(b) Empty bottles in alternative packs, full bottles on shrinkwrapped trays									
Revised Total 1 + 2 + 3	2333	823	819	4111	224	128	1370	208	10016
(c) Empty bottles in alternative packs, full bottles in cartons									
Revised Total 1 + 2 + 3	2576	904	862	4319	181	128	1537	389	10896

System 10

MATERIALS REQUIREMENTS IN kg/1000 containers FOR THE MAIN SYSTEM

Bauxite	17.148	Iron chromite	0.14	PET	—
Board	5.01	Iron ore	—	Polypropylene	136.26
Calcium sulphate	1.26	Lacquer	0.284	Sand	0.002
Carbon dioxide	0.67	Lead	—	Selenium	62.128
Cullet	1.26	Limestone	82.696	Sodium chloride	
Detergent	—	Lining compound	0.771	Steel scrap	—
Epoxy adhesive	—	Manganese	—	Sulphuric acid	4.84
Feldspar	46.8	Methanol	0.003	Tin	—
Fluorspar	3.557	Pallets/number	0.113	Toluene	0.004
Flux (solder)	—	Paper	0.456	Varnish	0.284
Glue	0.15	Petroleum coke	0.960	Wood	—
Ink	0.031	Polyethylene	1.38		

8.8 fl. oz. (0.25 litre) NON-RETURNABLE GLASS BOTTLE

CARBONATED SOFT DRINKS

Bottle mass = 0.176 kg

Main system
Empty bottle delivery : Bulk palletised
Closure employed : Large aluminium screw
Outer packaging : Shrinkwrapped trays
Trunking (return journey) : Empty return load
Retail distribution : Supermarket system
Retail sale : Supermarkets/Off-licences

Variations considered
Empty bottle delivery : Alternative packaging
Outer packaging : Cartons

SUMMARY OF TOTAL SYSTEM ENERGIES

All values in MJ/1000 containers

Main system
  (i) excluding retail sale and consumer use: Energy = 7262
  (ii) including retail sale and consumer use: Energy = 9797

Alternative systems
(a) Empty bottles bulk palletised, full bottles in cartons.
  (i) excluding retail sale and consumer use: Energy = 8142
  (ii) including retail sale and consumer use: Energy = 10677

(b) Empty bottles in alternative packaging, full bottles in shrinkwrapped trays.
  (i) excluding retail sale and consumer use: Energy = 7593
  (ii) including retail sale and consumer use: Energy = 10128

(c) Empty bottles in alternative packaging, full bottles in cartons.
  (i) excluding retail sale and consumer use: Energy = 8743
  (ii) including retail sale and consumer use: Energy = 11008

SYSTEM ENERGY VARIATIONS IN MJ/container – see Chapter 29
Minimum   7.88
Maximum   11.51
Average   9.80

SYSTEM ENERGY REQUIREMENTS PER 1000 CONTAINERS

Contribution	Electricity/MJ		Oil fuels/MJ			Other fuels/MJ			Total energy /MJ
	Fuel production and delivery energy	Energy content of fuel	Fuel production and delivery energy	Energy content of fuel	Feedstock energy	Fuel production and delivery energy	Energy content of fuel	Feedstock energy	
Main system									
Filling and packing	767	307	156	762	202	13	217	85	2509
Trunking	4	1	19	96	–	1	10	–	131
Retail delivery	3	1	13	64	–	1	7	–	89
Total 1	774	309	188	922	202	15	234	85	2729
Container manufacture	690	230	310	1520	–	110	1040	–	3900
Container delivery	4	1	18	91	–	1	10	–	125
Container packaging	106	35	20	99	28	–	36	65	389
Disposal	23	7	11	60	9	2	7	–	119
Total 2	823	273	359	1770	37	113	1093	65	4533
Retail sale	684	228	–	–	–	–	–	–	912
Consumer use	–	–	260	1363	–	–	–	–	1623
Total 3	684	228	260	1363	–	–	–	–	2535
Total 1 + 2	1597	582	547	2692	239	128	1327	150	7262
Total 1 + 2 + 3	2281	810	807	4055	239	128	1327	150	9797
Alternative systems									
(a) Empty bottles bulk palletised, full bottles in cartons									
Revised Total 1 + 2 + 3	2523	891	850	4262	197	129	1495	330	10677
(b) Empty bottles in alternative packs, full bottles in shrinkwrapped trays									
Revised Total 1 + 2 + 3	2371	840	824	4137	239	128	1380	209	10128
(c) Empty bottles in alternative packs, full bottles in cartons									
Revised Total 1 + 2 + 3	2613	921	867	4344	197	129	1548	389	11008

MATERIALS REQUIREMENTS IN kg/1000 containers FOR THE MAIN SYSTEM

Material	kg	Material	kg	Material	kg
Bauxite	19.307	Iron chromite	0.168	PET	—
Board	5.01	Iron ore	—	Polypropylene	—
Calcium sulphate	1.47	Lacquer	0.315	Sand	158.97
Carbon dioxide	0.67	Lead	—	Selenium	0.002
Cullet	1.47	Limestone	96.34	Sodium chloride	72.45
Detergent	—	Lining compound	0.854	Steel scrap	—
Epoxy adhesive	—	Manganese	—	Sulphuric acid	5.388
Feldspar	54.6	Methanol	0.003	Tin	—
Fluorspar	3.945	Pallets/number	0.113	Toluene	0.004
Flux (solder)	—	Paper	0.456	Varnish	0.315
Glue	0.15	Petroleum coke	1.065	Wood	—
Ink	0.031	Polyethylene	1.33		

System 12

8.8 fl. oz. (0.25 litre) NON-RETURNABLE GLASS BOTTLE

CARBONATED SOFT DRINKS

Bottle mass = 0.210 kg

Main system

Empty bottle delivery	: Bulk palletised
Closure employed	: Standard aluminium screw
Outer packaging	: Shrinkwrapped trays
Trunking (return journey)	: Empty return load
Retail distribution	: Supermarket system
Retail sale	: Supermarket/Off-licences

Variations considered

Empty bottle delivery	: Alternative packaging
Outer packaging	: Cartons

SUMMARY OF TOTAL SYSTEM ENERGIES

All values in MJ/1000 containers

Main system
- (i) excluding retail sale and consumer use: Energy = 7585
- (ii) including retail sale and consumer use: Energy = 10120

Alternative systems
- (a) Empty bottles bulk palletised, full bottles in cartons
  - (i) excluding retail sale and consumer use: Energy = 8465
  - (ii) including retail sale and consumer use: Energy = 11000

- (b) Empty bottles in alternative packaging, full bottles in shrinkwrapped trays.
  - (i) excluding retail sale and consumer use: Energy = 7916
  - (ii) including retail sale and consumer use: Energy = 10451

- (c) Empty bottles in alternative packaging, full bottles in cartons.
  - (i) excluding retail sale and consumer use: Energy = 8796
  - (ii) including retail sale and consumer use: Energy = 11331

SYSTEM ENERGY VARIATIONS IN MJ/container – see Chapter 29

Minimum	8.14
Maximum	11.89
Average	10.12

SYSTEM ENERGY REQUIREMENTS PER 1000 CONTAINERS

Contribution	Electricity/MJ		Oil fuels/MJ			Other fuels/MJ			Total energy /MJ
	Fuel production and delivery energy	Energy content of fuel	Fuel production and delivery energy	Energy content of fuel	Feedstock energy	Fuel production and delivery energy	Energy content of fuel	Feedstock energy	
Main system									
Filling and packing	619	238	136	659	141	10	174	84	2061
Trunking	4	1	19	96	–	1	10	–	131
Retail delivery	3	1	13	64	–	1	7	–	89
Total 1	626	240	168	819	141	12	191	84	2281
Container manufacture	820	270	370	1810	–	130	1250	–	4650
Container delivery	4	1	18	91	–	1	10	–	125
Container packaging	106	35	20	99	28	–	36	65	389
Disposal	27	8	13	71	11	2	8	–	140
Total 2	957	314	421	2071	39	133	1304	65	5304
Retail sale	684	228	–	–	–	–	–	–	912
Consumer use	–	–	260	1363	–	–	–	–	1623
Total 3	684	228	260	1363	–	–	–	–	2535
Total 1 + 2	1583	554	589	2890	180	145	1495	149	7585
Total 1 + 2 + 3	2267	782	849	4253	180	145	1495	149	10120
Alternative systems									
(a) Empty bottles bulk palletised, full bottles in cartons Revised Total 1 + 2 + 3	2510	863	891	4461	138	145	1663	329	11000
(b) Empty bottles in alternative packs, full bottles in shrinkwrapped trays Revised Total 1 + 2 + 3	2357	812	866	4335	180	145	1548	208	10451
(c) Empty bottles in alternative packs, full bottles in cartons Revised Total 1 + 2 + 3	2600	893	908	4543	138	145	1716	388	11331

MATERIALS REQUIREMENTS IN kg/1000 containers FOR THE MAIN SYSTEM

System 12

Material	System 12	Material	System 12	Material	System 12
Bauxite	14.1	Iron chromite	0.168	PET	—
Board	5.01	Iron ore	—	Polypropylene	—
Calcium sulphate	1.47	Lacquer	0.193	Sand	158.97
Carbon dioxide	0.67	Lead	—	Selenium	0.002
Cullet	1.47	Limestone	95.3	Sodium chloride	72.206
Detergent	—	Lining compound	0.523	Steel scrap	—
Epoxy adhesive	—	Manganese	—	Sulphuric acid	3.461
Feldspar	54.6	Methanol	0.003	Tin	—
Fluorspar	2.415	Pallets/number	0.113	Toluene	0.004
Flux (solder)	—	Paper	0.456	Varnish	0.193
Glue	0.15	Petroleum coke	0.652	Wood	—
Ink	0.031	Polyethylene	1.38		

9.5 fl. oz. (0.27 litre) RETURNABLE GLASS BOTTLE

CIDER

Bottle mass = 0.300 kg

Main system

Empty container delivery	: Bulk palletised
Closure employed	: Crown
Outer packaging	: Returnable crates
Trunking (return load)	: Carrying empty bottles, crates, etc.
Retail distribution	: Brewery system
Retail sale	: Public houses

Variations considered

Empty container delivery	: Customer crates
Retail sale	: Off-licences

SUMMARY OF TOTAL SYSTEM ENERGIES IN TERMS OF TRIPPAGE (*t*)

All values in MJ/1000 containers

Main system
   (i) excluding retail sale and consumer use: Energy = $2288 + 7390/t$
   (ii) including retail sale and consumer use: Energy = $5233 + 7390/t$

Alternative systems
(a) Empty bottles packed in customer crates, sales through public houses.
   (i) excluding retail sale and consumer use: Energy = $2288 + 7056/t$
   (ii) including retail sale and consumer use: Energy = $5233 + 7056/t$

(b) Empty bottles packed in customer crates, sales through off-licences.
   (i) excluding retail sale and consumer use: Energy = $2288 + 7056/t$
   (ii) including retail sale and consumer use: Energy = $4823 + 7056/t$

(c) Empty bottles bulk palletised, sales through off-licences.
   (i) excluding retail sale and consumer use: Energy = $2288 + 7390/t$
   (ii) including retail sale and consumer use: Energy = $4823 + 7390/t$

SYSTEM ENERGY VARIATIONS IN MJ/container – see Chapter 29
Minimum $4.97 + 5.52/t$
Maximum $5.49 + 11.20/t$
Average $5.23 + 7.39/t$

SYSTEM ENERGY REQUIREMENTS PER 1000 CONTAINERS

Contribution	Electricity/MJ		Oil fuels/MJ			Other fuels/MJ			Total energy /MJ
	Fuel production and delivery energy	Energy content of fuel	Fuel production and delivery energy	Energy content of fuel	Feedstock energy	Fuel production and delivery energy	Energy content of fuel	Feedstock energy	
Main system									
Filling and packing	540	181	133	641	76	30	298	43	1942
Trunking	6	2	29	145	–	1	15	–	198
Retail delivery	5	2	21	109	–	1	10	–	148
Total 1	551	185	183	895	76	32	323	43	2288
Container manufacturer	1180	390	530	2590	–	180	1780	–	6650
Container delivery	4	1	17	88	–	2	9	–	121
Container packaging	114	38	21	106	30	–	39	70	418
Disposal	39	12	18	102	15	3	12	–	201
Total 2	1337	441	586	2886	45	185	1840	70	7390
Retail sale	670	223	34	161	–	18	216	–	1322
Consumer use	–	–	260	1363	–	–	–	–	1623
Total 3	670	223	294	1524	–	18	216	–	2945
Total 1 + 2	1888	626	769	3781	121	217	2163	113	9678
Total 1 + 2 + 3	2558	849	1063	5305	121	235	2379	113	12623
Alternative systems									
(a) Empty bottles packed in customer crates									
Revised Total 2	1247	411	568	2794	15	185	1803	33	7056
(b) Retail sale through off-licences									
Revised Total 3	684	228	260	1363	–	–	–	–	2535

MATERIALS REQUIREMENTS IN kg/1000 containers FOR THE MAIN SYSTEM

Bauxite	8.4/t	Iron chromite	0.24/t
Board	2.25/t	Iron ore	6.971
Calcium sulphate	2.1/t	Lacquer	0.039
Carbon dioxide	0.73	Lead	–
Cullet	2.1/t	Limestone	1.287 + 133.8/t
Detergent	1.85	Lining compound	0.214
Epoxy adhesive	–	Manganese	0.056
Feldspar	78.0/t	Methanol	0.003
Fluorspar	–	Pallets/number	0.076 + 0.068/t
Flux (solder)	–	Paper	0.456
Glue	0.15	Petroleum coke	–
Ink	0.033	Polyethylene	0.59/t

PET	–
Polypropylene	1.17
Sand	227.1/t
Selenium	0.003/t
Sodium chloride	102.6/t
Steel scrap	1.231
Sulphuric acid	0.6/t
Tin	0.032
Toluene	0.004
Varnish	0.023
Wood	–

9.68 fl. oz (0.28 litre) RETURNABLE GLASS BOTTLE

BEER

Bottle mass = 0.300 kg

Main system
Empty container delivery : Bulk palletised
Closure employed   : Crown
Outer packaging   : Returnable crates
Trunking (return journey) : Carrying empty bottles, crates, etc.
Retail distribution   : Brewery system
Retail sale   : Public houses

Variations considered
Empty container delivery : Customer crates

SUMMARY OF TOTAL SYSTEM ENERGIES IN TERMS OF TRIPPAGE (t)

All values in MJ/1000 containers

Main system
   (i) excluding retail sale and consumer use: Energy = $2320 + 7390/t$
   (ii) including retail sale and consumer use: Energy = $5265 + 7390/t$

Alternative system
Empty bottles packed in customer crates.
   (i) excluding retail sale and consumer use: Energy = $2320 + 7056/t$
   (ii) including retail sale and consumer use: Energy = $5265 + 7056/t$

SYSTEM ENERGY VARIATIONS IN MJ/container – see Chapter 29
Minimum $5.00 + 5.52/t$
Maximum $5.53 + 11.34/t$
Average $5.27 + 7.39/t$

SYSTEM ENERGY REQUIREMENTS PER 1000 CONTAINERS

Contribution	Electricity/MJ		Oil fuels/MJ			Other fuels/MJ			Total energy /MJ
	Fuel production and delivery energy	Energy content of fuel	Fuel production and delivery energy	Energy content of fuel	Feedstock energy	Fuel production and delivery energy	Energy content of fuel	Feedstock energy	
Main system									
Filling and packing	540	181	133	641	76	30	298	43	1942
Trunking	7	2	33	169	–	1	18	–	230
Retail delivery	5	2	21	109	–	1	10	–	148
Total 1	552	185	187	919	76	32	326	43	2320
Container manufacture	1180	390	530	2590	–	180	1780	–	6650
Container delivery	4	1	17	88	–	2	9	–	121
Container packaging	114	38	21	106	30	–	39	70	418
Disposal	39	12	18	102	15	3	12	–	201
Total 2	1337	441	586	2886	45	185	1840	70	7390
Retail sale	670	223	34	161	–	18	216	–	1322
Consumer use	–	–	260	1363	–	–	–	–	1623
Total 3	670	223	294	1524	–	18	216	–	2945
Total 1 + 2	1889	626	773	3805	121	217	2166	113	9710
Total 1 + 2 + 3	2559	849	1067	5329	121	235	2382	113	12655
Alternative systems									
(a) Empty bottles packed in customer crates									
Revised Total 2	1247	411	568	2794	15	185	1803	33	7056

System 14

MATERIALS REQUIREMENTS IN kg/1000 containers FOR THE MAIN SYSTEM

Bauxite	8.4/t	Iron chromite	0.24/t	PET	–
Board	2.25/t	Iron ore	6.971	Polypropylene	1.17
Calcium sulphate	2.1/t	Lacquer	0.039	Sand	227.1/t
Carbon dioxide	0.73	Lead	–	Selenium	0.003/t
Cullet	2.1/t	Limestone	1.287 + 133.8/t	Sodium chloride	102.6/t
Detergent	1.85	Lining compound	0.214	Steel scrap	1.231
Epoxy adhesive	–	Manganese	0.056	Sulphuric acid	0.6/t
Feldspar	78.0/t	Methanol	0.003	Tin	0.032
Fluorspar	–	Pallets/number	0.076 + 0.068/t	Toluene	0.004
Flux (solder)	–	Paper	0.456	Varnish	0.028
Glue	0.15	Petroleum coke	–	Wood	–
Ink	0.033	Polyethylene	0.59/t		

9.68 fl. oz. (0.28 litre) RETURNABLE GLASS BOTTLE

CARBONATED SOFT DRINKS

Bottle mass = 0.355 kg

Main system

Empty bottle delivery   : Bulk palletised
Closure employed   : Crown
Outer packaging   : Returnable crates
Trunking (return journey) : Carrying empty bottles, crates, etc.
Retail distribution   : Brewery system
Retail sale   : Public houses

Variations considered

Empty bottle delivery   : Customer crates
Retail sale   : Off-licences

SUMMARY OF THE TOTAL SYSTEM ENERGIES IN TERMS OF TRIPPAGE (t)

All values in MJ/1000 containers

Main system

   (i) excluding retail sale and consumer use: Energy = $1826 + 8680/t$
  (ii) including retail sale and consumer use: Energy = $4771 + 8680/t$

Alternative systems

(a) Empty bottles packed in customer crates, sales through public houses.
   (i) excluding retail sale and consumer use: Energy = $1826 + 8346/t$
  (ii) including retail sale and consumer use: Energy = $4771 + 8346/t$

(b) Empty bottles packed in customer crates, sales through off-licences.
   (i) excluding retail sale and consumer use: Energy = $1826 + 8346/t$
  (ii) including retail sale and consumer use: Energy = $4361 + 8346/t$

(c) Empty bottles bulk palletised, sales through off-licences.
   (i) excluding retail sale and consumer use: Energy = $1826 + 8680/t$
  (ii) including retail sale and consumer use: Energy = $4361 + 8680/t$

SYSTEM ENERGY VARIATIONS IN MJ/container – see Chapter 29

Minimum $4.53 + 5.54/t$
Maximum $5.01 + 11.38/t$
Average $4.77 + 8.68/t$

Appendix 1

SYSTEM ENERGY REQUIREMENTS PER 1000 CONTAINERS

Contribution	Electricity/MJ		Oil fuels/MJ			Other fuels/MJ			Total energy /MJ
	Fuel production and delivery energy	Energy content of fuel	Fuel production and delivery energy	Energy content of fuel	Feedstock energy	Fuel production and delivery energy	Energy content of fuel	Feedstock energy	
Main system									
Filling and packing	387	130	117	565	76	17	145	43	1480
Trunking	6	2	29	145	–	1	15	–	198
Retail delivery	5	2	21	109	–	1	10	–	148
Total 1	398	134	167	819	76	19	170	43	1826
Container manufacture	1390	460	630	3060	–	220	2110	–	7870
Container delivery	5	1	22	112	–	2	12	–	154
Container packaging	114	38	21	106	30	–	39	70	418
Disposal	46	14	21	121	18	4	14	–	238
Total 2	1555	513	694	3399	48	226	2175	70	8680
Retail sale	670	223	34	161	–	18	216	–	1322
Consumer use	–	–	260	1363	–	–	216	–	1623
Total 3	670	223	294	1524	–	18	216	–	2945
Total 1 + 2	1953	647	861	4218	124	245	2345	113	10506
Total 1 + 2 + 3	2623	870	1155	5742	124	263	2561	113	13451
Alternative systems									
(a) Empty bottles delivered in customer crates									
Revised Total 2	1465	483	676	3307	18	226	2138	33	8346
(b) Retail sales through off-licences									
Revised Total 3	684	228	260	1363	–	–	–	–	2535

MATERIALS REQUIREMENTS IN kg/1000 containers FOR THE MAIN SYSTEM

Bauxite	$9.94/t$	Iron chromite	$0.284/t$
Board	$2.25/t$	Iron ore	6.971
Calcium sulphate	$2.485/t$	Lacquer	0.039
Carbon dioxide	0.73	Lead	—
Cullet	$2.485/t$	Limestone	$1.287 + 158.333/t$
Detergent	1.85	Lining compound	0.214
Epoxy adhesive	—	Manganese	0.056
Feldspar	$92.3/t$	Methanol	0.003
Fluorspar	—	Pallets/number	$0.076 + 0.068/t$
Flux (solder)	—	Paper	0.456
Glue	0.15	Petroleum coke	—
Ink	0.033	Polyethylene	$0.59/t$

PET	—
Polypropylene	1.17
Sand	$268.735/t$
Selenium	$0.004/t$
Sodium chloride	$121.41/t$
Steel scrap	1.231
Sulphuric acid	$0.71/t$
Tin	0.032
Toluene	0.004
Varnish	0.023
Wood	—

9.68 fl. oz. (0.28 litre) NON-RETURNABLE GLASS BOTTLE

BEER

Bottle mass = 0.180 kg

Main system
Empty container delivery : Bulk palletised
Closure employed   : Rip-cap
Outer packaging   : Shrinkwrapped multipacks
Trunking (return journey) : Empty return load
Retail distribution   : Supermarket system
Retail sale   : Supermarkets

Variations considered
None

SUMMARY OF TOTAL SYSTEM ENERGIES

All values in MJ/1000 containers

Main system
   (i) excluding retail sale and consumer use: Energy = 8292
   (ii) including retail sale and consumer use: Energy = 10827

SYSTEM ENERGY VARIATIONS IN MJ/container — see Chapter 29
Minimum   9.20
Maximum   14.00
Average   10.83

SYSTEM ENERGY REQUIREMENTS PER 1000 CONTAINERS

Contribution	Electricity/MJ		Oil fuels/MJ			Other fuels/MJ			Total energy /MJ
	Fuel production and delivery energy	Energy content of fuel	Fuel production and delivery energy	Energy content of fuel	Feedstock energy	Fuel production and delivery energy	Energy content of fuel	Feedstock energy	
Main system									
Filling and packing	937	359	197	960	210	24	409	155	3251
Trunking	7	2	36	182	–	2	19	–	248
Retail delivery	4	1	17	88	–	1	9	–	120
Total 1	948	362	250	1230	210	27	437	155	3619
Container manufacture	710	240	320	1550	–	110	1070	–	4000
Container delivery	4	1	19	99	–	2	10	–	135
Container packaging	114	38	21	106	30	–	39	70	418
Disposal	23	7	11	61	9	2	7	–	120
Total 2	851	286	371	1816	39	114	1126	70	4673
Retail sale	684	228	–	–	–	–	–	–	912
Consumer use	–	–	260	1363	–	–	–	–	1623
Total 3	684	228	260	1363	–	–	–	–	2535
Total 1 + 2	1799	648	621	3046	249	141	1563	225	8292
Total 1 + 2 + 3	2483	876	881	4409	249	141	1563	225	10827

System 16

MATERIALS REQUIREMENTS IN kg/1000 containers FOR THE MAIN SYSTEM

Bauxite	17.15	Iron Chromite	0.144	PET	—
Board	9.125	Iron ore	—	Polypropylene	—
Calcium sulphate	1.26	Lacquer	0.284	Sand	136.26
Carbon dioxide	0.73	Lead	—	Selenium	0.002
Cullet	1.26	Limestone	82.696	Sodium chloride	62.129
Detergent	—	Lining compound	0.77	Steel scrap	—
Epoxy adhesive	—	Manganese	—	Sulphuric acid	4.84
Feldspar	46.8	Methanol	0.017	Tin	—
Fluorspar	3.558	Pallets/Number	0.144	Toluene	0.018
Flux (solder)	—	Paper	0.456	Varnish	0.284
Glue	0.15	Petroleum coke	0.961	Wood	—
Ink	0.155	Polyethylene	1.760		

9.68 fl. oz. (0.28 litre) NON-RETURNABLE GLASS BOTTLE

BEER

Bottle mass = 0.200 kg

Main system
Empty container delivery : Bulk palletised
Closure type   : Crown
Outer packaging   : Shrinkwrapped trays
Trunking (return journey) : Empty return load
Retail distribution   : Supermarket system
Retail sale   : Supermarket/Off-licence

Variations considered
Closure type   : Standard aluminium screw
Outer packaging   : Cartons

SUMMARY OF TOTAL SYSTEM ENERGIES

All values in MJ/1000 containers

Main system
   (i) excluding retail sale and consumer use: Energy = 7399
  (ii) including retail sale and consumer use: Energy = 9934

Alternative systems
(a) Using aluminium screw closures and shrinkwrapped trays.
   (i) excluding retail sale and consumer use: Energy = 7886
  (ii) including retail sale and consumer use: Energy = 10421

(b) Using crown closures with full bottles packed in cartons.
   (i) excluding retail sale and consumer use: Energy = 8280
  (ii) including retail sale and consumer use: Energy = 10815

(c) Using aluminium screw closures and packaging in cartons.
   (i) excluding retail sale and consumer use: Energy = 8766
  (ii) including retail sale and consumer use: Energy = 11301

SYSTEM ENERGY VARIATIONS IN MJ/container – see Chapter 29
Minimum   8.44
Maximum   12.84
Average   9.93

SYSTEM ENERGY REQUIREMENTS PER 1000 CONTAINERS

Contribution	Electricity/MJ		Oil fuels/MJ			Other fuels/MJ			Total energy /MJ
	Fuel production and delivery energy	Energy content of fuel	Fuel production and delivery energy	Energy content of fuel	Feedstock energy	Fuel production and delivery energy	Energy content of fuel	Feedstock energy	
Main system									
Filling and packing	495	166	133	646	56	28	317	83	1924
Trunking	7	2	36	182	–	2	19	–	248
Retail delivery	4	1	17	88	–	1	9	–	120
Total 1	506	169	186	916	56	31	345	83	2292
Container manufacture	780	260	350	1720	–	120	1190	–	4420
Container delivery	4	1	19	99	–	2	10	–	135
Container packaging	114	38	21	106	30	–	39	70	418
Disposal	26	8	12	68	10	2	8	–	134
Total 2	924	307	402	1993	40	124	1247	70	5107
Retail sale	684	228	–	–	–	–	–	–	912
Consumer use	–	–	260	1363	–	–	–	–	1623
Total 3	684	228	260	1363	–	–	–	–	2535
Total 1 + 2	1430	476	588	2909	96	155	1592	153	7399
Total 1 + 2 + 3	2114	704	848	4272	96	155	1592	153	9934
Alternative systems									
(a) Using aluminium screw closures and shrinkwrapped trays									
Revised Total 1 + 2 + 3	2309	799	869	4376	181	148	1585	154	10421
(b) Using crown closures with filled bottles packed in cartons									
Revised Total 1 + 2 + 3	2357	785	891	4479	54	155	1760	334	10815
(c) Using aluminium screw closures, with filled bottles packed in cartons									
Revised Total 1 + 2 + 3	2551	880	912	4584	138	149	1753	334	11301

MATERIALS REQUIREMENTS IN kg/1000 containers FOR THE MAIN SYSTEM

Bauxite	5.6	Iron chromite	0.16	PET	—
Board	5.16	Iron ore	6.971	Polypropylene	—
Calcium sulphate	1.4	Lacquer	0.039	Sand	151.4
Carbon dioxide	0.73	Lead	—	Selenium	0.002
Cullet	1.4	Limestone	90.487	Sodium chloride	68.4
Detergent	—	Lining compound	0.214	Steel scrap	1.231
Epoxy adhesive	—	Manganese	0.056	Sulphuric acid	0.4
Feldspar	52.0	Methanol	0.003	Tin	0.032
Fluorspar	—	Pallets/number	0.118	Toluene	0.004
Flux (solder)	—	Paper	0.456	Varnish	0.023
Glue	0.15	Petroleum coke	—	Wood	—
Ink	0.033	Polyethylene	1.42		

System 18

9.68 fl. oz. (0.28 litre) METAL CAN

BEER

Can type : Three-piece tinplate

Main system

Outer packaging   : Hicone and trays
Trunking (return journey) : Empty return load
Retail distribution   : Supermarket system
Retail sale   : Supermarket/Off-licence

Variations considered

Outer packaging   : Cartons

SUMMARY OF TOTAL SYSTEM ENERGIES

All values in MJ/1000 containers

Main system
   (i) excluding retail sale and consumer use: Energy = 6222
  (ii) including retail sale and consumer use: Energy = 8757

Alternative system
Full cans packed in cartons.
   (i) excluding retail sale and consumer use: Energy = 7050
  (ii) including retail sale and consumer use: Energy = 9585

VARIATIONS IN SYSTEM ENERGY IN MJ/container – see Chapter 29
Minimum   8.05
Maximum   10.38
Average   8.76

SYSTEM ENERGY REQUIREMENTS PER 1000 CONTAINERS

Contribution	Electricity/MJ		Oil fuels/MJ			Other fuels/MJ			Total energy /MJ
	Fuel production and delivery energy	Energy content of fuel	Fuel production and delivery energy	Energy content of fuel	Feedstock energy	Fuel production and delivery energy	Energy content of fuel	Feedstock energy	
Main system									
Filling and packing	345	115	76	369	47	9	156	73	1190
Trunking	5	2	23	115	–	1	12	–	158
Retail delivery	2	1	10	49	–	–	5	–	67
Total 1	352	118	109	533	47	10	173	73	1415
Container manufacture	1242	542	173	848	227	176	1382	1	4591
Container delivery	2	1	7	38	–	1	4	–	53
Container packaging	50	17	4	17	–	2	26	18	134
Disposal	6	2	3	14	2	–	2	–	29
Total 2	1300	562	187	917	229	179	1414	19	4807
Retail sale	684	228	–	–	–	–	–	–	912
Consumer use	–	–	260	1363	–	–	–	–	1623
Total 3	684	228	260	1363	–	–	–	–	2535
Total 1 + 2	1652	680	296	1450	276	189	1587	92	6222
Total 1 + 2 + 3	2336	908	556	2813	276	189	1587	92	8757
Alternative system									
(a) Full containers packed in cartons									
Revised Total 1 + 2 + 3	2567	985	598	3015	229	189	1743	259	9585

System 18

MATERIALS REQUIREMENTS OF THE MAIN SYSTEM IN kg/1000 containers

Bauxite	31.73	Iron chromite	—	PET	—
Board	3.412	Iron ore	116.185	Polypropylene	—
Calcium sulphate	—	Lacquer	0.347	Sand	—
Carbon dioxide	0.73	Lead	0.638	Selenium	—
Cullet	—	Limestone	26.744	Sodium chloride	1.489
Detergent	—	Lining compound	0.161	Steel scrap	18.565
Epoxy adhesive	—	Manganese	0.926	Sulphuric acid	11.740
Feldspar	—	Methanol	—	Tin	0.481
Fluorspar	9.321	Pallets/number	0.576	Toluene	—
Flux (solder)	0.10	Paper	0.082	Varnish	0.136
Glue	—	Petroleum coke	2.516	Wood	0.123
Ink	0.067	Polyethylene	0.942		

11.6 fl. oz. (0.33 litre) METAL CAN

CARBONATED SOFT DRINKS

Can type: Three-piece tinplate

Main system
Outer packaging   : Shrinkwrapped trays
Trunking (return journey) : Empty return load
Retail distribution   : Supermarket system
Retail sale   : Supermarket/Off-licence

Variations considered
Outer packaging   : Cartons
Outer packaging   : Hicone and trays

SUMMARY OF TOTAL SYSTEM ENERGIES

All values in MJ/1000 containers

Main system
   (i) excluding retail sale and consumer use: Energy = 6272
   (ii) including retail sale and consumer use: Energy = 8807

Alternative systems
(a) Full cans packed in cartons.
   (i) excluding retail sale and consumer use: Energy = 7000
   (ii) including retail sale and consumer use: Energy = 9535

(b) Full cans packed in Hicone and trays.
   (i) excluding retail sale and consumer use: Energy = 5973
   (ii) including retail sale and consumer use: Energy = 8508

VARIATIONS IN SYSTEM ENERGY IN MJ/container – see Chapter 29
Minimum   8.08
Maximum   10.36
Average   8.81

SYSTEM ENERGY REQUIREMENTS PER 1000 CONTAINERS

Contribution	Electricity/MJ		Oil fuels/MJ			Other fuels/MJ			Total energy /MJ
	Fuel production and delivery energy	Energy content of fuel	Fuel production and delivery energy	Energy content of fuel	Feedstock energy	Fuel production and delivery energy	Energy content of fuel	Feedstock energy	
Main system									
Filling and packing	318	106	56	275	64	1	68	74	962
Trunking	4	1	21	104	–	1	11	–	142
Retail delivery	2	1	11	57	–	–	6	–	77
Total 1	324	108	88	436	64	2	85	74	1181
Container manufacture	1279	556	186	914	234	191	1498	1	4859
Container delivery	2	1	8	42	–	1	4	–	58
Container packaging	54	18	4	18	–	2	28	19	143
Disposal	6	2	3	16	2	–	2	–	31
Total 2	1341	577	201	990	236	194	1532	20	5091
Retail sale	684	228	–	1363	–	–	–	–	912
Consumer use	–	–	260	1363	–	–	–	–	1623
Total 3	684	228	260	1363	–	–	–	–	2535
Total 1 + 2	1665	685	289	1426	300	196	1617	94	6272
Total 1 + 2 + 3	2349	913	549	2789	300	196	1617	94	8807
Alternative systems									
(a) Full cans packed in cartons Revised Total 1 + 2 + 3	2551	981	583	2952	236	197	1773	262	9535
(b) Full cans packed in Hicone on trays Revised Total 1 + 2 + 3	2320	903	344	2750	283	196	1617	95	8508

MATERIALS REQUIREMENTS OF THE MAIN SYSTEM IN kg/1000 containers

Bauxite	31.73	Iron chromite	—	PET	—
Board	3.414	Iron ore	129.54	Polypropylene	—
Calcium sulphate	—	Lacquer	0.381	Sand	—
Carbon dioxide	0.88	Lead	—	Selenium	—
Cullet	—	Limestone	29.043	Sodium chloride	1.489
Detergent	—	Lining compound	0.161	Steel scrap	20.609
Epoxy adhesive	—	Manganese	1.032	Sulphuric acid	11.74
Feldspar	—	Methanol	—	Tin	0.966
Fluorspar	9.321	Pallets/number	0.634	Toluene	—
Flux (solder)	0.10	Paper	0.082	Varnish	0.153
Glue	—	Petroleum coke	2.516	Wood	0.135
Ink	0.075	Polyethylene	1.25		

11.6 fl. oz. (0.33 litre) METAL CAN

CARBONATED SOFT DRINKS

Can type: Three-piece tin-free steel

Main system
Outer packaging   : Shrinkwrapped trays
Trunking (return journey) : Empty return load
Retail distribution   : Supermarket system
Retail sale   : Supermarket/Off licence

Variation considered
Outer packaging   : Cartons

SUMMARY OF TOTAL SYSTEM ENERGIES

All values in MJ/1000 containers

Main system
   (i) excluding retail sale and consumer use: Energy = 6401
  (ii) including retail sale and consumer use: Energy = 8936

Alternative systems
(a) Full cans packed in cartons.
   (i) excluding retail sale and consumer use: Energy = 7129
  (ii) including retail sale and consumer use: Energy = 9664

(b) Full cans packed in Hicone and trays
   (i) excluding retail sale and consumer use: Energy = 6299
  (ii) including retail sale and consumer use: Energy = 8834

VARIATIONS IN SYSTEM ENERGY IN MJ/container – see Chapter 29
Minimum   8.18
Maximum   10.49
Average   8.94

SYSTEM ENERGY REQUIREMENTS PER 1000 CONTAINERS

Contribution	Electricity/MJ		Oil fuels/MJ			Other fuels/MJ			Total energy /MJ
	Fuel production and delivery energy	Energy content of fuel	Fuel production and delivery energy	Energy content of fuel	Feedstock energy	Fuel production and delivery energy	Energy content of fuel	Feedstock energy	
Main system									
Filling and packing	318	106	56	275	64	1	68	74	962
Trunking	4	1	21	104	–	1	11	–	142
Retail delivery	2	1	11	57	–	–	6	–	77
Total 1	324	108	88	436	64	2	85	74	1181
Container manufacture	1289	569	185	908	264	205	1567	1	4988
Container delivery	2	1	8	42	–	1	4	–	58
Container packaging	54	18	4	18	–	2	28	19	143
Disposal	6	2	3	16	2	–	2	–	31
Total 2	1351	590	200	984	266	208	1601	20	5220
Retail sale	684	228	–	–	–	–	–	–	912
Consumer use	–	–	260	1363	–	–	–	–	1623
Total 3	684	228	260	1363	–	–	–	–	2535
Total 1 + 2	1675	698	288	1420	330	210	1686	94	6401
Total 1 + 2 + 3	2359	926	548	2783	330	210	1686	94	8936
Alternative systems									
(a) Full containers packed in cartons									
Revised Total 1 + 2 + 3	2561	994	582	2946	266	211	1842	262	9664
(b) Full cans packed in Hicone on trays									
Revised Total 1 + 2 + 3	2330	916	540	2744	313	210	1686	95	8834

System 20

MATERIALS REQUIREMENTS OF THE MAIN SYSTEM IN kg/1000 containers

Bauxite	31.73	Iron chromite	–	PET	–
Board	3.414	Iron ore	111.207	Polypropylene	–
Calcium sulphate	–	Lacquer	0.49	Sand	–
Carbon dioxide	0.88	Lead	–	Selenium	–
Cullet	–	Limestone	23.518	Sodium chloride	1.489
Detergent	–	Lining compound	0.161	Steel scrap	23.992
Epoxy adhesive	0.176	Manganese	0.88	Sulphuric acid	11.74
Feldspar	–	Methanol	–	Tin	–
Fluorspar	9.321	Pallets/number	0.634	Toluene	–
Flux (solder)	–	Paper	0.082	Varnish	0.334
Glue	–	Petroleum coke	2.516	Wood	0.135
Ink	0.36	Polyethylene	1.25		

11.6 fl. oz. (0.33 litre) METAL CAN

CARBONATED SOFT DRINKS

Can Type: Two-piece tinplate

Main system
Outer packaging   : Shrinkwrapped trays
Trunking (return journey) : Empty return load
Retail distribution   : Supermarket system
Retail sale   : Supermarket/Off-licence

Variations considered
Outer packaging   : Cartons
                          : Hicone and shrinkwrapped trays

SUMMARY OF TOTAL SYSTEM ENERGIES

All values in MJ/1000 containers

Main system
   (i) excluding retail sale and consumer use: Energy = 6594
  (ii) including retail sale and consumer use: Energy = 9129

Alternative systems
(a) Full cans packed in cartons.
   (i) excluding retail sale and consumer use: Energy = 7322
  (ii) including retail sale and consumer use: Energy = 9857

(b) Full cans packed in Hicone and trays.
   (i) excluding retail sale and consumer use: Energy = 6492
  (ii) including retail sale and consumer use: Energy = 9027

VARIATIONS IN SYSTEM ENERGY IN MJ/container — see Chapter 29
Minimum   8.25
Maximum   10.82
Average   9.13

SYSTEM ENERGY REQUIREMENTS PER 1000 CONTAINERS

Contribution	Electricity/MJ		Oil fuels/MJ			Other fuels/MJ			Total energy /MJ
	Fuel production and delivery energy	Energy content of fuel	Fuel production and delivery energy	Energy content of fuel	Feedstock energy	Fuel production and delivery energy	Energy content of fuel	Feedstock energy	
Main system									
Filling and packing	318	106	56	275	64	1	68	74	962
Trunking	4	1	21	104	–	1	11	–	142
Retail delivery	2	1	11	57	–	–	6	–	77
Total 1	324	108	88	436	64	2	85	74	1181
Container manufacture	1467	626	200	977	235	178	1503	–	5186
Container delivery	2	1	8	42	–	1	4	19	58
Container packaging	54	18	4	18	–	2	28	–	143
Disposal	5	2	2	13	2	–	2	–	26
Total 2	1528	647	214	1050	237	181	1537	19	5413
Retail sale	684	228	–	–	–	–	–	–	912
Consumer use	–	–	260	1363	–	–	–	–	1623
Total 3	684	228	260	1363	–	–	–	–	2535
Total 1 + 2	1852	755	302	1486	301	183	1622	93	6594
Total 1 + 2 + 3	2536	983	562	2849	301	183	1622	93	9129
Alternative systems									
(a) Full cans packed in cartons Revised Total 1 + 2 + 3	2738	1051	596	3012	237	184	1778	261	9857
(b) Full cans packed in Hicone in trays Revised Total 1 + 2 + 3	2507	973	554	2810	284	183	1622	94	9027

MATERIALS REQUIREMENTS OF THE MAIN SYSTEM IN kg/1000 containers

Bauxite	31.73	Iron chromite	–
Board	3.414	Iron ore	104.292
Calcium sulphate	–	Lacquer	0.403
Carbon dioxide	0.88	Lead	–
Cullet	–	Limestone	25.732
Detergent	–	Lining compound	0.091
Epoxy adhesive	–	Manganese	0.831
Feldspar	–	Methanol	–
Fluorspar	9.321	Pallets/number	0.634
Flux (solder)	–	Paper	0.082
Glue	–	Petroleum coke	2.516
Ink	0.076	Polyethylene	1.25
PET	–		
Polypropylene	–		
Sand	–		
Selenium	–		
Sodium Chloride	1.489		
Steel scrap	18.663		
Sulphuric acid	11.74		
Tin	0.483		
Toluene	–		
Varnish	0.218		
Wood	0.135		

11.6 fl. oz. (0.33 litre) METAL CAN

CARBONATED SOFT DRINKS

Can type: Two-piece aluminium

Main system
Outer packaging   : Shrinkwrapped trays
Trunking (return journey) : Empty return load
Retail distribution   : Supermarket system
Retail sale   : Supermarket/Off-licence

Variations considered
Outer packaging   : Cartons
                           : Hicone and shrinkwrapped trays

SUMMARY OF TOTAL SYSTEM ENERGIES

All values in MJ/1000 containers

Main system
   (i) excluding retail sale and consumer use: Energy = 10750
   (ii) including retail sale and consumer use: Energy = 13285

Alternative systems
(a) Full cans packed in cartons.
   (i) excluding retail sale and consumer use: Energy = 11478
   (ii) including retail sale and consumer use: Energy = 14013

(b) Full cans packed in Hicone and shrinkwrapped trays.
   (i) excluding retail sale and consumer use: Energy = 10647
   (ii) including retail sale and consumer use: Energy = 13182

VARIATIONS IN SYSTEM ENERGY IN MJ/container — see Chapter 29
Minimum   11.94
Maximum   15.43
Average   13.29

SYSTEM ENERGY REQUIREMENTS PER 1000 CONTAINERS

Contribution	Electricity/MJ		Oil fuels/MJ			Other fuels/MJ			Total energy /MJ
	Fuel production and delivery energy	Energy content of fuel	Fuel production and delivery energy	Energy content of fuel	Feedstock energy	Fuel production and delivery energy	Energy content of fuel	Feedstock energy	
Main system									
Filling and packing	318	106	56	275	64	1	68	74	962
Trunking	4	1	21	104	–	1	11	–	142
Retail delivery	2	1	11	57	–	–	6	–	77
Total 1	324	108	88	436	64	2	85	74	1181
Container manufacture	3633	1716	335	1654	652	104	1260	–	9354
Container delivery	2	1	8	42	–	1	4	–	58
Container packaging	54	18	4	18	–	2	28	19	143
Disposal	3	1	1	7	1	–	1	–	14
Total 2	3692	1736	348	1721	653	107	1293	19	9569
Retail sale	684	228	–	–	–	–	–	–	912
Consumer use	–	–	260	1363	–	–	–	–	1623
Total 3	684	228	260	1363	–	–	–	–	2535
Total 1 + 2	4016	1844	436	2157	717	109	1378	93	10750
Total 1 + 2 + 3	4700	2072	696	3520	717	109	1378	93	13285
Alternative systems									
(a) Full cans packed in cartons Revised Total 1 + 2 + 3	4902	2140	730	3683	653	110	1534	261	14013
(b) Full cans packed in Hicone in trays Revised Total 1 + 2 + 3	4671	2062	688	3481	700	109	1378	93	13182

System 22

MATERIALS REQUIREMENTS OF THE MAIN SYSTEM IN kg/1000 containers

Material	Value	Material	Value	Material	Value
Bauxite	141.360	Iron chromite	—	PET	—
Board	3.414	Iron ore	0.41	Polypropylene	—
Calcium sulphate	—	Lacquer	0.202	Sand	—
Carbon dioxide	0.88	Lead	—	Selenium	—
Cullet	—	Limestone	28.209	Sodium chloride	6.635
Detergent	—	Lining compound	0.091	Steel scrap	0.073
Epoxy adhesive	—	Manganese	0.003	Sulphuric acid	52.302
Feldspar	—	Methanol	—	Tin	—
Fluorspar	41.527	Pallets/number	0.634	Toluene	—
Flux (solder)	—	Paper	0.081	Varnish	0.184
Glue	—	Petroleum coke	11.21	Wood	0.135
Ink	0.063	Polyethylene	1.25		

15.5 fl. oz. (0.44 litre) METAL CAN

BEER

Can type: Three-piece tinplate

Main system
Outer packaging   : Hicone and trays
Trunking (return journey) : Empty return load
Retail distribution   : Supermarket system
Retail sale   : Supermarket/Off-licence

Variations considered
Outer packaging   : Cartons

SUMMARY OF TOTAL SYSTEM ENERGIES

All values in MJ/1000 containers

Main system
   (i) excluding retail sale and consumer use: Energy = 7486
  (ii) including retail sale and consumer use: Energy = 10021

Alternative system
Full cans packed in cartons
   (i) excluding retail sale and consumer use: Energy = 8712
  (ii) including retail sale and consumer use: Energy = 11247

VARIATIONS IN SYSTEM ENERGY IN MJ/container – see Chapter 29
Minimum   9.20
Maximum   11.45
Average   10.02

SYSTEM ENERGY REQUIREMENTS PER 1000 CONTAINERS

Contribution	Electricity/MJ		Oil fuels/MJ			Other fuels/MJ			Total energy /MJ
	Fuel production and delivery energy	Energy content of fuel	Fuel production and delivery energy	Energy content of fuel	Feedstock energy	Fuel production and delivery energy	Energy content of fuel	Feedstock energy	
Main system									
Filling and packing	427	142	82	398	47	9	170	86	1361
Trunking	8	3	41	208	1	2	22	–	285
Retail delivery	3	1	15	75	–	1	8	–	103
Total 1	438	146	138	681	48	12	200	86	1749
Container manufacture	1363	587	207	1011	250	231	1782	1	5432
Container delivery	2	1	11	58	–	1	6	–	79
Container packaging	72	24	5	23	–	3	37	25	189
Disposal	7	2	3	19	3	1	2	–	37
Total 2	1444	614	226	1111	253	236	1827	26	5737
Retail sale	684	228	–	–	–	–	–	–	912
Consumer use	–	–	260	1363	–	–	–	–	1623
Total 3	684	228	260	1363	–	–	–	–	2535
Total 1 + 2	1882	760	364	1792	301	248	2027	112	7486
Total 1 + 2 + 3	2566	988	624	3155	301	248	2027	112	10021
Alternative systems									
(a) Full containers packed in cartons									
Revised Total 1 + 2 + 3	2904	1101	692	3454	254	249	2246	347	11247

MATERIALS REQUIREMENTS OF THE MAIN SYSTEM IN kg/1000 containers

Bauxite	31.73	Iron chromite	—	PET	—
Board	3.836	Iron ore	157.527	Polypropylene	—
Calcium sulphate	—	Lacquer	0.45	Sand	—
Carbon dioxide	1.17	Lead	0.908	Selenium	—
Cullet	—	Limestone	33.865	Sodium chloride	1.489
Detergent	—	Lining compound	0.161	Steel scrap	24.907
Epoxy adhesive	—	Manganese	1.255	Sulphuric acid	11.74
Feldspar	—	Methanol	—	Tin	0.645
Fluorspar	9.321	Pallets/number	0.896	Toluene	—
Flux (solder)	0.10	Paper	0.082	Varnish	0.188
Glue	—	Petroleum coke	2.516	Wood	0.193
Ink	0.105	Polyethylene	0.942		

15.5 fl. oz. (0.44 litre) METAL CAN

BEER

Can type: Three-piece tin-free steel

Main system
Outer packaging   : Hicone and trays
Trunking (return journey) : Empty return load
Retail distribution   : Supermarket system
Retail sale   : Supermarket/Off-licence

Variation considered
Outer packaging   : Cartons

SUMMARY OF TOTAL SYSTEM ENERGIES

All values in MJ/1000 containers

Main system
  (i) excluding retail sale and consumer use: Energy = 7525
 (ii) including retail sale and consumer use: Energy = 10060

Alternative systems
Full cans packed in cartons.
  (i) excluding retail sale and consumer use: Energy = 8745
 (ii) including retail sale and consumer use: Energy = 11280

VARIATIONS IN SYSTEM ENERGY IN MJ/container – see chapter 29
Minimum   9.23
Maximum   12.21
Average   10.06

SYSTEM ENERGY REQUIREMENTS PER 1000 CONTAINERS

Contribution	Electricity/MJ		Oil fuels/MJ			Other fuels/MJ			Total energy /MJ
	Fuel production and delivery energy	Energy content of fuel	Fuel production and delivery energy	Energy content of fuel	Feedstock energy	Fuel production and delivery energy	Energy content of fuel	Feedstock energy	
Main system									
Filling and packing	427	142	82	398	47	9	170	86	1361
Trunking	8	3	41	208	1	2	22	–	285
Retail delivery	3	1	15	75	–	1	8	–	103
Total 1	438	146	138	681	48	12	200	86	1749
Container manufacture	1351	593	206	1005	285	237	1794	1	5472
Container delivery	2	1	11	58	–	1	6	–	79
Container packaging	72	24	5	23	–	3	37	25	189
Disposal	7	2	3	18	3	1	2	–	36
Total 2	1432	620	225	1104	288	242	1839	26	5776
Retail sale	684	228	–	–	–	–	–	–	912
Consumer use	–	–	260	1363	–	–	–	–	1623
Total 3	684	228	260	1363	–	–	–	–	2535
Total 1 + 2	1870	766	363	1785	336	254	2039	112	7525
Total 1 + 2 + 3	2554	994	623	3148	336	254	2039	112	10060
Alternative systems									
(a) Packaging full cans in cartons									
Revised Total 1 + 2 + 3	2892	1107	685	3447	289	255	2258	347	11280

System 24

MATERIALS REQUIREMENTS OF THE MAIN SYSTEM IN kg/1000 containers

Bauxite	31.73	Iron chromite	—	PET	—
Board	3.836	Iron ore	111.318	Polypropylene	—
Calcium sulphate	—	Lacquer	0.49	Sand	—
Carbon dioxide	1.17	Lead	—	Selenium	—
Cullet	—	Limestone	23.538	Sodium chloride	1.489
Detergent	—	Lining compound	0.161	Steel scrap	24.012
Epoxy adhesive	0.176	Manganese	0.881	Sulphuric acid	11.74
Feldspar	—	Methanol	—	Tin	—
Fluorspar	9.321	Pallets/number	0.896	Toluene	—
Flux (solder)	—	Paper	0.082	Varnish	0.334
Glue	—	Petroleum coke	2.516	Wood	0.193
Ink	0.36	Polyethylene	0.942		

15.5 fl. oz. (0.44 litre) METAL CAN

BEER

Can type: Two-piece aluminium

Main system
Outer packaging	: Hicone and trays
Trunking (return journey)	: Empty return load
Retail distribution	: Supermarket system
Retail sale	: Supermarket/Off-licence

Variations considered
Outer packaging	: Cartons

SUMMARY OF TOTAL SYSTEM ENERGIES

All values in MJ/1000 containers

Main system
- (i) excluding retail sale and consumer use: Energy = 12050
- (ii) including retail sale and consumer use: Energy = 14585

Alternative system
Full cans in cartons.
- (i) excluding retail sale and consumer use: Energy = 13270
- (ii) including retail sale and consumer use: Energy = 15805

VARIATIONS IN SYSTEM ENERGY IN MJ/container — see Chapter 29
Minimum   13.19
Maximum   17.34
Average   14.59

SYSTEM ENERGY REQUIREMENTS PER 1000 CONTAINERS

Contribution	Electricity/MJ		Oil fuels/MJ			Other fuels/MJ			Total energy /MJ
	Fuel production and delivery energy	Energy content of fuel	Fuel production and delivery energy	Energy content of fuel	Feedstock energy	Fuel production and delivery energy	Energy content of fuel	Feedstock energy	
Main system									
Filling and packing	427	142	82	398	47	9	170	86	1361
Trunking	8	3	41	208	1	2	22	–	285
Retail delivery	3	1	15	75	–	1	8	–	103
Total 1	438	146	138	681	48	12	200	86	1749
Container manufacture	3891	1843	360	1778	711	110	1325	–	10018
Container delivery	2	1	11	58	–	1	6	–	79
Container packaging	72	24	5	23	–	3	37	25	189
Disposal	3	1	1	8	1	–	1	–	15
Total 2	3968	1869	377	1867	712	114	1369	25	10301
Retail sale	684	228	–	–	–	–	–	–	912
Consumer use	–	–	260	1363	–	–	–	–	1623
Total 3	684	228	260	1363	–	–	–	–	2535
Total 1 + 2	4406	2015	515	2548	760	126	1569	111	12050
Total 1 + 2 + 3	5090	2243	775	3911	760	126	1569	111	14585
Alternative systems									
(a) Full containers packed in cartons									
Revised Total 1 + 2 + 3	5428	2356	837	4210	713	127	1788	346	15805

MATERIALS REQUIREMENTS OF THE MAIN SYSTEM IN kg/1000 containers

Material	kg	Material	kg	Material	kg
Bauxite	142.471	Iron chromite	—	PET	—
Board	3.836	Iron ore	0.410	Polypropylene	—
Calcium sulphate	—	Lacquer	0.239	Sand	—
Carbon dioxide	1.17	Lead	—	Selenium	—
Cullet	—	Limestone	28.507	Sodium chloride	6.687
Detergent	—	Lining compound	0.091	Steel scrap	0.073
Epoxy adhesive	—	Manganese	0.003	Sulphuric acid	52.713
Feldspar	—	Methanol	—	Tin	—
Fluorspar	41.835	Pallets/number	0.896	Toluene	—
Flux (solder)	—	Paper	0.082	Varnish	0.227
Glue	—	Petroleum coke	11.299	Wood	0.193
Ink	0.079	Polyethylene	0.942		

15.5 fl. oz. (0.44 litre) METAL CAN

CARBONATED SOFT DRINKS

Can type: Three-piece tinplate

Main system

Outer packaging	: Shrinkwrapped trays
Trunking (return journey)	: Empty return load
Retail distribution	: Supermarket system
Retail sale	: Supermarket/Off-licence

Variations considered

Outer packaging	: Cartons
Outer packaging	: Hicone and trays

SUMMARY OF TOTAL SYSTEM ENERGIES

All values in MJ/1000 containers

Main system

    (i)   excluding retail sale and consumer use: Energy = 7245

    (ii) including retail sale and consumer use: Energy = 9780

Alternative systems

(a) Full cans packaged in cartons.

    (i)   excluding retail sale and consumer use: Energy = 8282

    (ii) including retail sale and consumer use: Energy = 10817

(b) Full cans packaged in Hicone and trays

    (i)   excluding retail sale and consumer use: Energy = 7064

    (ii) including retail sale and consumer use: Energy = 9599

VARIATIONS IN SYSTEM ENERGY IN MJ/container – see Chapter 29

Minimum	8.84
Maximum	11.59
Average	9.78

SYSTEM ENERGY REQUIREMENTS PER 1000 CONTAINERS

Contribution	Electricity/MJ		Oil fuels/MJ			Other fuels/MJ			Total energy /MJ
	Fuel production and delivery energy	Energy content of fuel	Fuel production and delivery energy	Energy content of fuel	Feedstock energy	Fuel production and delivery energy	Energy content of fuel	Feedstock energy	
Main system									
Filling and packing	363	121	66	326	85	2	77	86	1126
Trunking	8	3	41	208	–	2	22	–	284
Retail delivery	3	1	14	71	–	1	8	–	98
Total 1	374	125	121	605	85	5	107	86	1508
Container manufacture	1363	587	207	1011	250	231	1782	1	5432
Container delivery	2	1	11	58	–	1	6	–	79
Container packaging	72	24	5	23	–	3	37	25	189
Disposal	7	2	3	19	3	1	2	–	37
Total 2	1444	614	226	1111	253	236	1827	26	5737
Retail sale	684	228	–	–	–	–	–	–	912
Consumer use	–	–	260	1363	–	–	–	–	1623
Total 3	684	228	260	1363	–	–	–	–	2535
Total 1 + 2	1818	739	347	1716	338	241	1934	112	7245
Total 1 + 2 + 3	2502	967	607	3079	338	241	1934	112	9780
Alternative systems									
(a) Full cans packed in cartons Revised Total 1 + 2 + 3	2790	1063	656	3314	253	241	2153	347	10817
(b) Full cans packed in Hicone in trays Revised Total 1 + 2 + 3	2452	950	595	3015	300	241	1934	112	9599

System 26

MATERIALS REQUIREMENTS OF THE MAIN SYSTEM IN kg/1000 containers

Bauxite	31.73	Iron chromite	–	PET	–
Board	3.836	Iron ore	157.527	Polypropylene	–
Calcium sulphate	–	Lacquer	0.45	Sand	–
Carbon dioxide	0.88	Lead	–	Selenium	–
Cullet	–	Limestone	33.865	Sodium chloride	1.489
Detergent	–	Lining compound	0.161	Steel scrap	24.907
Epoxy adhesive	–	Manganese	1.255	Sulphuric acid	11.74
Feldspar	–	Methanol	–	Tin	1.232
Fluorspar	9.321	Pallets/number	0.896	Toluene	–
Flux (solder)	0.10	Paper	0.082	Varnish	0.188
Glue	–	Petroleum coke	2.516	Wood	0.193
Ink	0.105	Polyethylene	1.67		

15.5 fl. oz. (0.44 litre) METAL CAN

CARBONATED SOFT DRINKS

Can type: Three-piece tin-free steel

Main system

Outer packaging	: Shrinkwrapped trays
Trunking (return journey)	: Empty return load
Retail distribution	: Supermarket system
Retail sale	: Supermarket/Off-licence

Variations considered

Outer packaging	: Cartons
Outer packaging	: Hicone and trays

SUMMARY OF TOTAL SYSTEM ENERGIES

All values in MJ/1000 containers

Main system

  (i) excluding retail sale and consumer use: Energy = 7284

  (ii) including retail sale and consumer use: Energy = 9819

Alternative systems

(a) Full cans packaged in cartons.

  (i) excluding retail sale and consumer use: Energy = 8321

  (ii) including retail sale and consumer use: Energy = 10856

(b) Full cans packaged in Hicone and trays.

  (i) excluding retail sale and consumer use: Energy = 7103

  (ii) including retail sale and consumer use: Energy = 9638

VARIATIONS IN SYSTEM ENERGY IN MJ/container – see Chapter 29

Minimum	9.00
Maximum	11.76
Average	9.82

SYSTEM ENERGY REQUIREMENTS PER 1000 CONTAINERS

Contribution	Electricity/MJ		Oil fuels/MJ			Other fuels/MJ			Total energy /MJ
	Fuel production and delivery energy	Energy content of fuel	Fuel production and delivery energy	Energy content of fuel	Feedstock energy	Fuel production and delivery energy	Energy content of fuel	Feedstock energy	
Main system									
Filling and packing	363	121	66	326	85	2	77	86	1126
Trunking	8	3	41	208	–	2	22	–	284
Retail delivery	3	1	14	17	–	1	8	–	98
Total 1	374	125	121	605	85	5	107	86	1508
Container manufacture	1351	593	206	1005	285	237	1794	1	5472
Container delivery	2	1	11	58	–	1	6	–	79
Container packaging	72	24	5	23	–	3	37	25	189
Disposal	7	2	3	18	3	1	2	–	36
Total 2	1432	620	225	1104	288	242	1839	26	5776
Retail sale	684	228	–	–	–	–	–	–	912
Consumer use	–	–	260	1363	–	–	–	–	1623
Total 3	684	228	260	1363	–	–	–	–	2535
Total 1 + 2	1806	745	346	1709	373	247	1946	112	7284
Total 1 + 2 + 3	2490	973	606	3072	373	247	1946	112	9819
Alternative systems									
(a) Full bottles packed in Hicone and trays									
Revised Total 1 + 2 + 3	2440	956	594	3008	335	247	1946	112	9638
(b) Full cans packed in cartons									
Revised Total 1 + 2 + 3	2778	1069	655	3307	288	247	2165	347	10856

MATERIALS REQUIREMENTS OF THE MAIN SYSTEM IN kg/1000 containers

Bauxite	31.73	Iron chromite	–	PET	–
Board	3.836	Iron ore	94.786	Polypropylene	–
Calcium sulphate	–	Lacquer	0.448	Sand	–
Carbon dioxide	1.17	Lead	–	Selenium	–
Cullet	–	Limestone	20.985	Sodium chloride	1.489
Detergent	–	Lining compound	0.161	Steel scrap	20.443
Epoxy adhesive	0.132	Manganese	0.751	Sulphuric acid	11.74
Feldspar	–	Methanol	–	Tin	–
Fluorspar	9.321	Pallets/number	0.896	Toluene	–
Flux (solder)	–	Paper	0.082	Varnish	0.234
Glue	–	Petroleum coke	2.516	Wood	0.193
Ink	0.28	Polyethylene	1.67		

15.5 fl. oz. (0.44 litre) METAL CAN

CARBONATED SOFT DRINKS

Can type: Two-piece aluminium

Main system
Outer packaging : Shrinkwrapped trays
Trunking (return journey) : Empty return load
Retail distribution : Supermarket system
Retail sale : Supermarket/Off-licence

Variations considered
Outer packaging : Cartons
: Hicone and shrinkwrapped trays

SUMMARY OF TOTAL SYSTEM ENERGIES

All values in MJ/1000 containers

Main system
    (i) excluding retail sale and consumer use: Energy = 11809
    (ii) including retail sale and consumer use: Energy = 14344

Alternative systems
(a) Full cans packaged in cartons.
    (i) excluding retail sale and consumer use: Energy = 12846
    (ii) including retail sale and consumer use: Energy = 15381

(b) Full cans packaged in Hicone and trays.
    (i) excluding retail sale and consumer use: Energy = 11628
    (ii) including retail sale and consumer use: Energy = 14163

VARIATIONS IN SYSTEM ENERGY IN MJ/container — see Chapter 29
Minimum   12.96
Maximum   16.90
Average   14.34

SYSTEM ENERGY REQUIREMENTS PER 1000 CONTAINERS

Contribution	Electricity/MJ		Oil fuels/MJ			Other fuels/MJ			Total energy /MJ
	Fuel production and delivery energy	Energy content of fuel	Fuel production and delivery energy	Energy content of fuel	Feedstock energy	Fuel production and delivery energy	Energy content of fuel	Feedstock energy	
Main system									
Filling and packing	363	121	66	326	85	2	77	86	1126
Trunking	8	3	41	208	–	2	22	–	284
Retail delivery	3	1	14	71	–	1	8	–	98
Total 1	374	125	121	605	85	5	107	86	1508
Container manufacture	3891	1843	360	1778	711	110	1325	–	10018
Container delivery	2	1	11	58	–	1	6	–	79
Container packaging	72	24	5	23	–	3	37	25	189
Disposal	3	1	1	8	1	–	1	–	15
Total 2	3968	1869	377	1867	712	114	1369	25	10301
Retail sale	684	228	–	–	–	–	–	–	912
Consumer use	–	–	260	1363	–	–	–	–	1623
Total 3	684	228	260	1363	–	–	–	–	2535
Total 1 + 2	4342	1994	498	2472	797	119	1476	111	11809
Total 1 + 2 + 3	5026	2222	758	3835	797	119	1476	111	14344
Alternative systems									
(a) Full cans packaged in cartons									
Revised Total 1 + 2 + 3	5314	2318	807	4070	712	119	1695	346	15381
(b) Full cans packed in Hicone in trays									
Revised Total 1 + 2 + 3	4976	2205	746	3771	759	119	1476	111	14163

System 28

MATERIALS REQUIREMENTS FOR THE MAIN SYSTEM IN kg/1000 containers

Material	Value	Material	Value	Material	Value
Bauxite	142.471	Iron chromite	—	PET	—
Board	3.836	Iron ore	0.41	Polypropylene	—
Calcium sulphate	—	Lacquer	0.239	Sand	—
Carbon dioxide	1.17	Lead	—	Selenium	—
Cullet	—	Limestone	28.507	Sodium chloride	6.687
Detergent	—	Lining compound	0.091	Steel scrap	0.073
Epoxy adhesive	—	Manganese	0.003	Sulphuric acid	52.713
Feldspar	—	Methanol	—	Tin	—
Fluorspar	41.853	Pallets/number	0.896	Toluene	—
Flux (solder)	—	Paper	0.082	Varnish	0.227
Glue	—	Petroleum coke	11.299	Wood	0.193
Ink	0.079	Polyethylene	1.67		

17.6 fl. oz. (0.50 litre) NON-RETURNABLE GLASS BOTTLE

BEER

Bottle mass = 0.330 kg

Main system

Empty container delivery	: Bulk palletised
Closure employed	: Crown
Outer packaging	: Shrinkwrapped trays
Trunking (return journey)	: Empty return load
Retail distribution	: Supermarket system
Retail sale	: Supermarket/Off-licence

Variations considered

Closure employed	: Standard aluminium screw
Outer packaging	: Cartons

SUMMARY OF TOTAL SYSTEM ENERGIES

All values are in MJ/1000 containers

Main system
    (i) excluding retail sale and consumer use: Energy = 11688
    (ii) including retail sale and consumer use: Energy = 14223

Alternative system
(a) Using aluminium screw closures and shrinkwrapped trays.
    (i) excluding retail sale and consumer use: Energy = 12173
    (ii) including retail sale and consumer use: Energy = 14708

(b) Using crown closures and packaging in cartons.
    (i) excluding retail sale and consumer use: Energy = 13202
    (ii) including retail sale and consumer use: Energy = 15737

(c) Using aluminium screw closures and packaging in cartons.
    (i) excluding retail sale and consumer use: Energy = 13689
    (ii) including retail sale and consumer use: Energy = 16224

VARIATION IN SYSTEM ENERGY IN MJ/container — see Chapter 29
Minimum   11.89
Maximum   17.39
Average   14.22

SYSTEM ENERGY REQUIREMENTS PER 1000 CONTAINERS

Contribution	Electricity/MJ		Oil fuels/MJ			Other fuels/MJ			Total energy /MJ
	Fuel production and delivery energy	Energy content of fuel	Feedstock energy	Fuel production and delivery energy	Energy content of fuel	Fuel production and delivery energy	Energy content of fuel	Feedstock energy	
Main system									
Filling and packing	759	254	99	182	883	32	411	153	2773
Trunking	11	4	1	54	276	2	29	–	377
Retail delivery	5	2	–	23	119	1	13	–	163
Total 1	775	260	100	259	1278	35	453	153	3313
Container manufacture	1290	430	–	580	2850	200	1960	–	7310
Container delivery	7	2	–	31	156	3	16	–	215
Container packaging	171	57	47	32	160	–	57	105	629
Disposal	43	13	17	20	112	3	13	105	221
Total 2	1511	502	64	663	3278	206	2046	105	8375
Retail sale	684	228	–	–	–	–	–	–	912
Consumer use	–	228	–	260	1363	–	–	–	1623
Total 3	684	228	–	260	1363	–	–	–	2535
Total 1 + 2	2286	762	164	922	4556	241	2499	258	11688
Total 1 + 2 + 3	2970	990	164	1182	5919	241	2499	258	14223
Alternative systems									
(a) Using aluminium screw closures, full bottles in shrinkwrapped trays Revised Total 1 + 2 + 3	3164	1085	248	1203	6023	234	2492	259	14708
(b) Using crown closures and full bottles in cartons Revised Total 1 + 2 + 3	3388	1130	79	1255	6273	241	2795	576	15737
(c) Using aluminium screw closures and cartons for full bottles Revised Total 1 + 2 + 3	3583	1225	163	1276	6377	235	2788	577	16224

MATERIALS REQUIREMENTS FOR THE MAIN SYSTEM IN kg/1000 containers

Bauxite	9.24	Iron chromite	0.264	PET	—
Board	9.13	Iron ore	6.971	Polypropylene	—
Calcium sulphate	2.31	Lacquer	0.039	Sand	249.81
Carbon dioxide	1.33	Lead	—	Selenium	0.003
Cullet	2.31	Limestone	148.467	Sodium chloride	112.86
Detergent	—	Lining compound	0.214	Steel scrap	1.231
Epoxy adhesive	—	Manganese	0.056	Sulphuric acid	0.66
Feldspar	85.8	Methanol	0.003	Tin	0.032
Fluorspar	—	Pallets/number	0.196	Toluene	0.004
Flux (solder)	—	Paper	0.456	Varnish	0.023
Glue	0.15	Petroleum coke	—	Wood	—
Ink	0.033	Polyethylene	2.58		

17.6 fl. oz. (0.50 litre) NON-RETURNABLE GLASS BOTTLE

CARBONATED SOFT DRINKS

Bottle mass = 0.350 kg

Main system

Empty container delivery	: Bulk palletised
Closure employed	: Standard aluminium screw
Outer packaging	: Shrinkwrapped trays
Trunking (return journey)	: Empty return load
Retail distribution	: Supermarket system
Retail sale	: Supermarket/Off-licence

Variations considered

Empty bottle delivery	: Alternative packaging
Outer packaging	: Cartons

SUMMARY OF TOTAL SYSTEM ENERGIES

All values in MJ/1000 containers

Main system
- (i) excluding retail sale and consumer use: Energy = 12009
- (ii) including retail sale and consumer use: Energy = 14544

Alternative systems
- (a) Empty bottles bulk palletised, full bottles in cartons.
  - (i) excluding retail sale and consumer use: Energy = 13524
  - (ii) including retail sale and consumer use: Energy = 16059

- (b) Empty bottles packed in alternative packaging, full bottles in shrink-wrapped trays.
  - (i) excluding retail sale and consumer use: Energy = 12531
  - (ii) including retail sale and consumer use: Energy = 15066

- (c) Empty bottle packed in alternative packaging, full bottles in cartons.
  - (i) excluding retail sale and consumer use: Energy = 14046
  - (ii) including retail sale and consumer use: Energy = 16581

VARIATIONS IN SYSTEM ENERGY IN MJ/container – see Chapter 29
Minimum   11.45
Maximum   17.35
Average   14.54

SYSTEM ENERGY REQUIREMENTS PER 1000 CONTAINERS

Contribution	Electricity/MJ		Oil fuels/MJ			Other fuels/MJ			Total energy /MJ
	Fuel production and delivery energy	Energy content of fuel	Fuel production and delivery energy	Energy content of fuel	Feedstock energy	Fuel production and delivery energy	Energy content of fuel	Feedstock energy	
Main system									
Filling and packing	792	295	183	890	183	11	237	154	2745
Trunking	7	2	35	178	–	2	19	–	243
Retail delivery	5	2	26	133	–	1	14	–	181
Total 1	804	299	244	1201	183	14	270	154	3169
Container manufacture	1370	460	620	3020	–	210	2080	–	7760
Container delivery	7	2	31	156	–	3	16	–	215
Container packaging	171	57	32	160	47	–	57	105	629
Disposal	46	14	21	119	18	4	14	–	236
Total 2	1594	533	704	3455	65	217	2167	105	8840
Retail sale	684	228	–	–	–	–	–	–	912
Consumer use	–	–	260	1363	–	–	–	–	1623
Total 3	684	228	260	1363	–	–	–	–	2535
Total 1 + 2	2398	832	948	4656	248	231	2437	259	12009
Total 1 + 2 + 3	3082	1060	1208	6019	248	231	2437	259	14544
Alternative systems									
(a) Empty bottles bulk palletised, full bottles in cartons									
Revised Total 1 + 2 + 3	3500	1200	1281	6373	163	232	2733	577	16059
(b) Empty bottles in alternative packs, full bottles in shrinkwrapped trays									
Revised Total 1 + 2 + 3	3224	1107	1235	6148	248	232	2520	352	15066
(c) Empty bottles in alternative packs, full bottles in cartons									
Revised Total 1 + 2 + 3	3642	1247	1308	6502	163	233	2816	670	16581

MATERIALS REQUIREMENTS OF THE MAIN SYSTEM IN MJ/1000 containers

System 30

Bauxite	18.02	Iron chromite	0.28	PET	–
Board	9.13	Iron ore	–	Polypropylene	–
Calcium sulphate	2.45	Lacquer	0.193	Sand	264.95
Carbon dioxide	1.33	Lead	–	Selenium	0.004
Cullet	2.45	Limestone	157.74	Sodium chloride	120.086
Detergent	–	Lining compound	0.523	Steel scrap	–
Epoxy adhesive	–	Manganese	–	Sulphuric acid	3.741
Feldspar	91.0	Methanol	0.003	Tin	–
Fluorspar	2.415	Pallets/number	0.196	Toluene	0.004
Flux (solder)	–	Paper	0.456	Varnish	0.193
Glue	0.15	Petroleum coke	0.652	Wood	–
Ink	0.031	Polyethylene	2.58		

19.4 fl. oz. (0.55 litre) RETURNABLE GLASS BOTTLE

BEER

Bottle mass = 0.470 kg

Main system
Empty container delivery : Bulk palletised
Closure employed : Crown
Outer packaging : Returnable crates
Trunking (return journey) : Carrying empty bottles, crates, etc.
Retail distribution : Brewery system
Retail sale : Public houses

Variations considered
Empty container delivery : Customer crates

SUMMARY OF TOTAL SYSTEM ENERGIES AS A FUNCTION OF TRIPPAGE (t)

All values in MJ/1000 containers

Main system
   (i) excluding retail sale and consumer use: Energy = $3671 + 11611/t$
   (ii) including retail sale and consumer use: Energy = $6166 + 11611/t$

Alternative system
Empty bottles packed in customer crates.
   (i) excluding retail sale and consumer use: Energy = $3671 + 11074/t$
   (ii) including retail sale and consumer use: Energy = $6166 + 11074/t$

VARIATIONS IN SYSTEM ENERGY IN MJ/container – see Chapter 29
Minimum $6.29 + 8.36/t$
Maximum $6.95 + 15.17/t$
Average $6.62 + 11.61/t$

SYSTEM ENERGY REQUIREMENTS PER 1000 CONTAINERS

Contribution	Electricity/MJ		Oil fuels/MJ			Other fuels/MJ			Total energy /MJ
	Fuel production and delivery energy	Energy content of fuel	Fuel production and delivery energy	Energy content of fuel	Feedstock energy	Fuel production and delivery energy	Energy content of fuel	Feedstock energy	
Main system									
Filling and packing	860	288	191	923	137	38	397	57	2891
Trunking	14	5	71	363	1	3	38	–	495
Retail delivery	10	3	41	210	–	2	19	–	285
Total 1	884	296	303	1496	138	43	454	57	3671
Container manufacture	1840	610	830	4050	–	290	2790	–	10410
Container delivery	6	2	29	150	–	3	16	–	206
Container packaging	184	61	35	173	52	–	61	113	679
Disposal	61	19	28	160	24	5	19	–	316
Total 2	2091	692	922	4533	76	298	2886	113	11611
Retail sale	670	223	34	161	–	18	216	–	1322
Consumer use	–	–	260	1363	–	–	–	–	1623
Total 3	670	223	294	1524	–	18	216	–	2945
Total 1 + 2	2975	988	1225	6029	214	341	3340	170	15282
Total 1 + 2 + 3	3645	1211	1519	7553	214	359	3556	170	18227
Alternative systems									
(a) Empty bottles packed in customer crates									
Revised Total 2	1948	645	892	4383	24	298	2828	56	11074

MATERIALS REQUIREMENTS FOR THE MAIN SYSTEM IN kg/1000 containers

Bauxite	13.16/t	Iron chromite	0.376/t	PET	—
Board	3.5/t	Iron ore	6.971	Polypropylene	2.33
Calcium sulphate	3.29/t	Lacquer	0.039	Sand	355.79/t
Carbon dioxide	1.46	Lead	—	Selenium	0.005/t
Cullet	3.29/t	Limestone	1.287 + 209.62/t	Sodium chloride	160.74/t
Detergent	1.85	Lining compound	0.214	Steel scrap	1.231
Epoxy adhesive	—	Manganese	0.056	Sulphuric acid	0.94/t
Feldspar	122.2/t	Methanol	0.003	Tin	0.032
Fluorspar	—	Pallets/number	0.108 + 0.115/t	Toluene	0.004
Flux (solder)	—	Paper	0.456	Varnish	0.028
Glue	0.15	Petroleum coke	—	Wood	—
Ink	0.033	Polyethylene	1.01/t		

Appendix 1

19.4 fl. oz. (0.55 litre) NON-RETURNABLE GLASS BOTTLE

BEER

Bottle mass = 0.365 kg

Main system

Empty container delivery	: Bulk palletised
Closure employed	: Crown
Outer packaging	: Shrinkwrapped trays
Trunking (return journey)	: Empty return load
Retail distribution	: Supermarket system
Retail sale	: Supermarket/Off-licence

Variations considered

Closure employed	: Standard aluminium screw
Outer packaging	: Cartons

SUMMARY OF TOTAL SYSTEM ENERGIES

All values in MJ/1000 containers

Main system
(i) excluding retail sale and consumer use: Energy = 12922
(ii) including retail sale and consumer use: Energy = 15457

Variations considered
(a) Using aluminium screw closures and packaging in shrinkwrapped trays.
(i) excluding retail sale and consumer use: Energy = 13409
(ii) including retail sale and consumer use: Energy = 15944

(b) Using crown closures and packaging in cartons.
(i) excluding retail sale and consumer use: Energy = 14436
(ii) including retail sale and consumer use: Energy = 16971

(c) Using aluminium screw closures and packaging in cartons
(i) excluding retail sale and consumer use: Energy = 14921
(ii) including retail sale and consumer use: Energy = 17456

VARIATIONS IN SYSTEM ENERGY IN MJ/container – see **Chapter 29**
Minimum   12.89
Maximum   19.19
Average   15.46

SYSTEM ENERGY REQUIREMENTS PER 1000 CONTAINERS

Contribution	Electricity/MJ			Oil fuels/MJ			Other fuels/MJ			Total energy /MJ
	Fuel production and delivery energy	Energy content of fuel	Feedstock energy	Fuel production and delivery energy	Energy content of fuel	Feedstock energy	Fuel production and delivery energy	Energy content of fuel	Feedstock energy	
Main system										
Filling and packing	832	279	–	208	1006	99	36	464	153	3077
Trunking	11	4	–	56	287	1	3	30	–	392
Retail delivery	6	2	–	30	152	–	1	16	–	207
Total 1	849	285	–	294	1445	100	40	510	153	3676
Container manufacture	1430	480	–	650	3150	–	220	2160	–	8090
Container delivery	7	2	–	33	169	–	3	18	–	232
Container packaging	184	61	–	35	173	52	–	61	113	679
Disposal	47	15	–	22	124	18	4	15	–	245
Total 2	1668	558	–	740	3616	70	227	2254	113	9246
Retail sale	684	228	–	–	–	–	–	–	–	912
Consumer use	–	–	–	260	1363	–	–	–	–	1623
Total 3	684	228	–	260	1363	–	–	–	–	2535
Total 1 + 2	2517	843	–	1034	5061	170	267	2764	266	12922
Total 1 + 2 + 3	3201	1071	–	1294	6424	170	267	2764	266	15457

Alternative systems

(a) Using aluminium screw closures, packing full containers in shrink-wrapped trays

Revised Total 1 + 2 + 3	3396	1166	–	1315	6529	254	260	2757	267	15944

(b) Using crown closures, full bottles in cartons

Revised Total 1 + 2 + 3	3619	1210	–	1367	6778	85	268	3060	584	16971

(c) Using aluminium screw closures, full bottles in cartons

Revised Total 1 + 2 + 3	3814	1305	–	1388	6882	169	261	3052	585	17456

System 32

MATERIALS REQUIREMENTS FOR THE MAIN SYSTEM IN kg/100 containers

Bauxite	10.22	Iron chromite	0.292	PET	—
Board	9.33	Iron ore	6.971	Polypropylene	—
Calcium sulphate	2.555	Lacquer	0.039	Sand	276.305
Carbon dioxide	1.46	Lead	—	Selenium	0.004
Cullet	2.555	Limestone	164.077	Sodium chloride	124.83
Detergent	—	Lining compound	0.214	Steel scrap	1.231
Epoxy adhesive	—	Manganese	0.056	Sulphuric acid	0.73
Feldspar	94.9	Methanol	0.003	Tin	0.032
Fluorspar	—	Pallets/number	0.206	Toluene	0.004
Flux (solder)	—	Paper	0.456	Varnish	0.023
Glue	0.15	Petroleum coke	—	Wood	—
Ink	0.033	Polyethylene	2.67		

25 fl. oz. (0.71 litre) RETURNABLE GLASS BOTTLE

CARBONATED SOFT DRINKS

Bottle mass = 0.580 kg

Main system
Empty bottle delivery : Bulk palletised
Closure employed : Standard aluminium screw
Outer packaging : Returnable crates
Trunking (return load) : Carrying empty bottles, crates, etc.
Retail distribution : Brewery system
Retail sale : Public houses

Variations considered
Empty bottle delivery : Customer crates
Retail sale : Off-licences

SUMMARY OF TOTAL SYSTEM ENERGIES AS A FUNCTION OF TRIPPAGE (t)

All values in MJ/1000 containers

Main system
    (i) excluding retail sale and consumer use: Energy = $3261 + 14415/t$
    (ii) including retail sale and consumer use: Energy = $6206 + 14415/t$

Alternative systems
(a) Empty bottles packed in customer crates, sales through public houses.
    (i) excluding retail sale and consumer use: Energy = $3261 + 13750/t$
    (ii) including retail sale and consumer use: Energy = $6206 + 13750/t$

(b) Empty bottles packed in customer crates, sales through off-licences.
    (i) excluding retail sale and consumer use: Energy = $3261 + 13750/t$
    (ii) including retail sale and consumer use: Energy = $5796 + 13750/t$

(c) Empty bottles bulk palletised, sales through off-licences.
    (i) excluding retail sale and consumer use: Energy = $3261 + 14415/t$
    (ii) including retail sale and consumer use: Energy = $5796 + 14415/t$

VARIATIONS IN SYSTEM ENERGY IN MJ/container -- see Chapter 29
Minimum $5.90 + 9.78/t$
Maximum $6.52 + 20.68/t$
Average $6.21 + 14.42/t$

SYSTEM ENERGY REQUIRMENTS PER 1000 CONTAINERS

Contribution	Electricity/MJ		Oil fuels/MJ			Other fuels/MJ			Total energy /MJ
	Fuel production and delivery energy	Energy content of fuel	Fuel production and delivery energy	Energy content of fuel	Feedstock energy	Fuel production and delivery energy	Energy content of fuel	Feedstock energy	
Main system									
Filling and packing	770	288	176	859	221	12	158	92	2576
Trunking	11	4	55	279	1	2	30	–	382
Retail delivery	10	3	44	223	–	2	21	–	303
Total 1	791	295	275	1361	222	16	209	92	3261
Container manufacture	2270	760	1030	5000	–	360	3440	–	12860
Container delivery	10	3	47	240	1	4	25	–	330
Container packaging	227	76	43	213	63	–	75	140	837
Disposal	75	23	35	197	29	6	23	–	388
Total 2	2582	862	1155	5650	93	370	3563	140	14415
Retail sale	670	223	34	161	–	18	216	–	1322
Consumer use	–	–	260	1363	–	–	–	–	1623
Total 3	670	223	294	1524	–	18	216	–	2945
Total 1 + 2	3373	1157	1430	7011	315	386	3772	232	17676
Total 1 + 2 + 3	4043	1380	1724	8535	315	404	3988	232	20621
Alternative systems									
(a) Customer crates used to deliver empty bottles									
Revised Total 2	2405	803	1118	5465	30	370	3491	68	13750
(b) Retail sales through off-licences									
Revised Total 3	684	228	260	1363	–	–	–	–	2535

MATERIALS REQUIREMENTS OF THE MAIN SYSTEM IN kg/1000 containers

Bauxite	$8.22 + 16.24/t$	Iron chromite	$0.464/t$	PET	—
Board	$4.35/t$	Iron ore	—	Polypropylene	2.33
Calcium sulphate	$4.06/t$	Lacquer	0.193	Sand	$439.06/t$
Carbon dioxide	1.80	Lead	—	Selenium	$0.006/t$
Cullet	$4.06/t$	Limestone	$1.64 + 258.68/t$	Sodium chloride	$0.386 + 198.36/t$
Detergent	1.85	Lining compound	0.523	Steel scrap	—
Epoxy adhesive	—	Manganese	—	Sulphuric acid	$3.041 + 1.16/t$
Feldspar	$150.8/t$	Methanol	0.003	Tin	—
Fluorspar	2.415	Pallets/number	$0.189 + 0.141/t$	Toluene	0.004
Flux (solder)	—	Paper	0.456	Varnish	0.193
Glue	0.15	Petroleum coke	0.652	Wood	—
Ink	0.031	Polyethylene	$1.24/t$		

26.4 fl. oz. (0.75 litre) NON-RETURNABLE GLASS BOTTLE

CARBONATED SOFT DRINKS

Bottle mass = 0.500 kg

Main system
Empty container delivery   : Bulk palletised
Closure employed   : Standard aluminium screw
Outer packaging   : Shrinkwrapped trays
Trunking (return journey)   : Empty return load
Retail distribution   : Supermarket system
Retail sale   : Supermarket/Off-licence

Variations considered
Empty container delivery   : Alternative packaging
Outer packaging   : Cartons

SUMMARY OF TOTAL SYSTEM ENERGIES

All values in MJ/1000 containers

Main system
   (i) excluding retail sale and consumer use: Energy = 16426
  (ii) including retail sale and consumer use: Energy = 18961

Alternative systems
(a) Empty bottles bulk palletised, full bottles packed in cartons
   (i) excluding retail sale and consumer use: Energy = 17732
  (ii) including retail sale and consumer use: Energy = 20267

(b) Empty bottles packed in alternative packaging, full bottles packed in shrink-wrapped trays.
   (i) excluding retail sale and consumer use: Energy = 17145
  (ii) including retail sale and consumer use: Energy = 19680

(c) Empty bottles packed in alternative packaging, full bottles in cartons.
   (i) excluding retail sale and consumer use: Energy = 18451
  (ii) including retail sale and consumer use: Energy = 20986

VARIATIONS IN SYSTEM ENERGY IN MJ/container – see Chapter 29
Minimum   15.60
Maximum   26.58
Average   18.96

SYSTEM ENERGY REQUIREMENTS PER 1000 CONTAINERS

Contribution	Electricity/MJ		Oil fuels/MJ			Other fuels/MJ			Total energy /MJ
	Fuel production and delivery energy	Energy content of fuel	Fuel production and delivery energy	Energy content of fuel	Feedstock energy	Fuel production and delivery energy	Energy content of fuel	Feedstock energy	
Main system									
Filling and packing	968	354	213	1040	225	12	289	239	3340
Trunking	7	2	34	175	–	2	19	–	239
Retail delivery	7	3	37	188	–	2	20	–	257
Total 1	982	359	284	1403	225	16	328	239	3836
Container manufacture	1960	650	880	4310	1	310	2960	–	11070
Container delivery	9	3	43	221	–	4	23	–	304
Container packaging	239	80	45	224	67	–	79	147	881
Disposal	65	20	30	170	25	5	20	–	335
Total 2	2273	753	998	4925	93	319	3082	147	12590
Retail sale	684	228	–	–	–	–	–	–	912
Consumer use	–	–	260	1363	–	–	–	–	1623
Total 3	684	228	260	1363	–	–	–	–	2535
Total 1 + 2	3255	1112	1282	6328	318	335	3410	386	16426
Total 1 + 2 + 3	3939	1340	1542	7691	318	335	3410	386	18961
Alternative systems									
(a) Empty bottles bulk palletised, full bottles in cartons Revised Total 1 + 2 + 3	4303	1462	1603	7981	191	335	3697	695	20267
(b) Empty bottles in alternative packs, full bottles in shrinkwrapped trays Revised Total 1 + 2 + 3	4135	1405	1579	7869	318	336	3524	514	19680
(c) Empty bottles in alternative packs, full bottles in cartons Revised Total 1 + 2 + 3	4499	1527	1640	8159	191	336	3811	823	20986

System 34

MATERIALS REQUIREMENTS OF THE MAIN SYSTEM IN kg/1000 containers

Bauxite	22.22	Iron chromite	0.4	PET	—
Board	13.28	Iron ore	—	Polypropylene	—
Calcium sulphate	3.5	Lacquer	0.193	Sand	378.5
Carbon dioxide	1.99	Lead	—	Selenium	0.005
Cullet	3.5	Limestone	224.64	Sodium chloride	171.386
Detergent	—	Lining compound	0.523	Steel scrap	—
Epoxy adhesive	—	Manganese	—	Sulphuric acid	4.041
Feldspar	130	Methanol	0.003	Tin	—
Fluorspar	2.415	Pallets/number	0.149	Toluene	0.004
Flux (solder)	—	Paper	0.456	Varnish	0.193
Glue	0.15	Petroleum coke	0.652	Wood	—
Ink	0.031	Polyethylene	3.8		

35.2 fl. oz. (1.00 litre) RETURNABLE GLASS BOTTLE

CARBONATED SOFT DRINKS

Bottle mass = 0.705 kg

Main system
Empty bottle delivery   : Bulk palletised
Closure employed   : Standard aluminium screw
Outer packaging   : Returnable crates
Trunking (return journey) : Carrying empty bottles, crates, etc.
Retail distribution   : Brewery system
Retail sale   : Public houses

Variations considered
Empty bottle delivery   : Customer crates
Retail sale   : Off-licences

SUMMARY OF TOTAL SYSTEM ENERGIES AS A FUNCTION OF TRIPPAGE (t)

All values in MJ/1000 containers

Main system
    (i)   excluding retail sale and consumer use: Energy $= 3532 + 17702/t$
    (ii)   including retail sale and consumer use: Energy $= 6477 + 17702/t$

Alternative systems
(a)   Empty bottles packed in customer crates, sales through public houses.
    (i)   excluding retail sale and consumer use: Energy $= 3532 + 16776/t$
    (ii)   including retail sale and consumer use: Energy $= 6477 + 16776/t$

(b)   Empty bottles packed in customer crates, sales through off-licences.
    (i)   excluding retail sale and consumer use: Energy $= 3532 + 16776/t$
    (ii)   including retail sale and consumer use: Energy $= 6067 + 16776/t$

(c)   Empty bottles bulk palletised, sales through off-licences.
    (i)   excluding retail sale and consumer use: Energy $= 3532 + 17702/t$
    (ii)   including retail sale and consumer use: Energy $= 6067 + 17702/t$

VARIATIONS IN SYSTEM ENERGY IN MJ/container – see Chapter 29

Minimum $6.15 + 11.95/t$
Maximum $6.80 + 23.52/t$
Average $6.48 + 17.70/t$

SYSTEM ENERGY REQUIREMENTS PER 1000 CONTAINERS

Contribution	Electricity/MJ		Oil fuels/MJ			Other fuels/MJ			Total energy /MJ
	Fuel production and delivery energy	Energy content of fuel	Fuel production and delivery energy	Energy content of fuel	Feedstock energy	Fuel production and delivery energy	Energy content of fuel	Feedstock energy	
Main system									
Filling and packing	795	297	177	862	221	12	158	92	2614
Trunking	10	3	51	259	1	2	27	–	353
Retail delivery	19	6	81	416	1	3	39	–	565
Total 1	824	306	309	1537	223	17	224	92	3532
Container manufacture	2760	920	1250	6080	–	430	4180	–	15620
Container delivery	14	4	64	326	1	6	34	–	449
Container packaging	315	105	60	295	86	1	105	194	1161
Disposal	92	28	42	240	35	7	28	–	472
Total 2	3181	1057	1416	6941	122	444	4347	194	17702
Retail sale	670	223	34	161	–	18	216	–	1322
Consumer use	–	–	260	1363	–	–	–	–	1623
Total 3	670	223	294	1524	–	18	216	–	2945
Total 1 + 2	4005	1363	1725	8478	345	461	4571	286	21234
Total 1 + 2 + 3	4675	1586	2019	10002	345	479	4787	286	24179
Alternative systems									
(a) Empty bottles packed in customer crates									
Revised total 2	2934	975	1364	6685	36	443	4246	93	16776
(b) Retail sales through off-licences									
Revised Total 3	684	228	260	1363	–	–	–	–	2535

MATERIALS REQUIREMENTS OF THE MAIN SYSTEM IN kg/1000 containers

Bauxite	$8.22 + 19.74/t$	Iron chromite	$0.564/t$	PET	–
Board	$6.1/t$	Iron ore	–	Polypropylene	2.33
Calcium sulphate	$4.935/t$	Lacquer	0.193	Sand	$533.685/t$
Carbon dioxide	2.65	Lead	–	Selenium	$0.007/t$
Cullet	$4.935/t$	Limestone	$314.43/t$	Sodium chloride	$0.386 + 241.11/t$
Detergent	1.85	Lining compound	0.523	Steel scrap	–
Epoxy adhesive	–	Manganese	–	Sulphuric acid	$3.041 + 1.41/t$
Feldspar	$183.3/t$	Methanol	0.003	Tin	–
Fluorspar	2.415	Pallets/number	$1.807 + 0.193/t$	Toluene	0.004
Flux (solder)	–	Paper	0.456	Varnish	0.193
Glue	0.15	Petroleum coke	0.652	Wood	–
Ink	0.031	Polyethylene	$1.69/t$		

35.2 fl. oz. (1.00 litre) NON-RETURNABLE GLASS BOTTLE

BEER

Bottle mass = 0.600 kg

Main system

Empty bottle delivery	: Bulk palletised
Closure employed	: Crown
Outer packaging	: Shrinkwrapped trays
Trunking (return journey)	: Empty return load
Retail distribution	: Supermarket system
Retail sale	: Supermarket/Off-licence

Variations considered

Closure employed	: Standard aluminium screw
Outer packaging	: Cartons

SUMMARY OF TOTAL SYSTEM ENERGIES

All values in MJ/1000 containers

Main system
- (i) excluding retail sale and consumer use: Energy = 21511
- (ii) including retail sale and consumer use: Energy = 24046

Alternative systems
(a) Using aluminium screw closures and packaging in shrinkwrapped trays.
- (i) excluding retail sale and consumer use: Energy = 21997
- (ii) including retail sale and consumer use: Energy = 24532

(b) Using crown closures and packaging in cartons.
- (i) excluding retail sale and consumer use: Energy = 23439
- (ii) including retail sale and consumer use: Energy = 25974

(c) Using aluminium screw closures and packaging in cartons.
- (i) excluding retail sale and consumer use: Energy = 23925
- (ii) including retail sale and consumer use: Energy = 26460

VARIATIONS IN SYSTEM ENERGY IN MJ/container — see Chapter 29
Minimum   18.84
Maximum   33.61
Average   24.05

SYSTEM ENERGY REQUIREMENTS PER 1000 CONTAINERS

Contribution	Electricity/MJ		Oil fuels/MJ			Other fuels/MJ			Total energy /MJ
	Fuel production and delivery energy	Energy content of fuel	Fuel production and delivery energy	Energy content of fuel	Feedstock energy	Fuel production and delivery energy	Energy content of fuel	Feedstock energy	
Main system									
Filling and packing	1426	477	353	1713	183	52	752	291	5247
Trunking	18	6	89	453	1	4	48	–	619
Retail delivery	12	4	57	290	1	3	31	–	398
Total 1	1456	487	499	2456	185	59	831	291	6264
Container manufacture	2350	780	1060	5170	–	370	3560	–	13290
Container delivery	12	4	56	286	1	5	30	–	394
Container packaging	315	105	60	295	86	1	105	194	1161
Disposal	78	24	36	204	30	6	24	–	402
Total 2	2755	913	1212	5955	117	382	3719	194	15247
Retail sale	684	228	–	–	–	–	–	–	912
Consumer use	–	–	260	1363	–	–	–	–	1623
Total 3	684	228	260	1363	–	–	–	–	2535
Total 1 + 2	4211	1400	1711	8411	302	441	4550	485	21511
Total 1 + 2 + 3	4895	1628	1971	9774	302	441	4550	485	24046
Alternative systems									
(a) Using aluminium screw closures and shrinkwrapped trays Revised Total 1 + 2 + 3	5090	1723	1992	9878	387	434	4543	485	24532
(b) Using crown closures and cartons Revised Total 1 + 2 + 3	5432	1806	2062	10207	133	442	4963	929	25974
(c) Using aluminium screw closures and cartons Revised Total 1 + 2 + 3	5626	1902	2083	10311	217	435	4956	930	26460

MATERIALS REQUIREMENTS OF THE MAIN SYSTEM IN kg/1000 containers

System 36

Material	Value	Material	Value	Material	Value
Bauxite	16.8	Iron chromite	0.48	PET	—
Board	17.76	Iron ore	6.971	Polypropylene	—
Calcium sulphate	4.2	Lacquer	0.039	Sand	454.2
Carbon dioxide	2.65	Lead	—	Selenium	0.006
Cullet	4.2	Limestone	268.887	Sodium chloride	205.2
Detergent	—	Lining compound	0.214	Steel scrap	1.231
Epoxy adhesive	—	Manganese	0.056	Sulphuric acid	1.2
Feldspar	156.0	Methanol	0.003	Tin	0.032
Fluorspar	—	Pallets/number	0.360	Toluene	0.004
Flux (solder)	—	Paper	0.456	Varnish	0.023
Glue	0.15	Petroleum coke	—	Wood	—
Ink	0.033	Polyethylene	5.01		

35.2 fl. oz. (1.00 litre) NON-RETURNABLE GLASS BOTTLE

CIDER

Bottle mass = 0.600 kg

Main system

Empty bottle delivery	: Bulk palletised
Closure employed	: Standard aluminium screw
Outer packaging	: Shrinkwrapped trays
Trunking (return journey)	: Empty return load
Retail distribution	: Supermarket system
Retail sale	: Supermarket/Off-licence

Variations considered

Outer packaging	: Cartons

SUMMARY OF TOTAL SYSTEM ENERGIES

All values in MJ/1000 containers

Main system
  (i) excluding retail sale and consumer use: Energy = 21848
  (ii) including retail sale and consumer use: Energy = 24383

Alternative system
Full bottles packed in cartons
  (i) excluding retail sale and consumer use: Energy = 23776
  (ii) including retail sale and consumer use: Energy = 26311

VARIATIONS IN SYSTEM ENERGY IN MJ/container — see Chapter 29
Minimum   19.06
Maximum   32.49
Average   24.38

SYSTEM ENERGY REQUIREMENTS PER 1000 CONTAINERS

Contribution	Electricity/MJ		Oil fuels/MJ			Other fuels/MJ			Total energy /MJ
	Fuel production and delivery energy	Energy content of fuel	Fuel production and delivery energy	Energy content of fuel	Feedstock energy	Fuel production and delivery energy	Energy content of fuel	Feedstock energy	
Main system									
Filling and packing	1621	572	374	1817	268	45	745	291	5733
Trunking	15	5	74	378	1	3	40	–	516
Retail delivery	10	3	51	258	1	2	27	–	352
Total 1	1646	580	499	2453	270	50	812	291	6601
Container manufacture	2350	780	1060	5170	–	370	3560	–	13290
Container delivery	12	4	56	286	1	5	30	–	394
Container packaging	315	105	60	295	86	1	105	194	1161
Disposal	78	24	36	204	30	6	24	–	402
Total 2	2755	913	1212	5955	117	382	3719	194	15247
Retail sale	684	228	–	–	–	–	–	–	912
Consumer use	–	–	260	1363	–	–	–	–	1623
Total 3	684	228	260	1363	–	–	–	–	2535
Total 1 + 2	4401	1493	1711	8408	387	432	4531	485	21848
Total 1 + 2 + 3	5085	1721	1971	9771	387	432	4531	485	24383
Alternative system									
(a) Full bottles packed in cartons									
Revised Total 1 + 2 + 3	5621	1900	2062	10204	217	433	4944	930	26311

MATERIALS REQUIREMENTS OF THE MAIN SYSTEM IN kg/1000 containers

Material	Value	Material	Value	Material	Value
Bauxite	25.02	Iron chromite	0.48	PET	—
Board	17.76	Iron ore	—	Polypropylene	—
Calcium sulphate	4.2	Lacquer	0.193	Sand	454.2
Carbon dioxide	2.65	Lead	—	Selenium	0.006
Cullet	4.20	Limestone	269.24	Sodium chloride	205.586
Detergent	—	Lining compound	0.523	Steel scrap	—
Epoxy adhesive	—	Manganese	—	Sulphuric acid	4.241
Feldspar	156.0	Methanol	0.003	Tin	—
Fluorspar	2.415	Pallets/number	0.36	Toulene	0.004
Flux (solder)	—	Paper	0.456	Varnish	0.193
Glue	0.15	Petroleum coke	0.652	Wood	—
Ink	0.031	Polyethylene	5.01		

35.2 fl. oz. (1.00 litre) NON-RETURNABLE GLASS BOTTLE

CARBONATED SOFT DRINKS

Bottle mass = 0.630 kg

Main system

Empty bottle delivery	: Bulk palletised
Closure employed	: Standard aluminium screw
Outer packaging	: Shrinkwrapped trays
Trunking (return journey)	: Empty return load
Retail distribution	: Supermarket system
Retail sale	: Supermarkets/Off-licences

Variations considered

Empty container delivery	: Alternative packaging
Outer packaging	: Cartons

SUMMARY OF TOTAL SYSTEM ENERGIES

All values in MJ/1000 containers

Main system
  (i) excluding retail sale and consumer use: Energy = 20688
  (ii) including retail sale and consumer use: Energy = 23223

Alternative systems
(a) Empty bottles bulk palletised, full bottles in cartons.
  (i) excluding retail sale and consumer use: Energy = 22616
  (ii) including retail sale and consumer use: Energy = 25151

(b) Empty bottles packed in alternative packaging, full bottles in shrinkwrapped trays.
  (i) excluding retail sale and consumer use: Energy = 21650
  (ii) including retail sale and consumer use: Energy = 24185

(c) Empty bottles packed in alternative packaging, full bottles packed in cartons.
  (i) excluding retail sale and consumer use: Energy = 23578
  (ii) including retail sale and consumer use: Energy = 26113

VARIATIONS IN SYSTEM ENERGY IN MJ/container – see Chapter 29
Minimum   18.98
Maximum   30.94
Average   23.22

SYSTEM ENERGY REQUIREMENTS PER 1000 CONTAINERS

Contribution	Electricity/MJ		Oil fuels/MJ			Other fuels/MJ			Total energy /MJ
	Fuel production and delivery energy	Energy content of fuel	Fuel production and delivery energy	Energy content of fuel	Feedstock energy	Fuel production and delivery energy	Energy content of fuel	Feedstock energy	
Main system									
Filling and packing	1116	404	246	1203	268	12	343	291	3883
Trunking	15	5	74	378	1	3	40	–	516
Retail delivery	10	3	51	258	1	2	27	–	352
Total 1	1141	412	371	1839	270	17	410	291	4751
Container manufacture	2470	820	1110	5430	–	390	3740	–	13960
Container delivery	12	4	56	286	1	5	30	–	394
Container packaging	315	105	60	295	86	1	105	194	1161
Disposal	82	25	38	214	32	6	25	–	422
Total 2	2879	954	1264	6225	119	402	3900	194	15937
Retail sale	684	228	–	–	–	–	–	–	912
Consumer use	–	–	260	1363	–	–	–	–	1623
Total 3	684	228	260	1363	–	–	–	–	2535
Total 1 + 2	4020	1366	1635	8064	389	419	4310	485	20688
Total 1 + 2 + 3	4704	1594	1895	9427	389	419	4310	485	23223
Alternative systems									
(a) Empty bottles bulk palletised, full bottles in cartons									
Revised Total 1 + 2 + 3	5241	1772	1986	9860	219	420	4723	930	25151
(b) Empty bottles in alternative packs, full bottles in shrinkwrapped trays									
Revised Total 1 + 2 + 3	4967	1682	1943	9665	389	419	4463	657	24185
(c) Empty bottles in alternative packs, full bottles in cartons									
Revised Total 1 + 2 + 3	5504	1860	2034	10098	219	420	4876	1102	26113

System 38

MATERIALS REQUIREMENTS FOR THE MAIN SYSTEM IN kg/1000 containers

Material	Value	Material	Value	Material	Value
Bauxite	25.86	Iron chromite	0.504	PET	—
Board	17.76	Iron ore	—	Polypropylene	—
Calcium sulphate	4.41	Lacquer	0.193	Sand	476.91
Carbon dioxide	2.65	Lead	—	Selenium	0.006
Cullet	4.41	Limestone	282.62	Sodium chloride	215.846
Detergent	—	Lining compound	0.523	Steel scrap	—
Epoxy adhesive	—	Manganese	—	Sulphuric acid	4.301
Feldspar	163.8	Methanol	0.003	Tin	—
Fluorspar	2.415	Pallets/number	0.36	Toluene	0.004
Flux (solder)	—	Paper	0.456	Varnish	0.193
Glue	0.15	Petroleum coke	0.652	Wood	—
Ink	0.031	Polyethylene	5.01		

38 fl. oz. (1.08 litre) RETURNABLE GLASS BOTTLE

CIDER

Bottle mass = 0.675 kg

Main system
Empty bottle delivery	: Bulk palletised
Closure employed	: Standard aluminium screw
Outer packaging	: Returnable crates
Trunking (return journey)	: Carrying empty bottles, crates, etc.
Retail distribution	: Brewery system
Retail sale	: Public houses

Variations considered
Empty bottle delivery	: Customer crates
Retail sale	: Off-licences

SUMMARY OF TOTAL SYSTEM ENERGIES AS A FUNCTION OF TRIPPAGE (t)

All values in MJ/1000 containers

Main system
- (i) excluding retail sale and consumer use: Energy = $6241 + 17075/t$
- (ii) including retail sale and consumer use: Energy = $9186 + 17025/t$

Alternative systems

(a) Empty bottles packed in customer crates, sales through public houses.
- (i) excluding retail sale and consumer use: Energy = $6241 + 16048/t$
- (ii) including retail sale and consumer use: Energy = $9186 + 16048/t$

(b) Empty bottles packed in customer crates, sales through off-licences.
- (i) excluding retail sale and consumer use: Energy = $6241 + 16048/t$
- (ii) including retail sale and consumer use: Energy = $8776 + 16048/t$

(c) Empty bottles bulk palletised, sales through off-licences.
- (i) excluding retail sale and consumer use: Energy = $6241 + 17075/t$
- (ii) including retail sale and consumer use: Energy = $8776 + 17075/t$

VARIATIONS IN SYSTEM ENERGY IN MJ/container – see Chapter 29
Minimum	$8.73 + 11.57/t$
Maximum	$9.65 + 24.63/t$
Average	$9.19 + 17.03/t$

SYSTEM ENERGY REQUIREMENTS PER 1000 CONTAINERS

Contribution	Electricity/MJ		Oil fuels/MJ			Other fuels/MJ			Total energy /MJ
	Fuel production and delivery energy	Energy content of fuel	Fuel production and delivery energy	Energy content of fuel	Feedstock energy	Fuel production and delivery energy	Energy content of fuel	Feedstock energy	
Main system									
Filling and packing	1362	486	359	1738	298	54	663	92	5052
Trunking	18	6	90	457	1	4	48	–	624
Retail delivery	19	6	81	416	1	3	39	92	565
Total 1	1399	498	530	2611	300	61	750	92	6241
Container manufacture	2650	880	1190	5820	–	410	4000	–	14950
Container delivery	12	4	55	279	1	5	30	–	386
Container packaging	349	116	66	327	95	1	117	214	1285
Disposal	88	27	41	230	34	7	27	–	454
Total 2	3099	1027	1352	6656	130	423	4174	214	17075
Retail sale	670	223	34	161	–	18	216	–	1322
Consumer use	–	–	260	1363	–	–	–	–	1623
Total 3	670	223	294	1524	–	18	216	–	2945
Total 1 + 2	4498	1525	1882	9267	430	484	4924	306	23316
Total 1 + 2 + 3	5168	1748	2176	10791	430	502	5140	306	26261
Alternative systems									
(a) Empty bottles in customer crates									
Revised Total 2	2825	936	1295	6371	35	422	4062	102	16048
(b) Retail sales through off-licences									
Revised Total 3	684	228	260	1363	–	–	–	–	2535

MATERIALS REQUIREMENTS FOR THE MAIN SYSTEM IN kg/1000 containers

Bauxite	$8.22 + 18.9/t$	Iron chromite	$0.54/t$	PET	—
Board	$6.8/t$	Iron ore	—	Polypropylene	3.79
Calcium sulphate	$4.725/t$	Lacquer	0.193	Sand	$510.975/t$
Carbon dioxide	2.95	Lead	—	Selenium	$0.007/t$
Cullet	$4.725/t$	Limestone	$1.64 + 301.05/t$	Sodium chloride	$0.386 + 230.85/t$
Detergent	1.85	Lining compound	0.523	Steel scrap	—
Epoxy adhesive	—	Manganese	—	Sulphuric acid	$3.041 + 1.35/t$
Feldspar	$175.5/t$	Methanol	0.003	Tin	—
Fluorspar	2.415	Pallets/number	$0.189 + 0.211/t$	Toluene	0.004
Flux (solder)	—	Paper	0.456	Varnish	0.193
Glue	0.15	Petroleum coke	0.652	Wood	—
Ink	0.031	Polyethylene	$1.86/t$		

38.0 fl. oz. (1.08 litre) RETURNABLE GLASS BOTTLE

CARBONATED SOFT DRINKS

Bottle mass = 0.850 kg

Main system
Empty bottle delivery : Bulk palletised
Closure employed : Standard aluminium screw
Outer packaging : Returnable crates
Trunking (return journey) : Carrying empty bottles, crates, etc.
Retail distribution : Brewery system
Retail sale : Public houses

Variations considered
Empty bottle delivery : Customer crates
Retail sale : Off-licences

SUMMARY OF TOTAL SYSTEM ENERGIES AS A FUNCTION OF TRIPPAGE (t)
All values in MJ/1000 containers

Main system
(i) excluding retail sale and consumer use: Energy = $4044 + 21139/t$
(ii) including retail sale and consumer use: Energy = $6989 + 21139/t$

Alternative system
(a) Empty bottles in customer crates, sales through public houses.
(i) excluding retail sale and consumer use: Energy = $4044 + 20139/t$
(ii) including retail sale and consumer use: Energy = $6989 + 20139/t$

(b) Empty bottles packed in customer crates, sales through off-licences.
(i) excluding retail sale and consumer use: Energy = $4044 + 20139/t$
(ii) including retail sale and consumer use: Energy = $6579 + 20139/t$

(c) Empty bottles bulk palletised, sales through off-licences.
(i) excluding retail sale and consumer use: Energy = $4044 + 21139/t$
(ii) including retail sale and consumer use: Energy = $6579 + 21139/t$

VARIATIONS IN SYSTEM ENERGY IN MJ/container – see Chapter 29
Minimum $6.64 + 11.21/t$
Maximum $7.34 + 27.03/t$
Average $6.99 + 21.14/t$

SYSTEM ENERGY REQUIREMENTS PER 1000 CONTAINERS

Contribution	Electricity/MJ		Oil fuels/MJ			Other fuels/MJ			Total energy /MJ
	Fuel production and delivery energy	Energy content of fuel	Fuel production and delivery energy	Energy content of fuel	Feedstock energy	Fuel production and delivery energy	Energy content of fuel	Feedstock energy	
Main system									
Filling and packing	851	315	191	935	298	13	160	92	2855
Trunking	18	6	90	457	1	4	48	–	624
Retail delivery	19	6	81	416	1	3	39	–	565
Total 1	888	327	362	1808	300	20	247	92	4044
Container manufacture	3330	1110	1500	7330	–	520	5040	–	18830
Container delivery	15	5	69	350	1	6	37	–	483
Container packaging	341	114	64	319	93	1	114	209	1255
Disposal	111	34	51	289	43	9	34	–	571
Total 2	3797	1263	1684	8288	137	536	5225	209	21139
Retail sale	670	223	34	161	–	18	216	–	1322
Consumer use	–	–	260	1363	–	–	–	–	1623
Total 3	670	223	294	1524	–	18	216	–	2945
Total 1 + 2	4685	1590	2046	10096	437	556	5472	301	25183
Total 1 + 2 + 3	5355	1813	2340	11620	437	574	5688	301	28128
Alternative systems									
(a) Empty bottles packed in customer crates									
Revised Total 2	3530	1174	1629	8011	44	535	5116	100	20139
(b) Retail sales through off-licences									
Revised Total 3	684	228	260	1363	–	–	–	–	2535

System 40

MATERIALS REQUIREMENTS FOR THE MAIN SYSTEM IN kg/1000 containers

Bauxite	$8.22 + 23.8/t$	Iron chromite	$0.68/t$	PET	—
Board	$6.6/t$	Iron ore	—	Polypropylene	3.79
Calcium sulphate	$5.95/t$	Lacquer	0.193	Sand	$643.45/t$
Carbon dioxide	2.86	Lead	—	Selenium	$0.009/t$
Cullet	$5.95/t$	Limestone	$1.64 + 379.1/t$	Sodium chloride	$0.386 + 290.7/t$
Detergent	1.85	Lining compound	0.523	Steel scrap	—
Epoxy adhesive	—	Manganese	—	Sulphuric acid	$3.041 + 1.7/t$
Feldspar	$221.0/t$	Methanol	0.003	Tin	—
Fluorspar	2.415	Pallets/number	$0.189 + 0.208/t$	Toluene	0.004
Flux (solder)	—	Paper	0.456	Varnish	0.193
Glue	0.15	Petroleum coke	0.652	Wood	—
Ink	0.031	Polyethylene	$1.83/t$		

39.1 fl. oz. (1.11 litre) RETURNABLE GLASS BOTTLE

BEER

Bottle mass = 0.675 kg

Main system

Empty container delivery : Bulk palletised
Closure employed : Crown
Outer packaging : Returnable crates
Trunking (return load) : Carrying empty bottles, crates, etc.
Retail distribution : Brewery system
Retail sale : Public houses

Variations considered

Empty container delivery : Customer crates

SUMMARY OF TOTAL SYSTEM ENERGIES AS A FUNCTION OF TRIPPAGE (t)

All values in MJ/1000 containers

Main system

(i) excluding retail sale and consumer use: Energy = $5713 + 17075/t$
(ii) including retail sale and consumer use: Energy = $8658 + 17075/t$

Alternative system

Empty containers packed in customer crates.

(i) excluding retail sale and consumer use: Energy = $5713 + 16048/t$
(ii) including retail sale and consumer use: Energy = $8658 + 16048/t$

VARIATIONS IN SYSTEM ENERGY IN MJ/container — see Chapter 29

Minimum $8.23 + 11.57/t$
Maximum $9.09 + 23.64/t$
Average $8.66 + 17.08/t$

SYSTEM ENERGY REQUIREMENTS PER 1000 CONTAINERS

Contribution	Electricity/MJ		Oil fuels/MJ			Other fuels/MJ			Total energy /MJ
	Fuel production and delivery energy	Energy content of fuel	Fuel production and delivery energy	Energy content of fuel	Feedstock energy	Fuel production and delivery energy	Energy content of fuel	Feedstock energy	
Main system									
Filling and packing	1167	390	338	1634	214	61	670	92	4566
Trunking	14	5	72	364	1	3	39	–	498
Retail delivery	22	7	93	478	1	4	44	–	649
Total 1	1203	402	503	2476	216	68	753	92	5713
Container manufacture	2650	880	1190	5820	–	410	4000	–	14950
Container delivery	12	4	55	279	1	5	30	–	386
Container packaging	349	116	66	327	95	1	117	214	1285
Disposal	88	27	41	230	34	7	27	–	454
Total 2	3099	1027	1352	6656	130	423	4174	214	17075
Retail sale	670	223	34	161	–	18	216	–	1322
Consumer use	–	–	260	1363	–	–	–	–	1623
Total 3	670	223	294	1524	–	18	216	–	2945
Total 1 + 2	4302	1429	1855	9132	346	491	4927	306	22788
Total 1 + 2 + 3	4972	1652	2149	10656	346	509	5143	306	25733
Alternative systems									
(a) Empty bottles packed in customer crates									
Revised Total 2	2825	936	1295	6371	35	422	4062	102	16048

MATERIALS REQUIREMENTS FOR THE MAIN SYSTEM IN kg/1000 containers

Bauxite	18.9/t	Iron chromite	0.54/t	PET	—
Board	6.8/t	Iron ore	6.971	Polypropylene	3.79
Calcium sulphate	4.725/t	Lacquer	0.039	Sand	510.975/t
Carbon dioxide	2.95	Lead	—	Selenium	0.007/t
Cullet	4.725/t	Limestone	1.287 + 301.05/t	Sodium chloride	230.85/t
Detergent	1.85	Lining compound	0.214	Steel scrap	1.231
Epoxy adhesive	—	Manganese	0.056	Sulphuric acid	1.35/t
Feldspar	175.5/t	Methanol	0.003	Tin	0.032
Fluorspar	—	Pallets/number	0.189 + 0.211/t	Toluene	0.004
Flux (solder)	—	Paper	0.456	Varnish	0.023
Glue	0.15	Petroleum coke	—	Wood	—
Ink	0.033	Polyethylene	1.86/t		

52.8 fl. oz. (1.5 litre) NON-RETURNABLE PLASTIC (PET) BOTTLE

CARBONATED SOFT DRINKS

Bottle mass (PET) = **0.060** kg
Base mass (PET) = **0.005** kg

Main system

Empty container delivery	: Bulk palletised packs
Closure type	: Aluminium screw
Outer packaging	: Cartons
Trunking (return journey)	: Empty return load
Retail distribution	: Supermarket system
Retail sale	: Supermarkets

Variations considered

Outer packaging	: Shrinkwrapped trays

SUMMARY OF TOTAL SYSTEM ENERGIES

All values in MJ/1000 containers

Main system
  (i) excluding retail sale and consumer use: Energy = 22793
  (ii) including retail sale and consumer use: Energy = 25328

Alternative systems
(a) Filled bottles packed in shrinkwrapped trays
  (i) excluding retail sale and consumer use: Energy = 20942
  (ii) including retail sale and consumer use: Energy = 23477

SYSTEM ENERGY VARIATIONS IN MJ/container
Minimum = 21.6
Maximum = 25.3
Average = 25.3

SYSTEM ENERGY REQUIREMENTS PER 1000 CONTAINERS

Contribution	Electricity/MJ		Oil fuels/MJ			Other fuels/MJ			Total energy /MJ
	Fuel production and delivery energy	Energy content of fuel	Fuel production and delivery energy	Energy content of fuel	Feedstock energy	Fuel production and delivery energy	Energy content of fuel	Feedstock energy	
Main system									
Filling and packing	1850	648	376	1827	99	15	800	778	6393
Trunking	15	5	74	378	1	3	40	–	516
Retail delivery	10	3	51	258	1	2	27	–	352
Total 1	1875	656	501	2463	101	20	867	778	7261
Container manufacture	4549	1516	950	4629	3028	–	–	–	14672
Container delivery	6	2	27	137	–	2	15	–	189
Container packaging	171	57	33	161	45	–	59	105	631
Disposal	8	2	4	20	3	1	2	–	40
Total 2	4734	1577	1014	4947	3076	3	76	105	15532
Retail sale	684	228	–	–	–	–	–	–	912
Consumer use	–	–	260	1363	–	–	–	–	1623
Total 3	684	228	260	1363	–	–	–	–	2535
Total 1 + 2	6609	2233	1515	7410	3177	23	943	883	22793
Total 1 + 2 + 3	7293	2461	1775	8773	3177	23	943	883	25328
Alternative systems									
(a) Full bottles packed in shrinkwrapped trays									
Revised Total 1	1357	484	416	2059	304	19	446	325	5410

MATERIALS REQUIREMENTS FOR THE MAIN SYSTEM IN kg/1000 containers

System 42

Bauxite	8.54	Iron chromite	–	PET	64.85
Board	41.282	Iron ore	–	Polypropylene	–
Calcium sulphate	–	Lacquer	0.193	Sand	–
Carbon dioxide	3.99	Lead	–	Selenium	–
Cullet	–	Limestone	1.706	Sodium chloride	0.401
Detergent	–	Lining compound	0.523	Steel scrap	–
Epoxy adhesive	–	Manganese	–	Sulphuric acid	3.162
Feldspar	–	Methanol	0.003	Tin	–
Fluorspar	2.511	Pallets/number	0.308	Toluene	0.004
Flux (solder)	–	Paper	0.456	Varnish	0.193
Glue	0.15	Petroleum coke	0.678	Wood	–
Ink	0.031	Polyethylene	0.888		

70.4 fl. oz. (2.0 litre) NON-RETURNABLE PLASTIC (PET) BOTTLE

CARBONATED SOFT DRINKS

Bottle mass = 0.065 kg
Base mass = 0.015 kg

Main system

Empty container delivery	: Bulk palletised packs
Closure type	: Aluminium screw
Outer packaging	: Cartons
Trunking (return load)	: Empty return load
Retail distribution	: Supermarket system
Retail sale	: Supermarket

Variations considered

Outer packaging	: Shrinkwrapped trays

SUMMARY OF TOTAL SYSTEM ENERGIES

All values in MJ/1000 containers

Main system
    (i) excluding retail sale and consumer use: Energy = 26628
   (ii) including retail sale and consumer use: Energy = 29163

Alternative system
(a) Filled bottles packed in shrinkwrapped trays.
    (i) excluding retail sale and consumer use: Energy = 24637
   (ii) including retail sale and consumer use: Energy = 27172

SYSTEM ENERGY VARIATIONS IN MJ/container
Minimum = 25.00
Maximum = 29.20
Average = 29.20

SYSTEM ENERGY REQUIREMENTS PER 1000 CONTAINERS

Contribution	Electricity/MJ		Oil fuels/MJ			Other fuels/MJ			Total energy /MJ
	Fuel production and delivery energy	Energy content of fuel	Fuel production and delivery energy	Energy content of fuel	Feedstock energy	Fuel production and delivery energy	Energy content of fuel	Feedstock energy	
Main system									
Filling and packing	2376	822	475	2304	99	17	1078	1107	8278
Trunking and retail delivery	36	12	179	909	2	8	96	–	1242
Total 1	2412	834	654	3213	101	25	1174	1107	9520
Container manufacture	4771	1591	1037	5087	3618	–	–	–	16104
Container delivery	8	3	39	199	–	3	21	–	273
Container packaging	187	62	35	175	49	–	65	114	687
Disposal	9	3	4	22	3	–	3	–	44
Total 2	4975	1659	1115	5483	3670	3	89	114	17108
Retail sale	684	228	–	–	–	–	–	–	912
Consumer use	–	–	260	1363	–	–	–	–	1623
Total 3	684	228	260	1363	–	–	–	–	2535
Total 1 + 2	7387	2493	1769	8696	3771	28	1263	1221	26628
Total 1 + 2 + 3	8071	2721	2029	10059	3771	28	1263	1221	29163
Alternative system									
(a) Full bottles packed in shrinkwrapped trays									
Revised Total 1	1869	647	562	2774	321	24	717	615	7529

MATERIALS REQUIREMENTS OF THE MAIN SYSTEM IN kg/1000 containers

Bauxite	8.54	Iron chromite	–	PET	80.31
Board	57.08	Iron ore	0.193	Polypropylene	–
Calcium sulphate	–	Lacquer	0.193	Sand	–
Carbon dioxide	5.32	Lead	–	Selenium	–
Cullet	–	Limestone	1.706	Sodium chloride	0.401
Detergent	–	Lining compound	0.523	Steel scrap	–
Epoxy adhesive	–	Manganese	–	Sulphuric acid	3.162
Feldspar	–	Methanol	0.003	Tin	–
Fluorspar	2.511	Pallets/number	0.426	Toluene	0.004
Flux (solder)	–	Paper	0.456	Varnish	0.193
Glue	0.15	Petroleum coke	0.678	Wood	–
Ink	0.031	Polyethylene	80.31		

78 fl. oz. (2.22) litre) METAL CAN

BEER

Can type: Three-piece tinplate

Main system

Outer packaging	: Shrinkwrapped trays
Trunking (return journey)	: Empty return load
Retail distribution	: Supermarket system
Retail sale	: Supermarkets/Off licences

Variations considered

Outer packaging	: Cartons

SUMMARY OF TOTAL SYSTEM ENERGIES

All values in MJ/1000 containers

Main system
- (i) excluding retail sale and consumer use: Energy = 29092
- (ii) including retail sale and consumer use: Energy = 31627

Alternative system
Full containers packed in cartons.
- (i) excluding retail sale and consumer use: Energy = 29925
- (ii) including retail sale and consumer use: Energy = 32460

VARIATION IN SYSTEM ENERGY IN MJ/container — see Chapter 29
Minimum   29.13
Maximum   35.10
Average   31.63

SYSTEM ENERGY REQUIREMENTS PER 1000 CONTAINERS

Contribution	Electricity/MJ		Oil fuels/MJ			Other fuels/MJ			Total energy /MJ
	Fuel production and delivery energy	Energy content of fuel	Fuel production and delivery energy	Energy content of fuel	Feedstock energy	Fuel production and delivery energy	Energy content of fuel	Feedstock energy	
Main system									
Filling and packing	2979	993	674	3252	241	93	1318	328	9878
Trunking	36	12	177	900	2	8	95	–	1230
Retail delivery	15	5	76	385	1	3	41	–	526
Total 1	3030	1010	927	4537	244	104	1454	328	11634
Container manufacture	2478	953	682	3300	243	1014	7287	7	15964
Container delivery	12	4	55	283	1	5	30	–	390
Container packaging	358	119	23	115	–	14	184	125	938
Disposal	32	10	15	85	12	2	10	–	166
Total 1 + 2	2880	1086	775	3783	256	1035	7511	132	17458
Retail sale	684	228	–	–	–	–	–	–	912
Consumer use	–	–	260	1363	–	–	–	–	1623
Total 3	684	228	260	1363	–	–	–	–	2535
Total 1 + 2	5910	2096	1702	8320	500	1139	8965	460	29092
Total 1 + 2 + 3	6594	2324	1962	9683	500	1139	8965	460	31627
Alternative system									
(a) Full containers packed in cartons									
Revised Total 1 + 2 + 3	6837	2405	1993	9824	259	1140	9244	758	32460

Appendix 1

System 44

MATERIALS REQUIREMENTS FOR THE MAIN SYSTEM IN kg/1000 containers

Bauxite	—	Iron chromite	—	PET	—
Board	13.519	Iron ore	798.893	Polypropylene	—
Calcium sulphate	—	Lacquer	1.155	Sand	—
Carbon dioxide	5.88	Lead	—	Selenium	—
Cullet	—	Limestone	144.086	Sodium chloride	—
Detergent	—	Lining compound	0.268	Steel scrap	134.575
Epoxy adhesive	—	Manganese	6.366	Sulphuric acid	—
Feldspar	—	Methanol	—	Tin	3.487
Fluorspar	—	Pallets/number	4.359	Toluene	—
Flux (solder)	—	Paper	—	Varnish	0.34
Glue	—	Petroleum coke	—	Wood	0.931
Ink	0.263	Polyethylene	4.73		

98 fl. oz. (2.78 litre) METAL CAN

BEER

Can type: Three-piece tinplate

Main system

Outer packaging	: Shrinkwrapped trays
Trunking (return journey)	: Empty return load
Retail distribution	: Supermarket system
Retail sale	: Supermarkets/Off-licences

Variations considered

Outer packaging	: Cartons

SUMMARY OF TOTAL SYSTEM ENERGIES

All values in MJ/1000 containers

Main system
  (i) excluding retail sale and consumer use: Energy = 33425
  (ii) including retail sale and consumer use: Energy = 35960

Alternative system
  (i) excluding retail sale and consumer use: Energy = 34258
  (ii) including retail sale and consumer use: Energy = 36793

VARIATIONS IN SYSTEM ENERGY IN MJ/container
Minimum   33.07
Maximum   39.85
Average   35.96

SYSTEM ENERGY REQUIREMENTS PER 1000 CONTAINERS

Contribution	Electricity/MJ		Oil fuels/MJ			Other fuels/MJ			Total energy/MJ
	Fuel production and delivery energy	Energy content of fuel	Fuel production and delivery energy	Energy content of fuel	Feedstock energy	Fuel production and delivery energy	Energy content of fuel	Feedstock energy	
Main system									
Filling and packing	3904	1301	718	3463	241	101	1407	337	11472
Trunking	32	11	158	806	2	7	85	–	1101
Retail delivery	16	5	80	407	1	4	43	–	556
Total 1	3952	1317	956	4676	244	112	1535	337	13129
Container manufacture	2879	1106	787	3808	302	1184	8506	9	18581
Container delivery	14	4	66	337	1	6	36	–	464
Container packaging	416	139	26	127	–	17	214	142	1081
Disposal	33	10	15	86	13	3	10	151	170
Total 2	3342	1259	894	4358	316	1210	8766	151	20296
Retail sale	684	228	–	–	–	–	–	–	912
Consumer use	–	–	260	1363	–	–	–	–	1623
Total 3	684	228	260	1363	–	–	–	–	2535
Total 1 + 2	7294	2576	1850	9034	560	1322	10301	488	33425
Total 1 + 2 + 3	7978	2804	2110	10397	560	1322	10301	488	35960
Alternative system									
(a) Full cans packed in cartons Revised Total 1 + 2 + 3	8220	2885	2141	10538	319	1322	10580	788	36793

MATERIALS REQUIREMENTS OF THE MAIN SYSTEM IN kg/1000 containers

Bauxite	—	Iron chromite	—
Board	13.5	Iron ore	928.967
Calcium sulphate	—	Lacquer	1.43
Carbon dioxide	7.39	Lead	—
Cullet	—	Limestone	166.563
Detergent	—	Lining compound	0.268
Epoxy adhesive	—	Manganese	7.401
Feldspar	—	Methanol	—
Fluorspar	—	Pallets/number	5.195
Flux (solder)	—	Paper	—
Glue	—	Petroleum coke	—
Ink	0.351	Polyethylene	4.73

PET	—
Polypropylene	—
Sand	—
Selenium	—
Sodium chloride	—
Steel scrap	154.662
Sulphuric acid	—
Tin	4.006
Toluene	—
Varnish	0.483
Wood	1.118

Appendix 1

136 fl. oz. (3.86 litre) METAL CAN

BEER

Can type: Three-piece tinplate

Main system

Outer packaging	: Shrinkwrapped trays
Trunking (return journey)	: Empty return load
Retail distribution	: Supermarket system
Retail sale	: Supermarkets/Off-licences

Variations considered

Outer packaging	: Cartons

SUMMARY OF TOTAL SYSTEM ENERGIES

All values in MJ/1000 containers

Main system
  (i) excluding retail sale and consumer energy: Energy = 45076
  (ii) including retail sale and consumer energy: Energy = 47611

Alternative systems
Full containers packed in cartons.
  (i) excluding retail sale and consumer use: Energy = 46327
  (ii) including retail sale and consumer use: Energy = 48862

VARIATIONS IN SYSTEM ENERGY IN MJ/container — see Chapter 29
Minimum   43.69
Maximum   52.86
Average   47.61

SYSTEM ENERGY REQUIREMENTS PER 1000 CONTAINERS

Contribution	Electricity/MJ		Oil fuels/MJ			Other fuels/MJ			Total energy /MJ
	Fuel production and delivery energy	Energy content of fuel	Fuel production and delivery energy	Energy content of fuel	Feedstock energy	Fuel production and delivery energy	Energy content of fuel	Feedstock energy	
Main system									
Filling and packing	6037	2012	869	4196	362	116	1683	507	15782
Trunking	33	11	164	833	2	7	88	–	1138
Retail delivery	25	8	124	631	2	5	67	–	862
Total 1	6095	2031	1157	5660	366	128	1838	507	17782
Container manufacture	3755	1404	1069	5169	415	1590	11462	11	24875
Container delivery	20	6	93	475	1	8	51	–	654
Container packaging	582	194	35	174	–	24	298	197	1504
Disposal	51	16	23	132	19	4	16	–	261
Total 2	4408	1620	1220	5950	435	1626	11827	208	27294
Retail sale	684	228	–	–	–	–	–	–	912
Consumer use	–	–	260	1363	–	–	–	–	1623
Total 3	684	228	260	1363	–	–	–	–	2535
Total 1 + 2	10503	3651	2377	11610	801	1754	13665	715	45076
Total 1 + 2 + 3	11187	3879	2637	12973	801	1754	13665	715	47611
Alternative system									
(a) Full cans packed in cartons									
Revised Total 1 + 2 + 3	11551	4001	2683	13184	439	1755	14084	1165	48862

System 46

MATERIALS REQUIREMENTS OF THE MAIN SYSTEM IN kg/1000 containers

Bauxite	—	Iron chromite	—	PET	—
Board	19.88	Iron ore	1241.547	Polypropylene	—
Calcium sulphate	—	Lacquer	1.896	Sand	—
Carbon dioxide	10.26	Lead	—	Selenium	—
Cullet	—	Limestone	224.541	Sodium chloride	—
Detergent	—	Lining compound	0.336	Steel scrap	210.289
Epoxy adhesive	—	Manganese	9.892	Sulphuric acid	—
Feldspar	—	Methanol	—	Tin	6.706
Fluorspar	—	Pallets/number	7.386	Toluene	—
Flux (solder)	0.10	Paper	—	Varnish	0.743
Glue	—	Petroleum coke	—	Wood	1.583
Ink	0.527	Polyethylene	7.1		

Index